Fishponds in farming systems

Fishponds in farming systems

edited by:
A.J. van der Zijpp
J.A.J. Verreth
Le Quang Tri
M.E.F. van Mensvoort
R.H. Bosma
M.C.M. Beveridge

Wageningen Academic Publishers

ISBN-13: 978-90-8686-013-5

First published, 2007

Wageningen Academic Publishers
The-Netherlands, 2007

Contents

Chapter 1
Interdisciplinary research on fishponds

Programme for Optimisation of Nutrient Dynamics in integrated agriculture aquaculture farming systems

Introduction

During the next 25 years, the world population is expected to grow by a further 2 billion people to around 8 billion. This will lead to an increased demand for food. If an increase in food prices is to be prevented, food production needs to increase proportionally (Rosegrant *et al.*, 2001). However, since there is limited scope for further expansion of the area of arable land, there is a need to increase farm productivity from the existing land area. There are different ways to achieve this intensification of food production but most commonly it is done by importing nutrients, applying agrochemicals, and using genetically improved crop and animal species. While intensification seems necessary, there is a concern for ecological, economic and social sustainability. Importing nutrients, low nutrient use efficiencies, and applying agrochemicals may harm the environment. Genetically improved species and strains generally need higher levels of inputs, this means that farmers have to take more investment risk when applying these technologies. The poorer groups in society may not have access to the credit and knowledge necessary for implementing new technology. To secure sustainable food production for an expanding human population, nutrients must be used more efficiently.

The global yield from capture fisheries has levelled out (FAO, 2004) while the global human population continues to increase rapidly. Therefore, reliance upon aquaculture-based food supply to meet present and future needs has become greater than ever. Aquaculture is currently the fastest growing segment of food production in the world: it increased by an average annual rate of 8.8 percent from 1950 to 2004, and has quadrupled since 1990 from 15 to 60 million ton, expressing in some years an annual rate above 30% (FAO, 2006). At present aquaculture provides over 50% of the total fish consumption of 110 million tons. Developing countries provided more than 90% of the global aquaculture production, realising more than 70% of finfish and shrimp aquaculture production in non-aerated extensively managed earthen ponds. Thus small producers addressed by the program produce close to 30 million tons (live weight equivalent), with an export value of € 30 billion. Tilapia is the third most cultured fish, coming only after carps and salmonids. Tilapia has been the fish with the largest production expansion in recent years and the average annual growth of tilapia productions in the last decade was above 12%. The production of farmed tilapia has jumped from close to 380,000 metric tons in 1990 to over 1,500,000 metric tons in 2004 (FAO, 2006).

In parts of Asia, traditionally integrated systems with fish and pigs or poultry components are used. In view of their extensive use they might contribute to the sustainability of the farm. These systems were characterised by low external inputs and often relied on indigenous strains of fish, pigs and poultry. In such systems ponds can function as traps which accumulate runoff nutrients and recycle waste from the household, homestead gardens, surrounding embankments and livestock (Prein, 2002). An efficient use of accumulated pond sediment for fertilisation of surrounding land might contribute to the improvement of livelihoods.

Several studies indicate that farms with many links between the various farm components use nutrients more efficiently and can improve the livelihoods of the farm households (Edwards, 1998; Devendra and Thomas, 2002). Despite the fact that pond-dyke systems are widely used in China and also throughout South and Southeast Asia, few quantitative insights into nutrient cycling through pond-dyke systems exist.

Integrating an aquaculture component into a farming system may result in a more efficient and ecologically sound pond-dyke system which produces a range of products on a more sustained basis, but which may also lead to environmental pollution. As farming systems intensify, the relative importance of the various subsystems may also change. Depending on market demand, farmers may focus on one or two products and use genetically improved strains that yield the highest profits. Newly introduced genetically improved animals require high-quality feeds to perform optimally, which leads to an increase in production costs and more imports of nutrients into the system and possibly discharge of nutrients into the environment. This stresses the need to select fish breeds that are adapted optimally to the farming environment.

All the above mentioned matters, such as the need for more food, the importance of aquaculture and the role of small farmers' ponds in fish production, plus the question whether integrated farms with many components are more sustainable than specialised and/or intensified farms, lead to the formulation of two research programs called POND funded by the European Union and Wageningen University, respectively. The programs were designed to improve the impact of integrated aquaculture agriculture systems (IAAS), also referred to as pond-dyke systems, on the livelihoods of farming households. The aim is to contribute to more efficient utilisation of nutrients by using a genetically improved fish breed, by enhancing pond ecology, and by an increased on-farm recycling. The specific objectives were: (1) to assess the impact of pond aquaculture on the livelihoods of Asian farming households; (2) to investigate whether a specific tilapia breed for low-input environments needs to be selected; (3) to analyse the nutrient dynamics in the fishpond; (4) to enhance the role of aquaculture ponds in nutrient cycling on farms; (5) to create a model for nutrient cycling in pond-dyke systems; (6) to contribute to policies for enhancing the adoption of new aquaculture technologies. The livelihood assessment covered the availability of assets, the developmental history and the development policies related to agriculture, the agricultural practices and the livelihood strategies. The projects focused on tilapia given its' importance for farmers in developing countries and the wide range of culture environments encountered in tilapia aquaculture. The model, as mentioned under (5), should be the backbone of a support tool for decision-making on integrated agriculture-aquaculture (pond-dyke) systems to be used by policy makers and extension agents. For the decision-making support tool the most suitable combination of farm components contributing to improved systems resilience and sustainability of: income, food security and environment, needed to be identified. The new technologies should be based on the livelihoods of farm households and thus be created through participatory approaches

The POND programs started in 2001. The INREF-POND program was a collaboration of Wageningen University (WU) in the Netherlands, the World Fish Center, and Can Tho University (CTU) Vietnam. Four chair groups from Wageningen University contributed to the INREF-POND program: Aquaculture & Fisheries, Animal Production Systems, Animal

Breeding & Genetics and Soil Inventarisation and Land Evaluation. POND-LIVE regrouped Aquaculture & Fisheries of WU and of Stirling University, the Asian Institute of Technology (AIT) in Bangkok, Sisaket College of Agriculture and Technology, Bangladesh Agriculture University (BAU) in Mymensingh, and CTU. The activities with the World Fish Centre were mainly carried out at the Regional Centre for Africa and West Asia, Egypt. The specific partner at AIT was the School of Environment, Resources and Development of the Agricultural, Aquatic Systems and Engineering Program and at BAU the Department of Fisheries Management of the Faculty of Fisheries. In Can Tho, the College of Agriculture and Applied Biology, the College of Aquaculture & Fisheries and the Mekong Delta Development Research Institute contributed to the program.

Approach

The POND research was based upon a dual approach. The INREF-POND projects either did on-station research or monitored farming systems. The POND-LIVE projects followed the Participatory Learning in Action approach composed of six phases (Figure 1.1). The research was carried out within a livelihoods framework to ensure that the impact of pond-dyke systems was evaluated, not only in terms of aquatic production volume, but also in terms of employment opportunities, well-being, improved resilience and sustainability of the resource base. The presentation of intermediate results at progress and dissemination workshops,

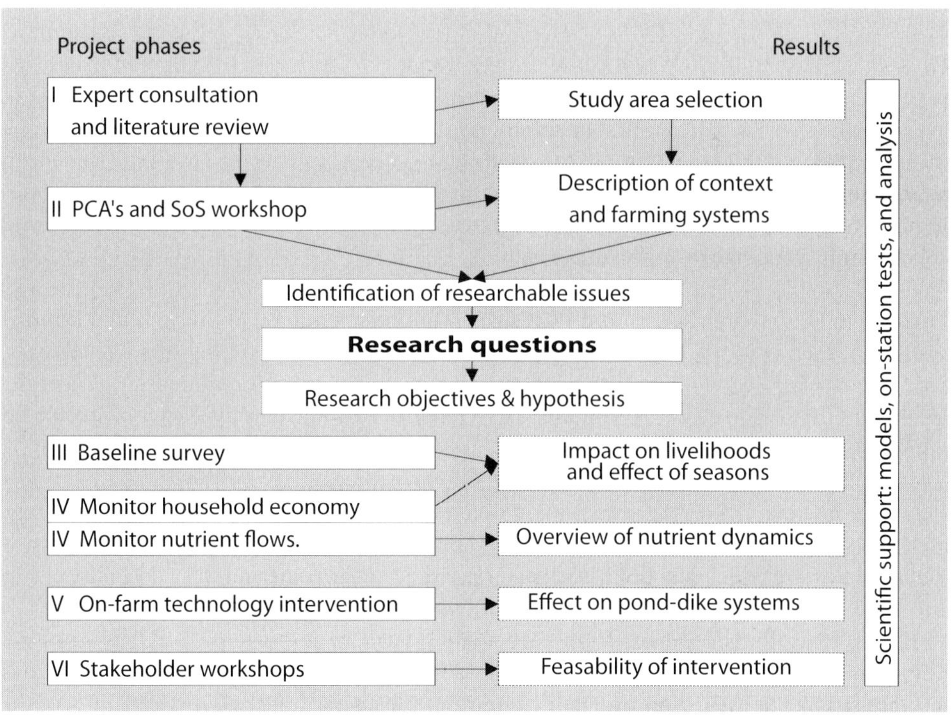

Figure 1.1. The POND research for development approach.

ensured that existing information on pond-dyke systems in the study areas were utilised, and that research results were placed in a wider conceptual framework.

The diagnostic studies and the on-farm experiments were integrated into one participatory approach of research for development. This participatory approach involving farmers, researchers, extension agents and policy makers should produce not only the new technology, but also institutional linkages and technical support that are better geared towards farmers' needs. The field studies were carried out in six phases.

Phase one: Secondary data were collected from literature, experts, local resource persons, and national policymakers. These data allowed the selection of three research areas with integrated pond-dyke systems.

Phase two: A Participatory Community Appraisal (PCA) was carried out in randomly selected villages of the research areas to analyse general context of the community, well-being, characteristics of farming systems, existing agricultural practices and constraints. The results of the PCA were presented in the State of the System (SOS) workshop and cross–checked by the participants. The SOS workshop involved four stakeholder groups: GOs, NGOs, fish traders and fish growers. The purpose of the SOS workshop was to:
1. get feedback from all stakeholders on findings of the PCA exercises;
2. identify problems and constraints of IAAS;
3. determine key research and development issues.

Phase three: During the SOS workshop it was felt that PCA provided information in relation to the community but did not provide in-depth information about the individual households' livelihoods, as well as farming systems. To develop a broader understanding of livelihoods at the household level and mapping out the pond-dyke systems baseline, surveys were carried out at farm household level. Most baseline surveys sampled proportionally among three categories of farmers: poor, intermediate and well-off ones, in three villages in each research area and collected a broad range of approximate data.

Phase four: The baseline survey and PCA helped to understand the livelihoods and farming systems in each country, but did not provide enough information on the seasonal changes. So, the household economics and pond nutrient dynamics were monitored longitudinally to identify the seasonality of livelihood characteristics and nutrient dynamics of integrated pond-dyke systems.

Phase five: The main objective of technological interventions was to assess the effect of change on the sustainability of pond-dyke systems and on the associated livelihoods. It was assumed that modification of existing practices could raise the farm income and the availability of fish and vegetables. During and after the on-farm trials, members of the project team organised workshops to collect feed-back from participating farmers and extension agents involved. They evaluated the final results in order to decide upon dissemination of research findings to other farmers in the region. The on-station research assessed the nutrient dynamics of the pond and the value of the pond sediment for agriculture on the dykes.

Phase six: The process of dissemination at local level was continuous in this approach, as extension officers and other farmers were involved in the process. For further dissemination the project organised policy workshops at district, provincial and regional level, and produced leaflets and other extension material. The context for promoting a technology is often problematic and complex. Therefore in some countries the institutional context was analysed to provide a broader illustration of the institutional constraints to adoption and adaptation of the technology.

The program consisted of several PhD projects (Figure 1.2), each formulated around two to three years of fieldwork focusing on monitoring, on-farm participatory trials or on-station experiments. Some PhD projects, funded through other sources, widened the overall context, broadening perceptions and insights. The monitoring and the participatory farm experiments with techniques for improving the nutrient efficiency of pond-dyke systems were carried out in Bangladesh, Thailand and Viet Nam. Extension workers and policy makers active in the three research sites were consulted and kept up-to-date on the research to ensure the approach was well embedded within the institutional framework. In each country three sites were selected for the monitoring and the on-farm trials. During the SOS-workshops the stake-holders identified the issues for the on-farm research.

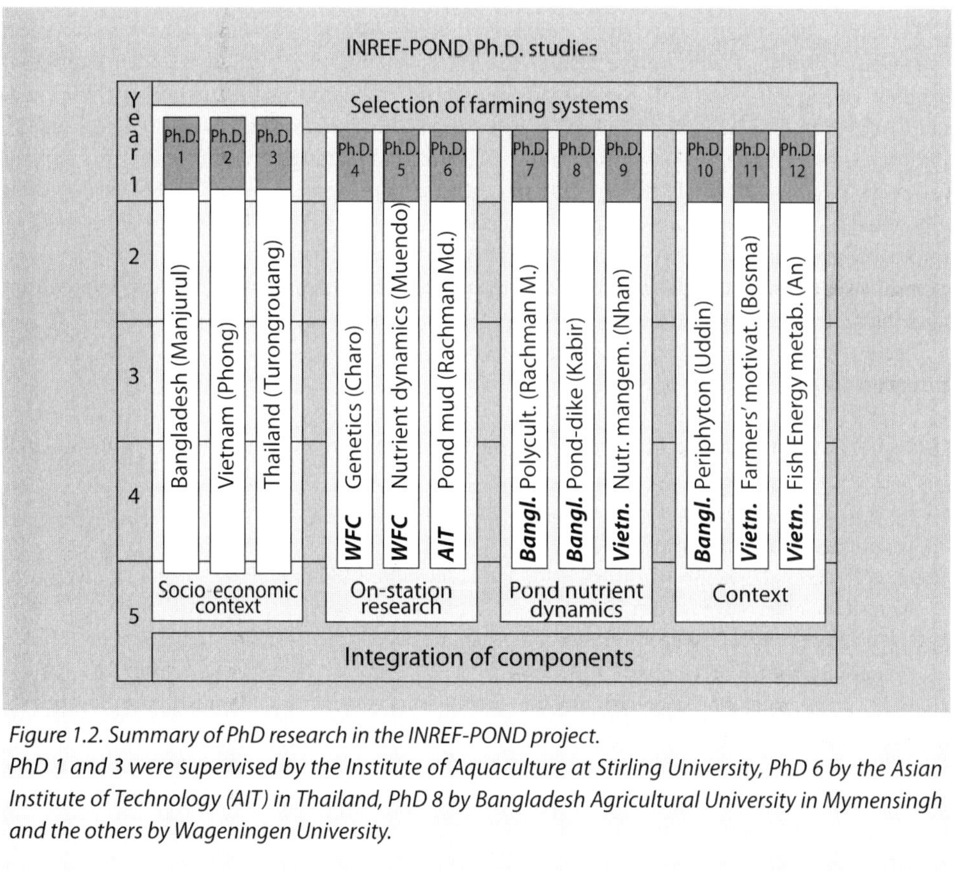

Figure 1.2. Summary of PhD research in the INREF-POND project.
PhD 1 and 3 were supervised by the Institute of Aquaculture at Stirling University, PhD 6 by the Asian Institute of Technology (AIT) in Thailand, PhD 8 by Bangladesh Agricultural University in Mymensingh and the others by Wageningen University.

Contents of the PhD studies

The on-station research at fish level focused on the genetic improvement of tilapia for aquaculture. It aimed to create and evaluate new tilapia breeds specifically selected to perform in high or low input environments. At the pond level, the research aimed at optimising the role of the pond in nutrient trapping and fish production to increase nutrient use efficiency in mixed farming system. Concurrently and subsequently the fertilising value of pond sediment for crop production (maize, morning glory, red pepper, and fruit trees) was investigated to assess the role of the fishpond in nutrient dynamics of IAAS. Different fish culture environments, i.e., low and high-input environments, were characterised in terms of nutrient use efficiency and nutrient accumulation in pond sediments. In addition the effect of polyculture on fish production and pond nutrient dynamics was studied to identify management options. Further, the possibilities to assess sustainability, to fine-tune the components of IAAS to reduce the nutrient flows to the environmental, and to extend research results to sub-Sahara Africa were investigated.

Agriculture development may affect the exploitation of natural resources on which the poorer groups in society depend for their livelihoods. Without sufficiently developed market infrastructure and supporting institutions, technological innovations are hardly adopted by farmers. These factors have to be considered when planning development (e.g. Udo and Cornelissen, 1998). Therefore the research results of POND were discussed with a range of experts. In the following chapters we present the POND results and the conclusions from the discussions during a symposium. In Chapter 2 the breeding of tilapia is discussed, in Chapter 3 the pond ecology and in Chapter 4 the farm nutrient balance. The contributions in Chapter 5 discuss sustainability aspects, present livelihood contributions, and places IAAS in the global market context. In Chapter 6 research and extension approaches are discussed, as well as the prospects for the contribution of IAAS to poverty alleviation. Finally, Chapter 7 is a general discussion on the global development perspective of integrated agriculture aquaculture farming systems.

References

Devendra, C. and D., Thomas, 2002. Smallholder farming systems in Asia. Agricultural Systems 71, 17-25.

Edwards, P., 1998. A system approach for the promotion of integrated aquaculture. Aquaculture, Economics and Management 2, 1-12.

FAO, 2004. The State Of the World Fisheries and Aquaculture (SOFIA). FAO Corporate Document Repository. http://www.fao.org

FAO, 2006. State of world aquaculture 2006. FAO Fisheries Department. FAO Fisheries Technical Paper. No. 500. Rome, FAO. 134 pp.

Prein, M., 2002. Integration of aquaculture into crop-animal systems in Asia. Agricultural Systems 71, 127-146.

Rosegrant, M.W., M.S. Paisner, S. Meijer and J. Witcover, 2001. 2020 Global Food Outlook: Trends, Alternatives, and Choices, International Food Policy Research Institute Washington D.C.

Udo H.M.J. and A.M.G. Cornelissen, 1998. Livestock in resource-poor farming systems. Outlook on Agriculture 27, 237-242.

Chapter 2

Tilapia breeding in low-input systems

Selection of fish for integrated agriculture-aquaculture systems

Hans Komen and Henk Bovenhuis
Animal Breeding & Genetics, Animal Sciences Group, Wageningen University, The Netherlands; Hans.Komen@wur.nl

Abstract

On-farm reproduction of tilapia reduces the growth potential. In low-input systems, the on-farm production of tilapia from on-station breeding schemes feeding protein-rich pellets might not be optimal due to genotype-environment interactions. In this paper we describe experiments to test this hypothesis; the results will be presented in the next chapters of this volume.

Keywords: reproduction, tilapia, inbreeding, GxE

Introduction

In parts of Asia, traditional integrated systems with fish and pigs or poultry are used to sustain the farm household. These systems are characterised by low external inputs and often rely on indigenous strains of fish, pigs and poultry. As farming systems intensify, the relative importance of the various subsystems change. Depending on market demand, farmers may focus on one or two products and use genetically improved strains that yield the highest profits. Newly introduced genetically improved animals require high-quality feeds to perform optimally, which leads to an increase in production costs and more imports of nutrients into the system and possibly discharge of nutrients into the environment (Wong, 1999). Introductions of productive livestock breeds from temperate climates to tropical countries are often unsuccessful because these strains can adapt insufficiently to the new culture environment, or are unsustainable because the technology needed to maintain high production levels (e.g. artificial insemination for crossbreeding, high quality feeds) cannot be afforded by farmers (Udo, 1982; Udo and Cornelissen, 1998; Kosgey *et al.*, 2006). The key question then, is whether one should produce one improved strain that serves all these different farming systems, or specialised strains for each particular environment. One strategy would be to select strains under local, sub-optimal farming conditions, especially when such strains prove to be more productive than commercially available strains which have been selected for growth under optimal conditions. Little is known about the consequences of these different selection schemes in farm animals or fish.

In aquaculture, the traditional way of increasing nutrient utilisation in ponds has been the polyculture of fish species that each utilises a different part of the pond food web. More recently, attempts have been made to apply selection programmes to fish for better performance in aquaculture systems (e.g. Eknath *et al.* 1993). However, integrated farming environments are highly variable, with a wide array of resources used as pond inputs, and ranging from low to high external nutrient input. It is unclear how selection for improved breeds and the nutrient management in these farms affect each other. Genetic selection for improved production characteristics is typically done in optimal environments where growth is maximal. However, such improved strains are not necessarily adapted to the conditions prevailing on many small-scale integrated farms. A fish strain selected for optimal performance with high-quality pelleted feeds is not necessarily good at converting the natural fish food produced in waste-fed ponds.

Genotype by environment interaction

The difference between measurements of a genotype or a population in two environments (e.g. optimal and sub-optimal) is called environmental sensitivity (represented by the slope of the lines in Figure 2.1). When genotypes differ in their sensitivity or 'reaction norm' to different environmental influences, a genotype by environment interaction (GxE) is said to occur (Falconer, 1990). There are several examples of genotype-environment interactions in pigs, poultry and cattle (Merks, 1988; Bourdon, 2000). In the presence of GxE, the genetic improvement obtained by selection in one environment may not be realised in other, different rearing environments. Using experimental evidence and mathematical models, Jinks and Connolly (1973, 1975) described the relationship between the selection environment and environmental sensitivity. They concluded that environmental sensitivity was reduced if selection pressure and environment effects were acting in opposite directions (antagonistic), and that sensitivity was increased if selection pressure and environment effects were acting in the same direction (synergistic). Falconer (1990) reviewed a number of experiments and

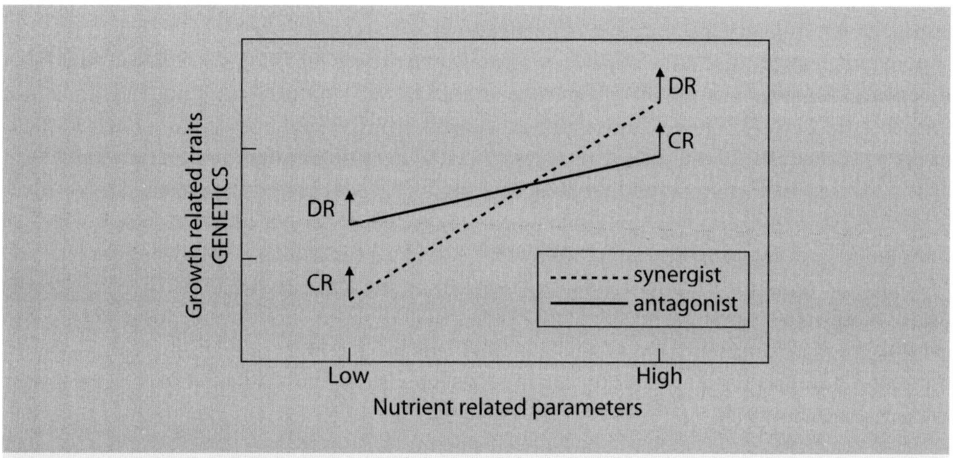

Figure 2.1. Environmental sensitivity in genetic selection.
DR = direct response to selection, CR = correlated response to selection. The slope of the lines is a measure of the difference between genotype performance in two environments (e.g., low and high nutrient input) and is called environmental sensitivity. Sensitivity is reduced when selection is done in the low environment ('antagonist') compared to selection in a high environment ('synergist'). The mean performance for a trait is better when selection is done in a low environment. For further explanation, see text.

concluded that to increase the mean performance of a genotype over a range of environments, selection should be done upwards in the less favourable environment.

According to FAO statistics, about 16 tilapiine species have been used for aquaculture production out of which ten species are commercially cultured. Since 1984, global tilapia production has been dominated by three species: the Nile tilapia (*Oreochromis niloticus* L.), the Mozambique tilapia (*Oreochromis mossambicus* Peters) and the blue tilapia (*Oreochromis aureus* Steindachner) (Rana, 1997). Nile tilapia is by far the most important cultured fish species among the tilapiines (El-Sayed, 2006). Due to the high cost of supplementary feeds, poor farmers, especially in sub-Saharan Africa, either grow Nile tilapia using organic fertilisation alone or with a wide variety of locally available farm resources (Liti *et al.*, 2005) making its farming conditions quite diverse. A number of selective breeding programmes to improve the growth of Nile tilapia have been initiated over the last 10-20 years (e.g. Hulata *et al.*, 1986; Eknath *et al.*, 1993, 1998; Bentsen *et al.*, 1998). These programmes have typically been carried out in relatively favourable environments receiving supplementary feed. One experiment with eight unselected strains of Nile tilapia in different culture environments suggested that there was little genotype by environment interaction (Bentsen *et al.* 1998). However, there are also reports that the gains of selection were lost when selected breeds were tested in less favourable environments (Macaranas *et al.*, 1997).

Experimental research

Given its importance for farmers in developing countries and the wide range of culture environments encountered in tilapia aquaculture, we decided to use Nile tilapia for a selection experiment in a low input environment. Two PhD projects were conceived to investigate in more detail the issues of genotype by environment interaction. The experimental work was done in collaboration with the World Fish Centre Regional Research Center for Africa and West Asia in Abbassa, Egypt (WFC-Egypt).

Fish for the selection experiment were obtained from a synthetic base population formed by the mating of four different strains of Nile tilapia from different parts of Egypt. The idea was to select fish from this base population divergently in two environments differing in source and input levels: (1) a favourable environment in which fish were fed 25% protein pellets by demand feeders and (2) a less favourable environment in which fish were fed on natural pond food induced by application of chicken manure (50 kg dry matter /ha/day). Heritability and response to selection for tilapia growth in low-input earthen ponds were estimated, and the growth performance and genetic correlation of selected lines in low and high-input environments were evaluated. The effect of selection for growth in low input ponds on survival, early development and body composition traits was measured. After two generations of selection, the fish were tested in five environments chosen to cover a wide range of the environments used in Africa and other resource poor regions. The following two chapters report on the results of these experiments.

References

Bentsen, H.B., A.E. Eknath, M.S. Palada-de Vera, J.C. Danting, H.L. Bolivar, R.A. Reyes, E.E. Dionisio, F.M. Longalong, A.V. Circa, M.M. Tayamen and B. Gjerde, 1998. Genetic improvement of farmed tilapias: growth performance in a complete diallel cross experiment with eight strains of *Oreochromis niloticus*. Aquaculture 160, 145-173.

Bourdon, R.M., 2000. Understanding animal breeding, 2nd Edition. Prentice Hall, Upper Saddle River, New Jersey, USA. 538 pp.

Eknath, A.E., M.M. Tayamen, M.S. Palada-de Vera, J.C. Danting, R.A. Reyes, E.E. Dionisio, J.B. Capili, H.L. Bolivar, T.A. Abella, A.V. Circa, H.B. Bentsen, B. Gjerde, T. Gjedrem and R.S.V.P. Pullin, 1993. Genetic improvement of farmed tilapias: growth performance of eight strains of *Oreochromis niloticus* tested in different farm environments. Aquaculture 111, 171-188.

Eknath, A.E., M.M. Dey, M. Rye, B. Gjerde, T.A. Abella, R. Sevilleja, M.M. Tayamen, R.A. Reyes and H.B. Bentsen, 1998. Selective breeding of Nile tilapia for Asia. In Proc. 6th World Cong. Genet. Appl. Livest. Prod. 27, 89-96.

El-Sayed, A.M., 2006. Tilapia culture. CABI publishing, Oxford, UK

Falconer, D.S., 1990. Selection in different environments. Genetical Research 56, 57-70.

Hulata, G., G.W. Wohlfarth and A. Halevy, 1986. Mass selection for growth rate in the Nile tilapia (*Oreochromis niloticus*). Aquaculture 57, 177-184.

Jinks, J.L., V. Connolly, 1973. Selection for specific and general response to environmental differences. Heredity 30, 33-40.

Jinks, J.L., V. Connolly, 1975. Determination of the environmental sensitivity of selection lines by the selection environment. Heredity 34, 401-406.

Liti, D., L. Cherop, J. Munguti and L. Chhorn, 2005. Growth and economic performance of Nile tilapia (*Oreochromis niloticus* L.) fed on two formulated diets and two locally available feeds in fertilized ponds. Aquac. Res. 36, 746-752.

Kosgey, I.S., R.L. Baker, H.M.J. Udo and J.A.M. van Arendonk, 2006. Successes and failures of small ruminant breeding programmes in the tropics: a review. Small ruminant research 61: 13-28.

Macaranas, J.M., P.B. Mather, S.N. Lal, T. Vereivalu, M. Lagibalavu and M.F. Capra, 1997. Genotype and environment: A comparative evaluation of four tilapia stocks in Fiji. Aquaculture 150, 11-24.

Merks, J.W.M., 1988. Genotype × environment interactions in pig breeding programmes. PhD-thesis, Wageningen University. 145 pp.

Rana, K.J., 1997. Status of global production and production trends. FAO Fish. Circ. No. 886. FAO, Rome.

Udo, H.M.J., 1982. Fattening of Large White, local Samoan and crossbred pigs under different management conditions. Tropical Animal Health and Production 18, 76-80.

Udo, H.M.J. and A.M.G. Cornelissen, 1998. Livestock in resource-poor farming systems. Outlook on Agriculture 27, 237-242.

Wong, A.C.Y., 1999. New developments in integrated dike-pond agriculture-aquaculture in the Zhujiang Delta, China: ecological implications. Ambio 28, 529-533.

Selecting Nile tilapia (*Oreochromis niloticus*) for growth in low-input environments

Harrison Charo-Karisa[1], Henk Bovenhuis[1], Mahmoud Rezk[2] and Hans Komen[1]
[1]Animal Breeding & Genetics, Animal Sciences Group, Wageningen University, The Netherlands; Hans.Komen@wur.nl
[2]Regional Center for Africa and West Asia, The World Fish Center, Abbassa, Egypt

Abstract

This chapter presents the results from a selective breeding experiment with Nile tilapia in a low-input pond environment. The environment was defined as low input as the pond only received chicken manure for fertilisation. No pellets were given to the fish during larval rearing or grow-out. The results show that good growth can be obtained under these conditions. Heritability for growth was high (0.6) and harvest body weight almost doubled over two generations of selection. A conclusion is that selecting Nile tilapia in low-input environments is cheap and sustainable and can be particularly relevant for resource-poor farmers in sub-Saharan Africa.

Keywords: breeding, tilapia, heritability, GxE, plasticity

Introduction

Farmed fish are produced in environments that vary widely depending on the culture site, the culture system and intensity, the levels of management applied, or in the combination of culture species used. To ensure the availability of consistently high quality genetic material for these wide-ranging environments (Little, 2004), it is crucial to design a good breeding scheme and to choose the best environmental conditions for selection. Selection for improved growth is typically done under favourable conditions where fish are fed protein-rich pelleted feed. This results in breeds that require more expensive feed in order to perform well, thus increasing the cost of production (Moav *et al.*, 1976). However, in resource-poor regions, Nile tilapia is typically reared in earthen ponds fertilised daily with organic fertilisers from livestock, occasionally supplemented with bran or leftover food. It is not clear whether fish selected under favourable conditions will perform equally well under low input conditions. In this chapter we will address this question and report on parameters that can be used to form guidelines for a breeding programme intended to produce improved broodstock and fish seed for resource-poor fish farmers. We will also discuss the prospects and requirements for establishing a practical breeding scheme for the small-scale farmer in resource-poor regions of the world, with a focus on Africa. We will begin by briefly reviewing the theory behind selecting for a wide range of low-input farming conditions, with a focus on phenotypic plasticity and the challenges associated with selecting for growth in low-input conditions.

Phenotypic plasticity

The ability of a single genotype to produce an array of phenotypes (or reaction norm) when exposed to a set of environments that differ in one or more aspects is referred to as phenotypic plasticity (Schlichting, 1986; Stearns, 1989). The benefit of plasticity lies in the ability to produce a better phenotype–environment match across many environments (Levins, 1968). Although plasticity cannot be measured on a single individual, there is considerable evidence that it is possible to select for plasticity (Scheiner, 2002). When an organism is faced with an environment which is less favourable in some aspects, trait values often change, either due to adaptive plasticity or to maladaptive stress response (Huey and Kingsolver, 1989; Newman, 1992). These responses are expressed in alternative forms of morphology, behaviour or physiological states (West-Ebehard, 1989). While it is generally agreed that natural selection should favour an organism that can produce an optimal phenotype in every environment it encounters, in reality, organisms fail to respond optimally in every environment because of genotype-by- environment interactions and evolutionary constraints or trade-offs (Gomulkiewicz and Kirkpatrick, 1992; Stearns, 1992).

The theoretical framework for the work described below was discussed in chapter 2.1, in which the Jinks–Connolly rule was introduced. This rule predicts that breeds selected under adverse (antagonistic or less favourable) conditions will be less sensitive to changes in the environment than breeds selected under favourable (agonistic) conditions. The Jinks–Connolly rule therefore predicts why plasticity would differ among different genotypes.

In the remainder of this section we will describe experiments designed to test this theory. The starting point was to select two lines from a single population of Nile tilapia in high- and low-input conditions for two generations. The high-input environment was defined as ponds in which fish were fed with 25% fish protein in the form of pellets. The low-input environment was defined as ponds fertilised only with chicken manure. The inputs in terms of available nitrogen were comparable in both systems (0.4–0.5 kg N/ha/day; Charo-Karisa *et al.*, 2006b). The only way the two environments differed was in the availability, form and digestibility of the protein source, and in differences in the pond environment as a result of the treatments.

Before this study, no information on the breeding of fish, nursing of fry, or the expected response to growth in low-input conditions was available in the literature. We therefore began by determining whether it was even possible to grow tilapia fry successfully in such low-input conditions.

Fry rearing in low- and high-input environments

In a pilot study (Charo-Karisa *et al.*, 2006a), we reared swim-up fry from 25 full-sib families separately for 42 days at 15 fry/m^2 density in 2 x 3 m hapas suspended in two earthen ponds The ponds were fertilised daily with 50 kg/ha/day chicken manure. In addition, fry in one row in each pond were fed twice daily with a 40% protein pelleted feed.

Surprisingly, dietary treatment effect was not significant, but there were large differences in growth between ponds: mean body weight at 42 days was 1.7 g in one pond and 0.4 g in the other. The heritability (h^2) of 42-day body weight estimated from the whole data set using a univariate model was 0.01±0.06. The bivariate heritability estimates were 0.59±0.19 in the first pond and 0.05±0.11 in the second pond. The common environmental/hapa (c^2) effects were estimated at 0.14±0.06 and 0.29±0.07 in the two respective ponds. Analysis of environmental variables showed that the two ponds differed significantly in dissolved oxygen. This observation combined with the low genetic correlation (r_g = -0.27±0.69) between body weights of fry in both ponds suggests the existence of genotype-by-environment interaction for tolerance to low dissolved oxygen in Nile tilapia juveniles.

Low-input environments are expected to result in slow growth rates. However, the results from this experiment show that the growth rates achieved in low-input conditions can be similar to growth rates obtained with pellet feeding. In other words, Nile tilapia juveniles can be grown at the same rate with or without supplementary feeds, provided that the pond in which they are growing is properly fertilised (Charo-Karisa *et al.*, 2006a). Although the rearing environment has a great influence on fish growth, this is mainly caused by factors not directly related to the type of inputs used.

Selecting Nile tilapia for growth in a low-input environment

The founder population (i.e. parents of the G_0 population) for the selection experiment was produced in spring of 2000 in a full diallel mating design among local Egyptian strains: Maryout, Zawia, Abbasa and Aswan (Rezk *et al.*, 2002, 2004). Eighty sires and 105 dams selected at random from the founder stock were used to produce the G_0. Each sire was mated to two dams and each dam mated to only one sire, thus generating full- and half-sib groups. For G_1, 50 sires and 87 dams were used, while for generation G_2, 54 sires and 104 dams were used. The females were removed when swim-up fry were sighted in the spawning hapas. Fry were given no supplementary feeds and were reared in the spawning hapas suspended in the earthen ponds until stocking. As soon as a family reached suitable tagging size, 24 fry chosen randomly fry from each full-sib family were individually tagged with Floy® tags and stocked in two 1000 m^2 fertilised earthen ponds for grow-out. Ponds were supplied daily with dry chicken manure from layer and broiler farms at the rate of 50 kg dm/ha, which corresponds to 0.4 kg nitrogen ha^{-1} day^{-1} and is enough to support yields of 4.3 kg fish BW ha^{-1} day^{-1} (calculated from Knud-Hansen *et al.*, 1991). After approximately 8 months of growth, which included 3–4 months of hibernation, fish were harvested by seine netting and breeding values were estimated using an animal model. Brooders were selected based on the rank of Estimated Breeding Values, separately for males and females. To minimise inbreeding, the number of individuals from a single full-sib or half-sib group was restricted; as a result, fish that were not ranked among the best may have been included as brooders.

The results of two generations of selection are presented in Table 2.1. In total, 6429 fully pedigreed experimental fish were included in the analysis. Survival to harvest was highly variable, ranging from 35 % to 77%, and was affected by initial weight, pond, and age effects. BW at harvest increased from a mean of 67.4 g in the founder population (G_0) to 129.5 g in G_2.

Table 2.1. Results from the selection experiments on Nile tilapia in 2 ponds over 3 generations (G_0, G_1, and G_2 in three consecutive years) with number of fish stocked, survival and tag loss in each pond, mean initial weight and mean body weight at harvest (BW), with standard deviations (SD) and coefficients of variation (CV), and actual final yield (RFY) and expected final yields (EFY).

	Stocking				Harvest						
		Initial weight (g)			Survival (%)	Tag loss (%)	BW (g)			Yield (kg/ha)	
	n	Mean	SD	CV			Mean	SD	CV	RFY	EFYφ
G_0 2002	1058	2.2	1.9	84	72	11	72	39	54	529	693
	1064	2.3	1.8	79	53	3	63	36	56	345	609
G_1 2003	1058	4.3	1.8	41	70	7	70	24	34	486	657
	1043	3.7	1.5	41	59	3	104	33	31	618	1007
G_2 2004	1101	4.3	3.1	72	77	2	128	42	33	1042	1238
	1105	4.5	3.8	84	35	2	131	40	30	493	1265

Adapted from Charo-Karisa *et al.*, 2006b.
φ Calculated from the difference between BW at harvest and initial body weight, assuming a stocking density of 1 fish/ m2 with no mortality.

Growth was significantly affected by initial weight, pond, sex and age effects. Generations were discrete and therefore genetic parameters were estimated separately for each year. Heritability estimates for BW ranged from 0.38 to 0.60, and the heritability for survival ranged from 0.03 to 0.14. The estimated selection response was 23.4 g between G_0 and G_1 and 13.0 g between G_1 and G_2 (Charo-Karisa *et al.*, 2006b).

Correlated responses

One of the major problems leading to reduction in growth rate in tilapia cultured in ponds is the tendency of females to mature and reproduce earlier and at small size. Consequently, it has been recommended that faster growth rate and delayed age at maturation both be included in the breeding goal for Nile tilapia (Bolivar *et al.*, 1993). Allocating resources to growth or reproduction is a fundamental trade-off in evolutionary life history and depends on environmental conditions (Stearns, 1992). Individuals in a population allocate their resources to somatic growth and reproduction depending on proximate environmental cues that are meant to confer fitness advantages in that particular environment. The use of life-history theory as a tool to predict reaction norms for age and size at maturity was originally suggested and developed by Stearns and his co-workers (Stearns, 1983; Stearns and Crandall, 1984, Stearns and Koella, 1986) as a deterministic process. Later, Heino *et al.* (2002) introduced the concept of reaction norms for age and size at maturation as a probabilistic process. These predictions sought to explain whether observed changes in age and size at maturity observed in organisms exposed to unavoidable growth stressors are adaptations that maximise fitness under given constraints. In sub-optimal or unpredictable environments it is expected that natural selection will favour alleles directing a greater part of the energetic resources toward early reproduction (Stearns, 1992).

In the present project we estimated phenotypic and genetic parameters for reproductive traits and gut length for Nile tilapia selected for growth in fertilised earthen ponds (Charo-Karrissa *et al.*, 2006c). The heritability estimate for maturity at harvest was 0.13, but we did not find evidence that selection in low-input environments resulted in earlier maturation. Although there was a tendency for females to mature earlier in later generations in this study, there was a low genetic correlation between growth performance and early maturity (Charo-Karisa *et al.*, 2006c). In contrast, Longalong *et al.* (1999) noted a high genetic correlation between early maturity and fast growth of Nile tilapia in an experiment carried out with high-protein fishmeal. In our study heritabilities for carcass traits of G_1 were low: 0.16 for gutted weight and 0.06 for dressing percentage. Phenotypic correlation between body weight and gutted weight was 0.84 and the genetic correlation was 0.20. Gut length increased with selection for body weight. The heritability estimate for gut length was 0.22. Moreover, gut length and body weight were genetically highly correlated. These results indicate that selection for growth in manure fed ponds will lead to specific adaptations of the digestive system.

Genotype-by- environment interaction

As explained in chapter 2.1 and our introduction, in the presence of genotype by environment interaction, genetic improvement obtained by selection in one environment may not be achieved in other environments. We tested the plasticity of the selected line by comparing the performance of the line selected in a low-input environment with a high line selected for growth in ponds where fish received 25% fish protein in the form of pellets (Charo-Karisa *et al.*, 2006c). The two lines originated from the same base population.

Fish were tested in five test environments: 40% protein pelleted feed (P200), 25% protein pellets (P100), 16% protein pellets (P50), 50 kg/ha chicken manure (M100) and 25 kg/ha chicken manure (M50). The test ponds were 200 m^2 and consisted of mud bottoms lined by concrete walls. The design was such that the nitrogen input was similar in the P50 and M50 environments and in the P100 and M100 treatments. The results are shown in Figure 2.2.

Survival from stocking to harvest ranged from 70% to 75 % in the high line and from 62% to 76% in the low line. Analyses revealed significant differences in growth performances of the two lines across test environments. The phenotypic mean body weight at harvest was highest for test environment P200 (123.4 g in the low line, and 131.7 g in the high line). The mean lowest phenotypic body weight was 92.1g (test environment M50) in the low line and 82.4g (test environment P100) in the high line. Although the high line performed better in more test environments, there was a significant line-by-test-environment interaction, indicating that both lines were sensitive to the environment. Family-by-test-environment interaction was significant only in the low line. This genotype–environment interaction was related to the response of the families to nitrogen and dissolved oxygen in the ponds (Charo-Karisa *et al.*, 2006d). It can be concluded from this study that from a genetic point of view there are no strong arguments against selecting in a resource-poor environment.

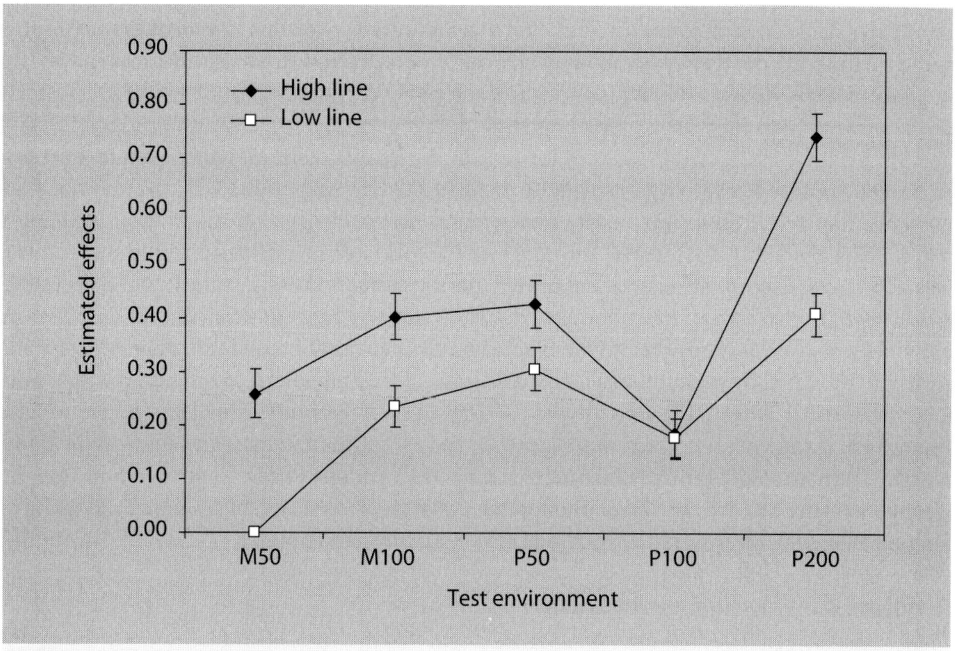

Figure 2.2. Results from the GxE experiment showing the estimated effects for line by test environment interaction for log body weight (standard errors indicated by error bars). See text for explanation of test environments. (Source: Charo-Karisa et al.*, 2006c).*

Establishment of breeding schemes for resource poor regions

Most global aquaculture production (approximately 90%) is currently in developing countries (Hishamunda and Ridler, 2002), with Africa contributing approximately 1% of the global production (Jamu and Brummet, 2004). Given the under-utilised water resources, the availability and inexpensiveness of labour, the high demand for fish, and the suitable climate, there is much potential for improving aquaculture in Africa. However, African aquaculture has had a long history of failure (Machena and Moehl, 2001), which has been blamed on lack of clear policy and technological know-how, inadequate infrastructure, poor extension support, lack of fish seed and feed, and poor financing (Hempel, 1993; Machena and Moehl, 2001).

About 95% of aquaculture production in Africa comes from small-scale farmers, with tilapia accounting for about 40% of total production. Small-scale production systems are extensive/semi-intensive utility-oriented pond systems (100–500 m^2) operated by households and integrated with other agricultural activities (Machena and Moehl, 2001). In highly productive agricultural areas where customary land inheritance has led to small plots of land, small-scale aquaculture is the only option left. Targeting small-scale systems for improved production is therefore synonymous with increasing total production of the aquaculture sector in Africa. Currently, these systems are characterised by low yields and small fish size at marketing. Given that tilapia prices in most cases depend on fish size and quality, farmed fish should meet acceptable standards (Adesulu, 2000; Afolabi *et al.*, 2000). Hence, fish farming must compete with other agricultural activities for labour, water, and farm inputs at the farm level, as well as with fish from capture fisheries in the market place.

A major set-back to aquaculture production in Africa has been the dependence on donor funding, which has made aquaculture unsustainable (Machena and Moehl, 2001). Kosgey *et al.* (2006) have reviewed several breeding programmes for small ruminants among resource poor-farmers in the tropics and highlighted issues determining their success or failure. Important in terms of causing failure are top–down approaches in setting breeding goals and insufficient involvement of farmers; the latter leads to inappropriate breeding goals being set (i.e. goals unsuitable for the target conditions) and to a lack of continuous monitoring of animals at the multiplier level. These issues may apply in selective breeding programmes for fish breeding as well. The management practices and breeding goals in resource-poor regions may differ from those in favourable environments. The selection strategy chosen will only be appropriate if it takes account of farmer needs, the production environments, and climatic conditions.

Environment of selection

Whereas the highly intensive culture practised in developed countries relies heavily on protein-rich pelleted supplemental feeds, in most developing countries there are limited protein sources. In fertilised ponds like those used in our study, the diet mainly consists of phytoplankton. This diet has limited protein, tends to reduce assimilation efficiency, and is generally less favourable to fish growth (Bowen *et al.*, 1995). Fertilising ponds with manure also promotes the development of benthos such as chironomids and oligochaetes (Friday, 1987; Kullberg and Peterson, 1987) which not only playing a significant role in the release of nutrients but are also eaten by tilapia (Egna *et al.*, 1997; Teichert-Coddington, *et al.*, 1997). The provision of substrates in the fertilised ponds can increase the growth of periphyton, thereby significantly improving fish yields (van Dam *et al.*, 2002); this indicates that there is scope for improving the productivity of low-input fertilised ponds.

Feed accounts for more than 50% of the operation costs of intensive tilapia culture (El-Sayed, 1998; Wu *et al.*, 1999). For nursing of fry, 40% protein pellets are used. Apart from being expensive, these high protein pellets are not readily available in resource-poor regions. This means that a hatchery that uses manure instead of pellets will spend less and thus produce cheaper seed and broodstock. Similarly, a farmer who uses manure in the grow-out period spends less than a farmer using pellets. To illustrate this we compare the input amounts used for a 200 m^2 pond for an environment receiving 50 kg/ha/day chicken manure with an environment where fish are fed 25% protein pellets. On average, a manured pond receives 1 kg chicken manure daily while a pellet-fed pond receives 0.25 kg pellets per day. Chicken manure in Egypt is sold at 250 EGP per ton or 0.25 EGP per kg, and 25% protein pellets are sold at 2 EGP per kg. For a grow-out period of 180 days, the farmer/breeder using chicken manure spends 45 EGP per pond while the farmer/breeder using pellets spends twice as much (90 EGP[1]).

[1] EGP = Egyptian Pound; 7 EGP = 1 €.

A two-pronged approach to improve tilapia yields at only a small extra cost would consist of simultaneously improving natural pond productivity and using fish selected for growth in this environment. Integrating fish farming with poultry or other livestock increases the opportunities for saving and is thus more cost-effective.

Delivering quality seed to farmers: a participatory approach to farmer involvement

A major setback to farming of tilapia in resource-poor regions has been the lack of quality fish seed (Machena and Moehl, 2001; Little *et al.*, 2001; Little, 2004). The production and dissemination of fish seed may be centralised or decentralised. In the centralised approach, the breeding programme is located in a central place and farmers get the fry through multiplier centres. The advantage of this approach is that genetic quality can easily be maintained. The disadvantage is that the rural communities remain dependent on a distant source which may become unreliable (Little and Hulata, 2000). In the decentralised approach, the juveniles are produced close to the grow-out areas, allowing farmers to have a greater control of the seed availability. Because of the problematic infrastructure in most of Africa, the decentralised approach seems a better option. It has the advantage of cutting transport costs and has been used successfully in Asia (Little, 2004). However, breeding programmes would also need to be decentralised. This is only possible if the breeding programmes are simple, cheap and small scale.

In establishing a sustainable breeding programme it is recommended to involve local fish farmers from inception of the breeding programme onwards. This could help tackle the perception that aquaculture is technically complex and difficult to handle (Machena and Moehl, 2001). A participatory approach (i.e. farmer involvement) will enable farmers to perceive the breeding programme as their own, and if aquaculture extension officers are involved, can simultaneously be used to train farmers in general aquaculture practices. By allowing farmers to define the breeding goals at the onset, the breeding programme ensures that the resultant breeds match local conditions and needs, thus ensuring programme sustainability. Ambali and Malekano (2004) have pointed out that selective breeding programmes are likely to be abandoned in Africa because they are expensive, and because it takes a long time to improve a strain through selective breeding. However, as we have shown, in our study there was a progressive increase in size of the fry produced across only two generations of selection (Charo-Karrissa *et al.*, 2006b). We have further demonstrated that a breeding programme does not necessarily have to be expensive. What is required is the commitment of farmers to participate. Nile tilapia farmers, who are accustomed to using unimproved seed, are likely to be encouraged to participate in the breeding program as they will use improved seed after every generation of selection.

In aquaculture, some traits of major economic importance include growth rate and meat yield, feed conversion efficiency, disease resistance and survival, meat quality, and age at sexual maturation (Gjedrem, 2000). In small-scale farms, the most important breeding goal is improved growth. In fertilised ponds, survival is one of the most important traits for improvement and is substantially heritable (Charo-Karisa *et al.*, 2006c). Shultz (1986) recommends minimising the number of traits to be improved, in order to minimise the costs of trait measurement and to keep the breeding programme focused. This may be particularly relevant in resource-poor regions. Therefore, body weight and survival should be the first target traits for improvement. Given the high genetic correlation between body weight and body measurements (Charo-Karisa *et al.*, 2006c), body measurements, which are easier and more convenient to measure, could be used as the traits of selection instead of body weight.

Conclusions

There are good prospects for setting up sustainable breeding programmes for resource-poor tilapia farming conditions without requiring expensive supplementary protein pellets. Because poverty alleviation and food security are primary goals in the developing world, the initiation and implementation of cheaper breeding programmes will ensure that the genetically improved material is accessible to the rural fish farmer. A high response can be achieved within a few generations in Nile tilapia, and because of the short generation times, fish farmers can reap the benefits of genetic improvement programmes without delay, thus reducing the risk of failure of the breeding programme. Appropriate breeding goals for these breeding schemes need to be set up with the involvement of the local farmers, preferably for each agro-ecological zone (but within the financial capabilities of the community), to ensure that the breed meets the requirement of local farmers. The establishment of multiplication centres where farmers can collect fry or a decentralised breeding programme could ensure that the improved fish seed reaches the farm when required. The full benefit of the breeding programme will be achieved by improving husbandry practices that enhance water quality and improve nutrition, such as the use of substrates on which periphtyon can grow. It is important to train researchers and aquaculture extension officers how to keep records, because although record-keeping is a tool for monitoring progress, it has not been common practice for fish farmers to keep proper records. Record-keeping is a necessary step for the maintenance of domesticated strains. The initiation and implementation of a breeding programme as described can still be expensive for the rural poor; however, donor dependence should be discouraged. Instead, the role of governments in formulating policies that make it easier for fish farmers to access credit should be enhanced. In this way, fish farmers can also learn how to manage their own finances or to manage corporate breeding schemes.

References

Adesulu, E.A., 2000. Current status of tilapia in Nigerian aquaculture. In: K. Fitzsimmons (ed.) Tilapia Aquaculture: Proceedings from the Fourth International Symposium on Tilapia in Aquaculture, Volume 2. North-eastern Regional Agricultural Engineering Service, Ithaca, New York, pp. 577-583.

Afolabi, J.A., P.B. Imoudu and A.O. Fagbenro, 2000. Economic and technical viability of tilapia culture in homestead concrete tanks in peri-urban Nigeria. In: K. Fitzsimmons (ed.) Tilapia Aquaculture: Proceedings from the Fourth International Symposium on Tilapia in Aquaculture, Volume 2. North-eastern Regional Agricultural Engineering Service, Ithaca, New York, pp. 575-581.

Ambali, A.J.D. and L.B. Malekano, 2004. Genetic improvement with specific reference to tilapia genetic resources in Africa and their use in aquaculture-potential benefits and risks. In: Gupta, M.V., Bartley, D.M., Acosta, B.O. (Eds.). Use of Genetically Improved and Alien Species for Aquaculture and Conservation of Aquatic Biodiversity in Africa. WorldFish Center Conference Proceedings 68, pp. 10-15.

Bentsen, H.B., A.E. Eknath, M.S. Palada de Vera, J.C. Danting, H.L. Bolivar, R.A. Reyes, E.E. Dionisio, F.M. Longalong, A.V. Circa, M.M. Tayamen and B. Gjerde, 1998. Genetic improvement of farmed tilapias: growth performance in a complete diallel cross experiment with eight strains of *Oreochromis niloticus*. Aquaculture 160, 145-173.

Bolivar, R.B., A.E. Eknath, H.L. Bolivar and T.A. Abella, 1993. Growth and reproduction of individually tagged Nile tilapia *Oreochromis niloticus*.of different strains. Aquaculture 111, 159–160.

Bowen, S.H., E.V. Lutz and M.O. Ahlgren, 1995. Dietary protein and energy as determinants of food quality:trophic strategies compared. Ecology 76, 899-907.

Charo-Karisa, H., Komen, J., Reynolds, S., Rezk, M.A., Ponzoni R.W. and Bovenhuis, H., 2006a. Genetic and environmental factors affecting growth of Nile tilapia (*Oreochromis niloticus*) juveniles: modelling spatial correlations between hapas. Aquaculture 255, 586-596.

Charo-Karisa, H., Komen, J., Rezk, M.A., Ponzoni R.W., van Arendonk, J.A.M., and Bovenhuis, H., 2006b. Heritability estimates and response to selection for growth of Nile tilapia (*Oreochromis niloticus*) in low-input earthen ponds. 2006. Aquaculture, in press.

Charo-Karisa, H., Bovenhuis, H., Rezk, M.A., Ponzoni R.W., van Arendonk, J.A.M., and Komen, J., 2006c. Phenotypic and genetic parameters for body measurements, reproductive traits and gut length of Nile tilapia (*Oreochromis niloticus*) selected for growth in low-input earthen ponds. In: Charo-Karisa, H., Selection for growth of Nile tilapia (*Oreochromis niloticus* L.) in low-input environments. PH-D thesis, Wageningen University, 2006. pp. 91-107.

Charo-Karisa, H., H. Bovenhuis, M.A. Rezk, R.W. Ponzoni, J.A.M. van Arendonk and H. Komen, 2006d. Genotype by environment interaction in two lines of Nile tilapia (*Oreochromis niloticus* L.) divergently selected for growth in different pond environments. In: Charo-Karisa, H., Selection for growth of Nile tilapia (*Oreochromis niloticus* L.) in low-input environments. PhD thesis, Wageningen University, 2006, pp. 109-128.

Egna, H.S., Boyd, C. E. and Burke, D.A., 1997. Introduction. In: Egna, H.S. and Boyd, C.E (Eds.), Dynamics of Pond Aquaculture. CKRC Press, Boca Raton, New York., pp. 1-14.

El-Sayed, A-F.M., 1998. Total replacement of fish meal with animal protein resources in Nile tilapia, *Oreochromis niloticus* (L.), feeds. Aquac. Res. 29, 275-280.

Friday, L.E., 1987. The diversity of microinvertebrate and macrophyte community in ponds. Freshwater Biology 18, 87-104.

Gjedrem, T., 2000. Genetic improvement of cold-water species. Aquac. Res. 31, 25-33.

Gomulkiewicz, R. and M. Kirkpatrick, 1992. Quantitative genetics and the evolution of reaction norms. Evolution 46, 390-411.

Heino, M., U. Dieckmann and O.R. Godø, 2002. Measuring probabilistic reaction norms for age and size at maturation. Evolution 56, 669–678.

Hempel, E., 1993. Constraints and possibilities for developing aquaculture. Aquaculture International 1, 2-19.

Hishamunda, N. and N.B. Ridler, 2002. Macro policies to promote sustainable commercial aquaculture. Aquaculture International 10, 491-505.

Huang, C.M. and L.C. Liao, 1990. Response to mass selection for growth rate in *Oreochromis niloticus*. Aquaculture 85, 199-205.

Huey, R.B. and J. Kingsolver, 1989. Evolution of thermal sensitivity of ectotherm performance. Trends in Ecology and Evolution 4, 131–135.

Jamu, D. and R. Brummett, 2004. Opportunities and Challenges for African Aquaculture. In: Gupta, M.V., Bartley, D.M. and Acosta, B.O. (Eds.), Use of Genetically Improved and Alien Species for Aquaculture and Conservation of Aquatic Biodiversity in Africa. WorldFish Center Conference Proceedings 68, pp. 1-9.

Knud-Hansen, C.F., T.R. Batterson, C.D. McNabb, I.S. Harahat, K. Sumantadinata and H.M. Eidman, 1991. Nitrogen input, primary productivity and fish yield in fertilized freshwater ponds in Indonesia. Aquaculture 94, 49-63.

Kosgey, I.S., R.L. Baker, H.M.J. Udo and J.A.M. Van Arendonk, 2006. Successes and failures of small ruminant breeding programmes in the tropics: a review. Small Ruminant Research 61, 13-28.

Kullberg, A. and R.C. Peterson, 1987. Dissolved organic carbon, seston and macroinvertebrate drift in an acidified and limed humic stream. Freshwater Biology, 17: 553-564.

Levins, R., 1968. Evolution in changing environments. Princeton University Press, Princeton, New Jersey, USA.

Little, D.C., 2004. Delivering better quality seed to farmers. In: Bolivar, R., Mair, G. and Fitzsimmons, K., (Eds.), New Dimensions in Farmed Tilapia. Proceedings of the 6th International Symposium on Tilapia in Aquaculture, Manila, pp. 3-17.

Little, D.C. and G. Hulata, 2000. Strategies for tilapia seed production. In: Beveridge, M.C.M. and McAndrew, B.J. (Eds.), Tilapias: Biology and Exploitation. Fish and Fisheries Series, vol. 25. Kluwer Academic Publishing, Dordrecht, pp. 267– 326.

Little, D.C., A. Satapornvanit and P. Edwards, 2001. Freshwater fish seed quality in Asia. In: Edwards, P., Little, D.C. and Demaine, H. (Eds.), Rural Aquaculture. CABI Publishing, Wallingford, UK., pp. 185-196.

Longalong, F.M., A.E. Eknath and H.B. Bentsen, 1999. Response to bi-directional selection for frequency of early maturing females in Nile tilapia (*Oreochromis niloticus*). Aquaculture 178, 13–25.

Machena, C. and J. Moehl, 2001. Sub-Saharan African aquaculture: regional summary. In: Subasinghe, R.P., Bueno, P., Phillips, M.J., Hough, C., McGladdery, S.E., Arthur, J.R., (Eds.), Aquaculture in the Third Millennium. Technical Proceedings of the Conference on Aquaculture in the Third Millennium, Bangkok, Thailand, 20-25 February 2000. NACA, Bangkok and FAO, Rome, pp. 341-355.

Moav, R., M. Soller, G. Hulata and G. Wohlfarth, 1976. Genetic aspects of the transition from traditional to modern fish farming. Theoretical and Applied Genetics 47, 285- 290.

Newman, R.A., 1992. Adaptive plasticity in amphibian metamorphosis. Bioscience 42, 671–678.

Rezk, M.A., E.A. Kamel, T.I. Dawood and R.A. Dunham, 2004. Heritability estimates for some important traits in Egyptian Nile tilapia, *Oreochromis niloticus*. In: Aquaculture-An Ecologically sustainable and Profitable venture, World Aquaculture Society Conference Proceedings, 1-5 March 2004, Honolulu, Hawaii, USA. Aquaculture 2004 Book of Abstracts, 708 pp.

Rezk, M.A., E.A. Kamel, A.A. Ramadan and R.A. Dunham, 2002. Comparative growth of Egyptian tilapias in response to declining water temperature. Aquaculture 207, 239-247.

Scheiner, S.M., 2002. Selection experiments and the study of phenotypic plasticity. J. Evol. Biol. 15, 889-898.

Schlichting, C.D.,1986. The evolution of phenotypic plasticity in plants. Ann. Rev. Ecol. Syst. 17, 667-693.

Shultz, F.T., 1986. Developing a commercial breeding program. Aquaculture 57, 65-76.

Stearns, S.C., 1983. A natural experiment in life-history evolution: Field data on the introduction of Mosquitofish (*Gambusia affinis*) to Hawaii. Evolution 37, 601-617.

Stearns, S.C., 1989. The evolutionary significance of phenotypic plasticity. Bioscience 39, 436–445.

Stearns, S.C., 1992. The Evolution of Life Histories. Oxford: Oxford University Press.

Stearns, S.C. and R.E. Crandall, 1984. Plasticity for age and size at sexual maturity: a life-history response to unavoidable stress. In: Potts, G. and Wootton, R.J. (Eds.), Fish Reproduction. Academic Press, London., pp.13-33.

Stearns, S.C. and J.C. Koella, 1986. The evolution of phenotypic plasticity in life-history traits: predictions of reaction norms for age and size at maturity. Evolution 40, 893-913.

Teichert-Coddington, D.R., T.J. Popma and L.L. Lovshin,1997. Attributes of tropical pond-cultured fish. In: Egna, H.S. and Boyd, C.E (Eds.), Dynamics of Pond Aquaculture. CKRC Press, Boca Raton, New York., pp. 183-197.

van Dam, A.A., M.C.M. Beveridge, M.E. Azim and M.C.J. Verdegem, 20002. The potential of fish production based on periphyton. Rev. Fish Biol. Fisher. 12, 1–31.

West-Eberhard, M.J., 1989. Phenotypic plasticity and the origins of diversity. Annual Review of Ecology and Systematics 20, 249–278.

Wu, Y.V., K.W. Tudor, P.B. Brown and R.R. Rosati, 1999. Substitution of plant proteins or meat and bone meal for fish meal in diets of Nile tilapia. North American Journal of Aquaculture 61, 58-63.

Mass spawning of *Oreochromis niloticus* for fry production: threats and opportunities for small-scale aquaculture and selective breeding

Yonas Fessehaye[1,2], Henk Bovenhuis[1], Mahmoud Rezk[3], Raul Ponzoni[3] and Hans Komen[1]
[1]Animal Breeding and Genetics Group, Wageningen University, P.O. Box 338, 6700 AH, Wageningen, The Netherlands
[2]Aquaculture and Fisheries group, Wageningen University, P.O. Box 338, 6700 AH, Wageningen, The Netherlands
[3]Regional Center for Africa and West Asia, The World Fish Center, Abbassa, Egypt

Abstract

Mass selection based on mass spawning can present a feasible and low-cost selective breeding scheme. In mass spawning, however, mating is not controlled and a small number of breeding individuals may account for a large proportion of the next generation, leading to higher inbreeding, decrease in performance of stocks and no response to selection. To understand the mating systems, reproductive and population parameters under mass spawning, we conducted mass spawnings with female to male sex ratios of 1:1, 2:1 and 3:1. We analysed parentage, established the effective number of parents contributing to the next generation and calculated reproductive parameters and the associated level of inbreeding in an experimental breeding unit. Considering the breeding units as basic building blocks, mass selection breeding programmes based on mass spawning can be implemented by extrapolating the number of breeding units depending on the size of the nucleus and the intensity of selection. Such schemes require minimum investments in infrastructure and can be adopted as a model for low-cost and effective selective breeding schemes that will produce the desired response without compromising the long-term fitness of culture populations due to inbreeding.

Keywords: *Oreochromis niloticus*, natural mating, mass selection, inbreeding

Introduction

An essential component of aquaculture breeding programmes is the controlled production of sufficient seeds of desired quality. In Nile tilapia, broodstock are often stocked in small and confined units such as aquaria, tanks and hapas at a sex ratio of 2-5 females to one male (Bautista *et al.*, 1988). Stocking two females and one male, for example, will enable controlled mating of selected parents and equal contribution of families; however, confinement in small units can also result in aggression and fighting between males, thereby reducing successful reproduction and seed production (Behrends *et al.*, 1993). Alternatively, fry can be produced in group (mass) spawning: typically achieved by stocking groups of males and females in ponds, tanks or in large hapas suspended in fertilised ponds at different female to male ratios (Little *et al.*, 1995). Mass spawning can be a cheaper option for mass selection breeding programme and multiplier units in the breeding plan to provide sufficient and superior fish for grow-out. However, breeding schemes using mass spawning can have two constraints: (1) Mate selection and mating of the animals is not controlled and hence a small number of breeding individuals may account for a large proportion of the subsequent generation; and (2) high fecundity in most fish allows the use of small number of parents and selection intensities are high. These two factors are generally recognised as the main sources of inbreeding in hatchery-produced fish populations (Bentsen and Gjerde, 1994; Gjedrem, 1998). High rates of inbreeding are known to affect survival, reproduction, growth, the expression of genetic defects and the level of genetic variability (Kincaid, 1983).

The mating system and other reproductive parameters may play a vital role in the genetic diversity of populations. Therefore, in order to develop a successful mass selection breeding scheme that employs mass spawning, it is crucial to thoroughly understand natural mating systems. In the past, the emphasis has been on increasing the quantity of fry produced rather than on the genetic quality of fry. In this paper we discuss the mating systems, reproductive and population parameters in mass spawnings of *O. niloticus* under semi-natural conditions and discuss the threats and opportunities of developing a mass selection breeding scheme that uses mass spawning.

Reproductive success, effective population size and inbreeding

Reproductive success and effective population size

In an idealised population, mating is random i.e. reproductively active individuals have an equal opportunity to mate and contribute equally to the gene pool. However, theory predicts that differences between the sexes in potential reproductive rate and a sex ratio biased strongly towards males should result in intense male competition, polygynous mating and high variance in male reproductive success (Kvarnemo and Ahnesjo, 1996). In order to determine the reproductive success of *O. niloticus* males, we conducted parentage analysis of mass spawning experiments at female to male sex ratios of 1:1, 2:1 and 3:1 (henceforth referred to as sex ratios 1, 2 and 3). The set-up of the spawning experiment is detailed in Fessehaye *et al.* (2006a, c). Briefly, the three female to male sex ratios studied were 18 females and 18 males (ratio 1), 25 females and 12 males (ratio 2) and 27 females and 9 males (ratio 3). The fish were stocked in an 8 x 2x 1 m^3 (L x W x H) hapa enclosure suspended in a pond. Broodstock were allowed to spawn naturally for ten days, after which fertilised eggs were collected from brooding females, transferred to artificial incubators and reared until the stage of yolk sac fry. Ten yolk sac fry per female from sex ratio 2 were genotyped for 11 microsatellite markers; for sex ratios 1 and 3, 20 yolk sac fry per female were genotyped for seven microsatellites. Pedigree was reconstructed by assigning offspring to their putative parents using the software PAPA 2.0 (Package for Analysis of Parental Allocation). Overall, 760 offspring (sex ratio 2: 10 offspring *76 spawns) and 1120 offspring (sex ratios 1 and 3: 20 offspring*56 spawns) were assigned parentage with high level of confidence. The relative contribution of males in siring offspring is depicted in Figure 2.3. In all three sex ratios there was clear evidence of male dominance in siring offspring. In sex ratio 2, one third of males (4 out of 12 males) sired 75% offspring

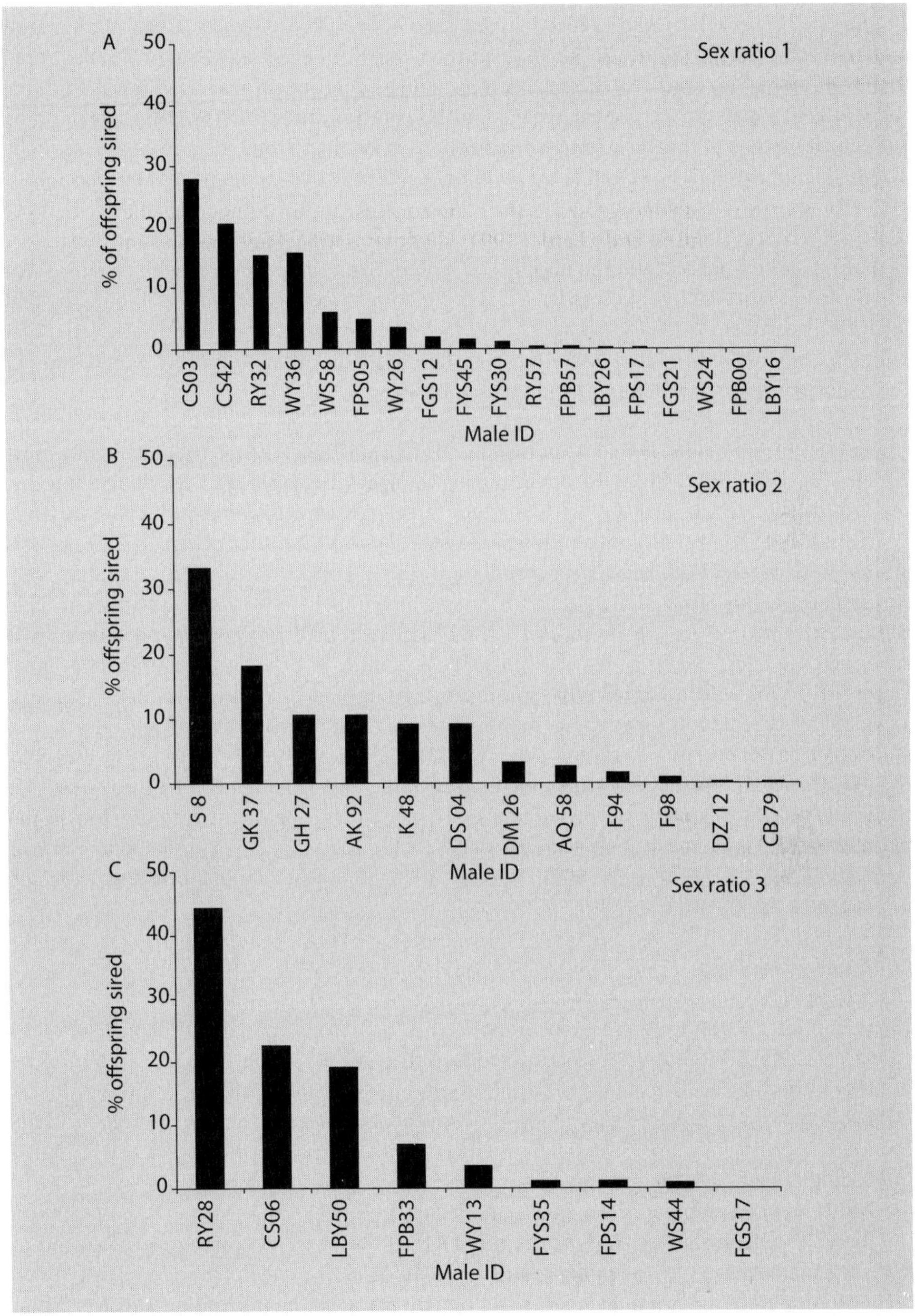

Figure 2.3. Percent of offspring sired by males at different female to male sex ratios in mass spawnings under semi-natural conditions (Fessehaye et al., *2006a, c).*

(Figure 2.3B) (Fessehaye *et al.*, 2006a). In sex ratio 1, one third of males (6 out of 18 males) sired 89% of offspring (Figure 2.3A). In sex ratio 3, one third of males (3 out of 9 males) sired 93% of offspring (Figure 2.3C) (Fessehaye *et al.*, 2006b). The operational sex ratio (the ratio of sexually active males to fertilisable females) in all sex ratios was biased towards males, as only few females spawn at the same time (asynchronous spawning) while most males are sexually active. Such male-biased sex ratios led to intense reproductive competition between males, with stronger male–male competition in sex ratio 1 compared to sex ratio 3 as there were more males in sex ratio 1. Reproductive success was more skewed in sex ratio 3 than in sex ratio 1 (Table 2.2), which agrees with the findings of Widemo and Owens (1995) who reported that the degree of reproductive success for a lekking mating system varies with demography and is inversely correlated with the number of competing males. In general, it can be inferred that Nile tilapia exhibit reproductive skew, with certain individual males dominating paternity in the majority of spawns produced over a spawning season.

In contrast to the males, the contribution of females was much more balanced, with most females (70–92%) spawning at least once during the spawning period. As shown in Table 2.2, male reproductive variance ($\sigma_{km}{}^2$) was much greater than female reproductive variance ($\sigma_{kf}{}^2$) indicating that male contribution was more skewed than that of females. There was a large disparity between the census population size (*N*) and the effective population size, N_e (Table 2.2). This reduced magnitude of N_e was attributable to the large variances in male reproductive success ($\sigma_{km}{}^2$), plus the small number of contributing males. Matocq (2004) reported that a decrease in N_e may be caused by a number of behavioural and demographic factors operating within populations. Specifically, strong skew in male reproductive success, as would be expected in polygynous mating systems, could cause a substantial decrease in N_e relative to the census population size. A N_e/N ratio of between 0.30 and 0.44 indicates that less than 50% of the parents are contributing to the next generation, which suggests that the population is experiencing higher rates of genetic drift, the drift being greatest in the population with a female to male sex ratio of 3:1 (Table 2.2). This low N_e equated to a rate of inbreeding of 3.0–4.4% per generation, which is 2–2.5 times the rate of inbreeding expected in an idealised population.

Effects of inbreeding

Harmful effects of inbreeding have been noticed in numerous animal and plant species. In an earlier study, inbreeding significantly reduced the phenotypic performance of a number of traits important in tilapia aquaculture (Fessehaye *et al.*, 2007, Table 2.3). Taking the ΔF estimate of sex ratio 3 (Table 2.2), the decline in phenotypic performance of different traits can be extrapolated for breeding schemes that employ natural mating at that sex ratio. An inbreeding rate of 4.4% per generation can cause fry survival to decline by 2.4% (or ≈ 4% of mean survival), the number of hatched fry to decline by 42 (11% of the mean), and fecundity to fall by 33 eggs per female (5% of the mean). Such levels of inbreeding can also lead to male reproductive success being depressed by as much as 18%, the depression being much higher under more intense male–male competition (sex ratio 1). Inbreeding affects traits at early stages; furthermore, inbred animals that survive to adulthood may still suffer reduced fitness in terms of poor performance in mating competition and reduced fecundity. The

Table 2.2. Census number (N), reproductive variance (σ^2_k), effective population size, predicted (PN_e) and idealised total population size (IN_e) and predicted ($P\Delta F$) and idealised rate of inbreeding ($I\Delta F$) for the mass spawnings with varying sex ratios. Adapted from Fessehaye et al. *(2006a,b).*

Year	N^1	♀/♂	σ^2_{km}	σ^2_{kf}	N_{em}	N_{ef}	$PN_e{}^2$	$IN_e{}^3$	$P\Delta F^4$	$I\Delta F^5$	N_e/N
2003	37	2:1	2,094	149	5.0	17.8	15.6	32	0.0309	0.0156	0.43
2003	37	2:1	725	86	5.4	15.8	16.1	32	0.0301	0.0156	0.45
2005	36	3:1	10,165	380	3.2	17.1	10.8	27	0.0437	0.0185	0.30
2005	36	1:1	1,309	460	5.5	10.2	14.3	36	0.0341	0.0139	0.40

N_{em} and N_{ef} are the effective numbers of males and females respectively; σ^2_{km} and σ^2_{kf} are reproductive variances for males and females respectively.
[1]Census size is the total number of males and females (*N*).
[2, 3]Both predicted and idealised N_e, N_{em} and N_{ef} were calculated following Lande and Barrowclough (1987). Predicted total N_e (PN_e), and idealised total N_e (IN_e) were calculated using the formula.
[4]For predicted ΔF the contribution of parents was obtained from parentage analysis.
[5]For idealised ΔF an equal contribution of parents was assumed, and using the fractional contribution of each parent (=c^2) and the average fractional contribution of females ($\bar{C}_f$) and males ($\bar{C}_m$), the ΔF was calculated after Brown *et al.* (2005):

$$\Delta F = \frac{1}{2}\sum_{Parents} c_i^2 - \frac{1}{4}(\bar{c}_m)^2 - \frac{1}{4}(\bar{c}_f)^2$$ (For details see Fessehaye *et al.*, 2006a).

Table 2.3. Mean ± SD of trait, regression coefficient of inbreeding (F %) and genetic depression in phenotypic performance per 4.4% increase in the level of inbreeding.

Trait	Mean ± SD	Regression coefficient (F%)	Inbreeding depression per 4.4% of F[1]
Fry survival (%)	63.6 ± 14.7[a]	-0.55	2.4
Initial weight (g)	2.1 ± 1.7[a]	-0.14	0.6
Total n of hatched fry	384 ± 277[a]	-9.6	42.2
Female fecundity	674 ± 423[b]	-7.4	32.6
Male RS* (proportion)	0.10 ± 0.24[b]	-0.04	0.18 (or 18%)

Mean and regression coefficients were taken from [a](Fessehaye *et al.*, 2007) and [b](Fessehaye *et al.*, 2006b), and extrapolated to calculate the per generation inbreeding depression of the traits in mass spawning at a female to male ratio of 3:1.
*RS = reproductive success was calculated as the proportion of offspring sired by that male.
[1]Decrease in phenotypic performance per 4.4% increase in inbreeding.

lower reproductive success of inbred adults under natural conditions suggests the presence of a selective pressure that favours outbred adults and counterbalances the accumulation of inbreeding in natural mating systems (Fessehaye *et al.*, 2006b). When tilapia aquaculture relies on mass spawning for fry production, the accumulation of inbreeding and its actual impact on different traits might be less than predicted. However, as a precaution, it remains essential for breeding programmes to manage ΔF, because a loss of genetic variation will adversely influence both the short- and long-term limits of response.

Breeding programmes using mass spawning

Most aquaculture stocks currently used in developing countries are genetically similar or even inferior to wild, undomesticated stocks. There is anecdotal and empirical evidence that inbreeding and inadvertent selection are responsible for substantial (up to 40%) declines in growth among *Oreochromis* captive populations in both large and small-scale fish farms in Africa (Brummett *et al.*, 2004). The experience of selective breeding in tilapias has been reviewed by Penman and McAndrew (2000), who concluded that the low heritability for growth observed in many trials was the outcome of the low genetic variation typical of the cultured stocks used. Genetic diversity is crucial in breeding programmes because it is critical for both the short- and long-term limits of response to selection (Falconer, 1989). Therefore, capturing of high genetic diversity in the founder populations and minimising the loss of alleles are crucial strategies to ensure higher probability of success for breeding programs. A breeding scheme using mass spawning to produce fry can achieve these objectives by (1) increasing the effective population size and (2) maintaining separate lines of breeding stock and occasionally exchanging breeders between these separate lines. These two strategies will be discussed below.

Increasing effective population size (Ne)

Effective population size defines the effective number of individuals contributing to the next generation and helps to predict the fixation probability of favourable and deleterious alleles (Crow and Kimura, 1970). It therefore determines the effect of both inbreeding depression and genetic drift (Falconer, 1989; Caballero, 1994). Genetic drift influences the level of genetic variation. In both spawning experiments of 2003 and 2005 (Table 2.2), the effective population size was approximately half of what would be expected in an idealised population under random mating ($N_e/N < 0.5$); suggesting the potential for rapid losses of genetic diversity. These low N_e values are equivalent to an inbreeding rate of 3.0– 4.4%, which is up to four times the level of inbreeding acceptable for breeding programmes (1%: Bijma, 2000). This confirms the need to increase the N_e in the population, in order to avoid the risks of inbreeding in selection programmes relying on mass spawning for fry production. Several methods can be employed to increase the effective population size. For a fixed number of parents, the most effective way to increase the N_e would be to reduce the variance in reproductive success by equalising the contribution of males and females to the next generation. In practice, this might be difficult in mass spawning conditions where there is no control on mate selection or mating of the animals and where there is no marker information to reconstruct pedigree. *O. niloticus* females spawn asynchronously, which leads to a higher male–male competition and lower N_e. Synchronising the females, therefore, might lead to less male–male competition, which in turn might increase the participation of more males in fertilising eggs. Synchronisation of the ovarian cycle in *O. niloticus* females is difficult but the problem can be alleviated through some manipulations. Little *et al.* (1993) have suggested that conditioning of tilapia females at high density prior to transferring them to spawning tanks improves spawning synchrony and intensity. The spawning data from the females in our study also showed that some females spawned up to four times, whereas others did not spawn during the spawning period. This, together with the wide range of fecundity among females, can lead to unequal contribution and hence might cause a reduction in effective population sizes. This can be prevented by removing brooding females from breeding tanks and by taking equal number of fry from each brooding female. The number of progeny that could be sampled from each brooding female can be calculated by adopting the method in Komen *et al.* (2006), who determined the number of offspring that can be sampled per family by constraining the level of inbreeding to 1%. Though this intervention strategy might result in fewer fry being produced, it is feasible, given that the primary objective for a breeding nucleus is to capture more genetic variation rather than produce more fry.

Another practice used routinely to conserve diversity is to increase the effective population size by spawning as many fish as possible. This option is more expensive because it requires more space, increased logistics, and the purchase and maintenance of broodstock, but it pays off in terms of capturing a higher level of genetic diversity which will in turn result in greater genetic progress. Gjerde and Rye (1998) reported that the high fecundity and higher broodstock number make it possible to have a broad long-term breeding objective in the nucleus, while at the same time applying selection for a single or a few traits when producing grow-out animals and thus reducing the conflict between short and long-term breeding objectives as seen in breeding programmes for livestock species.

Given a fixed total number of individuals, the factor with the greatest potential to affect the effective population size and over which we have most control is the sex ratio. Theoretically, N_e is maximum when sex ratio is 1:1 ($N_e = 4N_{em}N_{ef}/ (N_{em}N_{ef})$). It will be recalled that we conducted mass spawnings with three different sex ratios; the predicted level of inbreeding in each sex ratio is shown in Table 2.2. Sex ratio 1 was expected to result in a lower level of inbreeding, but as the table shows, it resulted in even more inbreeding than sex ratio 2. We attribute this to the high male density and greater male-male competition in sex ratio 1 leading to relatively less male participation and more reproductive variance. In terms of the seed production and the associated level of inbreeding, a sex ratio of 2 females to one male seems to be a more viable option for a breeding scheme based on mass spawning.

Apart from the traditional manipulations of increasing effective population size mentioned above, another way to increase the N_e of small populations entails using genetic markers. Several approaches such as frequency-dependent selection and selection for heterozygosity at marker loci have been proposed as strategies to increase N_e (Wang, 2001). These methods have been rather inefficient, however, because marker information is limited for many fish species (Wang, 2001). A second-generation genetic linkage map has been constructed for *Oreochromis spp.*, which will enable mapping and selective breeding of quantitative traits (Lee *et al.*, 2005).

Marker information could also be used in parentage analysis; knowledge on parentage could be employed in implementation of a two stage selection scheme that combines walk-back and optimum contribution selection proposed by Sonesson (2005), which makes it possible to achieve high genetic gains at a constrained rate of inbreeding. The first step involves selection of a batch of phenotypically superior fish; in the second stage optimum contribution selection is used. By implementing such a scheme, the unbalanced contribution and higher rates of inbreeding in tilapia mass selection schemes can be prevented and thus result in higher genetic gains. Genotyping two batches of 50 fish were often necessary and could capture 76–92% of the overall genetic level achieved by typing all fish. There are two added advantages of such a scheme: first, only a small number of animals need to be typed and this will lower the genotyping costs; secondly, there is no common environmental effect, as fish are communally reared in one big holding facility (Sonesson, 2005).

Population subdivision and exchange of breeders

In addition to strategies designed to increase effective population size, genetic diversity could be maintained by subdividing a captive population into subpopulations (Kimura and Crow, 1963). Each subpopulation will lose genetic variation (i.e. heterozygosity and alleles) at a rate proportional to the effective size of that subpopulation, but different subpopulations may be fixed for different alleles. Crossing of groups and transfer of animals between such groups mitigates genetic drift and will result in maintenance of more genetic diversity in the total captive population (Hedrick and Gilpin, 1997). It has been shown that pooling a group of isolated subpopulations maintains more genetic diversity and increases reproductive fitness compared with a single large population of the same total size (Margan *et al.*, 1998). The creation of sub-lines and selection within each sub-line has also been successfully implemented in dairy cattle breeding programmes: Terawaki *et al.* (1998) simulated genetic selection in a closed nucleus herd subdivided into two lines and with varying degree of crossing between the two lines and found that after ten generations of selection, the scheme that crossed the two sub-lines every generation resulted in the lowest level of inbreeding, albeit with a slight decrease in genetic gain. Such schemes could also be adopted in the selective breeding of tilapia.

Given the information on the number of parents leaving offspring to the next generation, the strategy 'selection within cohorts and exchange of breeders' as suggested by Eknath (1991) for Indian carp and implemented by McPhee *et al.* (2004) for weight selection in redclaw crayfish (*Cherax quadricarinatus*) could also be implemented in tilapia breeding programmes. If the number of parents contributing to the next generation is known, the rate of inbreeding per generation can be calculated using the formula of Falconer and Mackay (1996), in which ΔF is the average rate of inbreeding per generation; f and m are the number of females and males leaving progeny to the next generation:

$$\Delta F = \frac{1}{8f} + \frac{1}{8m}$$

Taking the mass spawning we conducted in 2003 as an example, the effective numbers of males and females contributing to the next generation were as follows: hapa A ($N_{em} = 5$, $N_{ef} = 18$) and hapa B ($N_{em} = 5$, $N_{ef} = 16$) (Table 2.2). These two experimental breeding hapas can be considered as model broodstock units, and the level of inbreeding per generation for different number of breeding units can be calculated (Figure 2.4). For example, assuming a similar number of contributing males and females as in the experimental breeding units, three units of breeders will result in a level of inbreeding of 1.06% per generation, which is very close to the acceptable level of inbreeding for most breeding programmes. For a small-scale farmer-based breeding programme, using three breeding units of similar size as our experimental spawning stock would have a census population size of 111 adults (= 37*3; 25 females and 12 males per cohort) and an N_e of 56 (= 111*0.5) spawning adults should be sufficient to meet the inbreeding limit of 1%. In conservation management, on the other hand, it is a generally accepted view that a minimum effective population size of 50 spawning individuals is required to capture genetic diversity that will sustain the short-term fitness of a population (Allendorf and Ryman, 2002). Using a $N_e/N \approx 0.5$ from our study, this translates into a census population size of 100. Therefore, from both the rate of inbreeding per generation and the conservation rule of thumb, three breeding units of broodstock can be recommended as a founder population for a small-scale breeding scheme run by an individual farmer. Nucleus breeding schemes offer the potential for greater genetic progress than would be achieved by traditional within-farm selection (Smith, 1988). The rate of genetic progress depends among other things on the size of the nucleus, selection intensity and the effectiveness of selection within the nucleus (Weigel, 2001). The experimental breeding units can be considered as building blocks and the reproductive parameters might be extrapolated to design bigger schemes. In general, to accelerate rates of genetic progress in a nucleus it is necessary to increase the population size

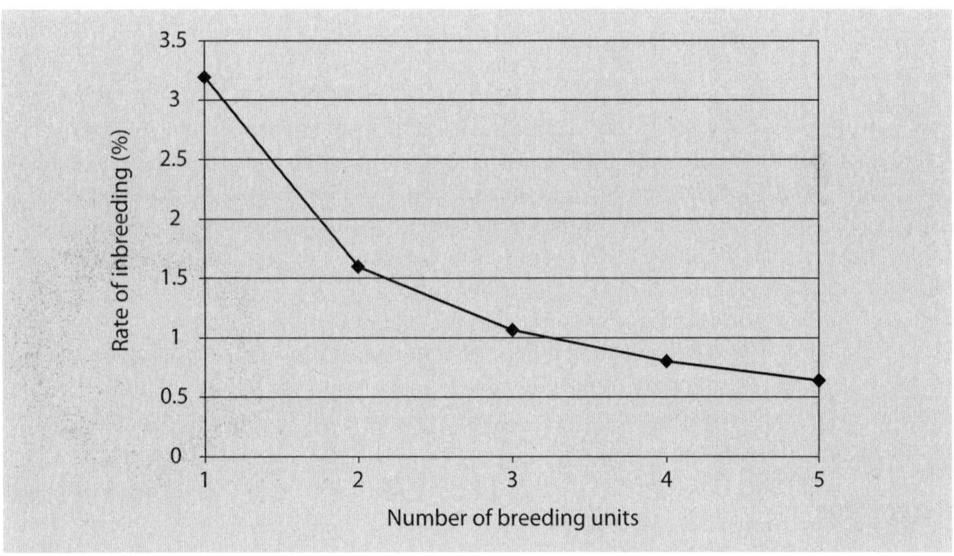

Figure 2.4. Number of breeding units (cohorts) and associated level of inbreeding per generation. Rate of inbreeding was calculated based on the effective number of males and females contributing in an experimental mass spawning with 25 females and 12 males. One breeding unit had a census population size of 37 (25 females and 12 males). Effective numbers of males and females contributing in a breeding unit were determined using parentage analysis of offspring (Fessehaye et al., *2006a).*

and intensify selection. If a bigger population in a nucleus were to pose an economic challenge, structured exchange between smaller nucleuses could still be effective – provided that the schemes have similar population size, selection criteria and goals.

Exchange of breeders between breeding units in a farm or between nucleus breeding schemes can be achieved by selecting the best males in one unit and then transferring them to another cohort in a cyclical mating system as described by Cockerham (1970). Cyclical mating systems have been used successfully in Asian breeds of Nile tilapia through within-family selection; the results have shown that the average rates of inbreeding of 1.4% per generation in the stocks selected for twelve generations were less than the rates typically observed in cultured populations (Bolivar and Newkirk, 2000). In contrast to single pair matings, selection within cohorts with exchange of breeders between cohorts following a prescribed pattern is a feasible design, even if resources are limited. Field personnel find it easy to follow and can thus adhere to instructions provided (Ponzoni *et al.*, 2006). In populations under selection, avoiding the loss of desirable genetic diversity may be impossible, but the problem of inbreeding in aquaculture stocks can be forestalled by exchange of breeders or by keeping large effective population sizes. It should be emphasised that subdividing the population or exchange of breeders does nothing to reduce the level of inbreeding in the base population, but it serves only to reduce the rate at which further inbreeding is accumulated. Therefore, it is essential for the starting broodstock to have a relatively high genetic diversity.

Conclusions

Breeding schemes relying on mass spawning for fry production present a straightforward and economically feasible option for low-cost genetic improvement programmes. However, many mass selection schemes in fish have not resulted in the desired response; this has been partly attributed to poor genetic management, including inadvertent selection, inbreeding and founder effects. The results presented above have established the effective number of parents contributing to the next generation and the associated level of inbreeding per generation in experimental mass-spawning broodstock units (breeding units). The information can be used to develop practical scenarios for mass selection breeding schemes that maximise selection response while at the same time managing inbreeding. If the breeding units are considered to be basic building blocks, mass selection breeding programmes based on mass spawning can be implemented. The number of breeding units can be extrapolated depending on the size of the nucleus and the intensity of selection. Such schemes require minimum infrastructural investment and can be adopted as a model for low-cost and effective selective breeding schemes for resource-poor regions.

References

Allendorf, F.W., and N. Ryman, 2002. The role of genetics in population viability analysis. In: McCullough and Beissinger, S.R. (Eds.). Population viability analysis. University of Chicago press, Chicago.

Bautista, A.M. and M.H. Carlos, A.I. San Antonio, 1988. Hatchery production of *Oreochromis niloticus* L. at different sex ratio and stocking densities. Aquaculture 73, 85–95.

Behrends, L.L., J.B. Kingsley and A.H. Price, 1993. Hatchery production of blue tilapia, *Oreochromis aureus* (Steindachner) in small suspended hapa nets. Aquacult. Fish. Manage. 24, 237–243.

Bentsen, H.B. and B. Gjerde, 1994. Design of fish breeding programs. Proceedings of the 5th World Congress Genet. Appl. Livest. Prod. 19, pp. 353 -359.

Bijma, P., 2000. Long term genetic contribution. Prediction of rates of inbreeding and genetic gain in selected populations. PhD thesis, Animal Breeding and Genetics Group, Wageningen University, The Netherlands.

Bolivar, R.B., and G.F. Newkirk, 2000. Response to selection for body weight of Nile tilapia (*Oreochromis niloticus*) in different culture environments. In: Fitzsimmons, K. and Carvalho, J.F (eds). Tilapia Aquaculture in the 21st century. Proceedings from the fifth International Symposium on Tilapia in Aquaculture, Vol. 1, Rio de Janeiro, Brazil pp.12-23.

Brown, R.C., J.A. Woolliams and B.J. McAndrew, 2005. Factors influencing effective population size in commercial populations of gilthead seabream, Sparus aurata. Aquaculture 247, 219-225

Brummett, R.E, D.E. Angoni and V. Pouomogne, 2004. On-farm and on-station comparison of wild and domesticated Cameroonian populations of *Oreochromis niloticus*. Aquaculture 242, 157-164.

Caballero, A., 1994. Developments in the prediction of effective population size. Heredity 73, 657-679.

Cockerham, C.C., 1970. Avoidance and rate of inbreeding. In: Kojima, K (Ed.) Mathematical topics in population genetics. Springer Verlag, Berlin. pp. 104-127.

Crow, J.F., and M. Kimura, 1970. An introduction to Population Genetics Theory. Harper and Row, New York, 591 pp.

Eknath, A.E., 1991. Simple broodstock management to control indirect selection and inbreeding. Indian carp example. NAGA, The ICLARM quarterly 738, 13-14.

Falconer, D.S., 1989. Introduction to Quantitative Genetics. Third edition. Longman, England.

Falconer, D.S. and T.F.C. Mackay, 1996. Introduction to Quantitative Genetics, 4th edition Longman, Essex, England.

Fessehaye, Y., Z. El-bialy, M.A. Rezk, R. Crooijmans, H. Bovenhuis and H. Komen, 2006a. Mating systems and male reproductive success in Nile tilapia (*Oreochromis niloticus*) in breeding hapas: A microsatellite analysis. Aquaculture 256, 148-158.

Fessehaye, Y., H. Komen, M.A. Rezk, J.A.M, Arendonk and H. Bovenhuis, 2007. Effects of inbreeding on survival, body weight and fluctuating asymmetry (FA) in Nile tilapia, *Oreochromis niloticus* . Aquaculture (in press).

Fessehaye, Y., H. Bovenhuis, M.A. Rezk, R. Crooijmans, J.A.M. Arendonk and H. Komen, 2006b. Effects of inbreeding and sex ratio on mating systems and reproductive success in mass-spawning of Nile tilapia, *Oreochromis niloticus. In*: Fessehaye, Y., 2006. Natural mating in Nile tilapia (*Oreochromis niloticus*): Implications for reproductive success, inbreeding and cannibalism. PhD thesis, Wageningen University, The Netherlands.

Gjedrem, T., 1998. Developments in fish breeding and genetics. Acta Agricultura Scandinavica A 28 (Suppl.), 19-26.

Gjerde B., and M. Rye, 1998. Design of breeding programmes in aquaculture species: Possibilities and constraints. In: Bartley, D.M. and Basurco B. (ed.). Genetics and breeding of Mediterranean aquaculture species. Zaragoza: CIHEAM-IAMZ, 1998. Seminar of the CIHEAM Network on Technology of Aquaculture in the Mediterranean (TECAM), 1997/04/28-29, Zaragoza (Spain), pp. 181-192.

Hedrick, P.W., and M. Gilpin, 1997. Metapopulations genetics: effective population size. In: Hanski, I. and M. Gilpin, (ed.) Metapopulation dynamics: ecology, genetics, and evolution. New York, Academic press, pp. 165-181.

Kimura, M., and J.F. Crow, 1963. On the maximum avoidance of inbreeding. Genet. Res. 4, 399-415.

Kincaid, H.L., 1983. Inbreeding in fish populations used for aquaculture. Aquaculture 33, 215-227.

Komen, H., H. Bovenhuis and J.A.M. van Arendonk, 2006. Consequences of reproductive characteristics for fish breeding schemes. Paper presented at the 8th World Congress Genet. Appl. Livest. Prod. August 13-18, 2006, Brazil.

Kvarnemo, C. and I. Ahnesjo, 1996. The dynamics of operational sex ratios and competition for mates. Trends in Ecology and Evolution 11, 404–408.

Lande, R. and G.F. Barrowclough, 1987. Effective population size, genetic variation, and their use in population management. In: Soule, M.E. (Ed.) Viable populations for conservation. Cambridge University Press, Cambridge, UK, pp. 87-123.

Lee, B.Y., W.J. Lee, J.T. Streelman, K.L. Carleton, A.E. Howe, G. Hulata, A. Slettan, A., J.E. Stern, Y. Terai and T.D. Kocher, 2005. A second-generation genetic linkage map of Tilapia (*Oreochromis Spp.*). Genetics 170, 237-244.

Little, D.C., D.J. Macintosh and P. Edwards, 1993. Improving spawning synchrony in the Nile tilapia, *Oreochromis niloticus* (L.). Aquacult. Fish. Manage. 24, 399–405.

Little, D.C., C.K. Lin and W.A. Turner, 1995. Commercial scale tilapia fry production in Thailand. World Aquaculture 26, 20–24.

Margan, S.H., R.K. Nurthen, M.E. Montgomery, L.M. Woodworth, E.H. Lowe, D.A. Briscoe and R. Frankham, 1998. Single large or several small? Population fragmentation in the captive management of endangered species. Zoo Biol. 17, 467-480.

Matocq, M.D., 2004. Reproductive success and effective population size in woodrats (*Neotoma macrotis*). Molecular Ecology 13, 1635-1642.

McPhee, C.P., C.M. Jones and S.A. Shanks, 2004. Selection for increased weight at 9 months in redclaw crayfish (*Cherax quadricarinatus*). Aquaculture 237, 131-140.

Penman, D.J., and B.J. McAndrew, 2000. Genetics for the management and improvement of cultured tilapias. Pp 227-266. In: Beveridge, M.C.M and McAndrew, B.J., (eds.). Tilapias: Biology and Exploitation. Fisheries Series 25 Kluwer, Dordrecht, The Netherlands.

Ponzoni, R.W., N.H. Nguyen and H.L. Khaw, 2006. Importance and implementation of simple and advanced selective breeding programs for aquaculture species in developing countries. Paper presented at the 8th World Congress Genet. Appl. Livest. Prod, August 13-18, 2006, Brazil.

Smith, C., 1988. Genetic improvement of livestock in developing countries using nucleus breeding units. World Animal Review 65, 2-10.

Sonesson, A.K., 2005. A combination of walk-back and optimum contribution selection in fish: a simulation study. Genet. Sel. Evol. 37, 587-599.

Terawaki, Y., M. Suzuki, Y. and Fukuri, Y., 1998. Genetic response and inbreeding in a sub-population mating system for diary cattle. Proc. Sixth World Congr. Genet. Appl. Livest. Prod. 25, 471-474.

Wang, J., 2001. Optimal marker-assisted selection to increase the effective size of small populations. Genetics 157, 867-874.

Weigel, K.A., 2001. Controlling inbreeding in modern breeding programs. Journal of Dairy Science 84 (E. Suppl.), 177-184.

Widemo, F., and I.P.F. Owens, 1995. Lek size, male mating skew and the evolution of lekking. Nature 373, 148-151.

Decentralised Nile tilapia seed production

David C. Little[1], Benoy K. Barman[1,2], Mohammad M. Haque[1,3] and M. Abdul Wahab[4]
[1] Institute of Aquaculture, University of Stirling, Stirling, FK9 4LA, United Kingdom
[2] WorldFish Center, Dhaka, Bangladesh and South Asia Office, House 22B, Road 7, Block F, Banani, Dhaka 1213, Bangladesh
[3] Department of Aquaculture, Bangladesh Agricultural University, Mymensingh, Bangladesh
[4] Department of Fisheries Management, Bangladesh Agricultural University, Mymensingh, Bangladesh

Abstract

Local production of mixed sex Nile tilapia in irrigated rice fields has been introduced, established and then spread through farmer-to farmer contact in Northwest Bangladesh benefiting poor households in a number of ways. Food fish farmers have improved access to high quality seed at the time of peak demand early in the monsoon season. The seed producers benefit through small but strategic cash flows but also improved production of their fish for their own consumption, both as large fingerlings and fish after further grow on. A range of social benefits emerged during qualitative assessments. Initially introduced through on-farm research and then Farmer Field Schools' adopters starting with common carp and Nile tilapia have adapted the basic concept, tending to expand to a certain level and increasing the number of species produced. Difficulty in delivery of sufficient quality broodfish to rural areas is a major impediment to broader adoption of decentralised production. Capacity among local promoters to support initial broodfish and knowledge supply was important to the development of decentralised seed networks. Confusion over the perceived qualities and constraints of mixed and mono-sex tilapias respectively can undermine acceptance of decentralised seed production strategies among policy makers.

Keywords: tilapia, decentralised, seed, rice fields, smallholder

Introduction

Decentralised approaches to freshwater fish seed production have been developed in Bangladesh resulting in high quality mixed sex Nile tilapia seed being produced and used in rural areas. Lack of timely and quality seed availability remains a constraint to adoption of smallholder aquaculture (Edwards *et al.*, 1991) throughout Asia, despite the widespread development of hatcheries. Demand for quality Nile tilapia is high (Dey, 2000) but their availability remains a particular challenge as its reproductive biology and behaviour constrain large-scale production (Little and Hulata, 2000). Together with common carp, Nile tilapia can be produced without sophisticated knowledge and technology in rural areas but questions remain as to the scale that such approaches can be promoted effectively and the level of sustained benefits achieved.

Breeding and overpopulation problems associated with mixed sex tilapias are widely perceived as being largely overcome with the increasing availability of mono-sex and improved strain tilapias (ADB, 2004) but the reality is that mono-sex production still remains highly limited in most contexts to serving commercially and urban-demand orientated producers. Even in Central Thailand, almost two decades after its introduction, mono-sex Nile tilapia is estimated to be around 50% of seed marketed (Belton, 2006). Furthermore there is evidence that mixed sex seed, if homogeneous in terms of age and size, perform to a level that can meet the demands of producers and consumers (Dan and Little, 2000; Little *et al.*, 2003; Little and Edwards, 2004). In rural areas where mono-sex Nile tilapia production have been promoted, continued interest in stocking high quality mixed sex fish continues, for a variety of reasons (AIT/DOF, 2000). Mixed sex tilapia production remains important even in areas of Central Thailand where mono-sex fish are readily available, suggesting that the nature of demand for food fish, and thus seed, is diverse. The importance of relatively small-sized tilapias (<200 g) to household food security is becoming better understood through detailed household level studies (e.g. Barman *et al.*, 2004) and market research (Faruque *et al.*, 2006; Kundu *et al.*, 2006; B.K. Barman, personal communication).

Demand for freshwater fish seed is highly skewed towards availability of rainfall in much of Asia, with the major demand peak occurring after the onset of the monsoons, a period that generally follows a dry, cooler season. Decentralised Nile tilapia and common carp seed production in the irrigated spring rice crop results in large fingerlings being available during this major period of demand, contrasting with alternative, even resource intensive approaches, that typically struggle to meet such concentrated demand (Bhujel *et al.*, 2001). Timely and adequate supplies of good quality seed has been a precondition in all regions both for scaling up production and adoption of aquaculture by new entrants (World Bank, 2006).

Interest in the potential roles of Nile tilapia in Northwest Bangladesh (which is considered the country's most marginal agricultural region) is long established. Barman *et al.* (2002) found that farmers in more than 20 communities villages ranked it as the second most important cultured species in terms of its overall value, second only to silver carp; this assessment was made prior to the introduction of any improved strains. The species was considered a valuable species for 'seasonal' ponds, and promoted as part of polycultures by a Regional aquaculture project (Northwest Fisheries Extension Project) beginning in 1989. This project also introduced the concept of rice-field based common carp seed production (Gregory and Kamp, 1999) that subsequently became the focus of major development interventions throughout the Region.

This paper presents a historical perspective of the development of decentralised Nile tilapia seed production in Northwest Bangladesh. Some comparative insights are given from northern Vietnam which has a similar agro-ecology and dependence on carp culture and where the introduction of decentralised approaches has been less successful. The paper also offers some preliminary analysis regarding the impacts of decentralised seed production on rural communities and approaches to its promotion.

Methodology

Following the technical success of Nile tilapia seed production in spring rice fields in Vietnam (Nguyen *et al.*, 1997) the concept was trialled in Bangladesh among rice-common carp farmers. A total of 19 households, assessed as being 'poor', piloted the approach in two villages where rice field based common carp seed production had been established for a number of years (Barman and Little, 2006). Nile tilapia broodfish (GIFT strain) were limited and each household received a total of only 18 mature fish. Production advice was given and performance was monitored at the producer household level.

After a successful pilot phase, the project contributed to the development of Farmer Field School Curricula of a large-scale development project (GO-Interfish) that was promoting fish seed production in rice fields in the Region. The importance of stocking fish in rice fields as part of a basket of Integrated Pest Management options had been established earlier as a key instrument in the reduction of pesticide use. The impacts of including information derived from the trial on how to use Nile tilapia within the options for farmers led to this knowledge being spread widely in the Region. Subsequent impact and evaluation studies have been conducted (2002-2006) to assess longer term outcomes of the research at household, community and institutional levels.

Outcomes

Analysis of the initial trial results suggested that introduction of Nile tilapia changed the management and outcomes of rice-field based seed fish production dramatically. Previously common carp seed production was a low level, subsistence orientated activity. After introduction of Nile tilapia, more than 80% of participating households sold some of the large-sized seed in addition to using it to restock their own systems and for direct consumption (Haque *et al.*, 2006; Figure 2.5). Production of seed from irrigated rice fields matched with the amount of rainfall and the major window of demand for fish seed in the area (Figure 2.6).

An initial assessment of the sustained uptake and further spread of Nile tilapia seed production from its points of introduction suggested a rapid 'organic' spread from the two communities (Barman et al, 2004). Three years later 120 farmers in a total of twenty communities had begun the practice without any further extension effort. Movement of seed and broodfish by traders and better-off farmers with perennial water appear to have been critical for this take-up by other households in the village and further afield. The practice has been characterised by farmers gradually increasing plot size, including further species and expanding their production over consecutive seasons (Barman *et al.*, 2004; Figure 2.7) until reaching 1,500-6,000 fingerlings crop^{-1} household^{-1}.

Adoption of decentralised seed production

Adoption analysis conducted since (2003-2005) indicates that rural households across a broad socio-economic spectrum benefit from decentralised seed production in rice fields (Haque *et al.*, 2006). The majority of the poorer households tend to use production for cash rather than subsistence, and analysis of seasonal cash flow reveals that this reduces vulnerability during stressful periods of the year. The major impact is through increased productivity of household food fish production; those producing seed in the rice field increased their yields of fish by 60% over control groups (Figure 2.8). Inter-seasonal differences play a critical role in the livelihoods

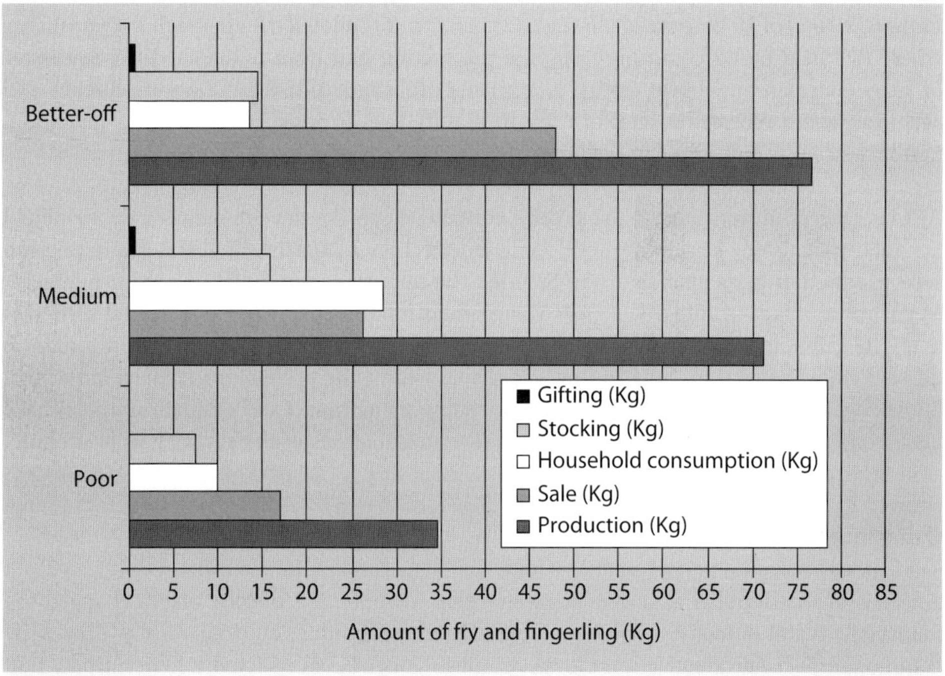

Figure 2.5. Use of fish seed produced in irrigated rice fields by households of different socio-economic status, including 'Gifting' = giving away seed to other households, 'Stocking' use of seed for food fish production on-farm, household consumption = fish from rice fields consumed directly within the household, 'Sale' -sold as juveniles; production-total seed produced (Haque et al., *2006).*

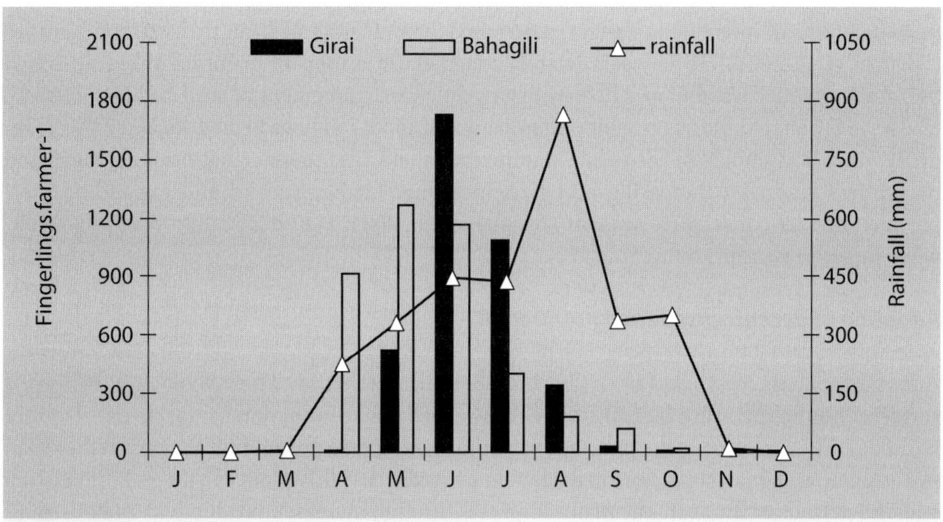

*Figure 2.6. Monthly variations in fingerlings production in relation to annual rainfall in the study areas (*Girai *and* Bahagili*) (Barman and Little, 2006).*

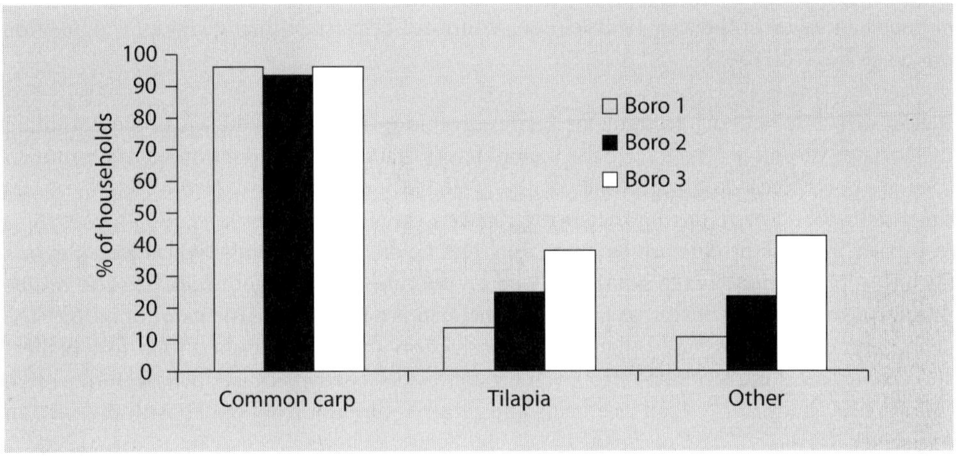

Figure 2.7. Fish species used for fingerling production in the rice fields of farming households over three consecutive irrigated rice production seasons (= Boro) in the GO-Interfish project area (Barman et al.*, 2004).*

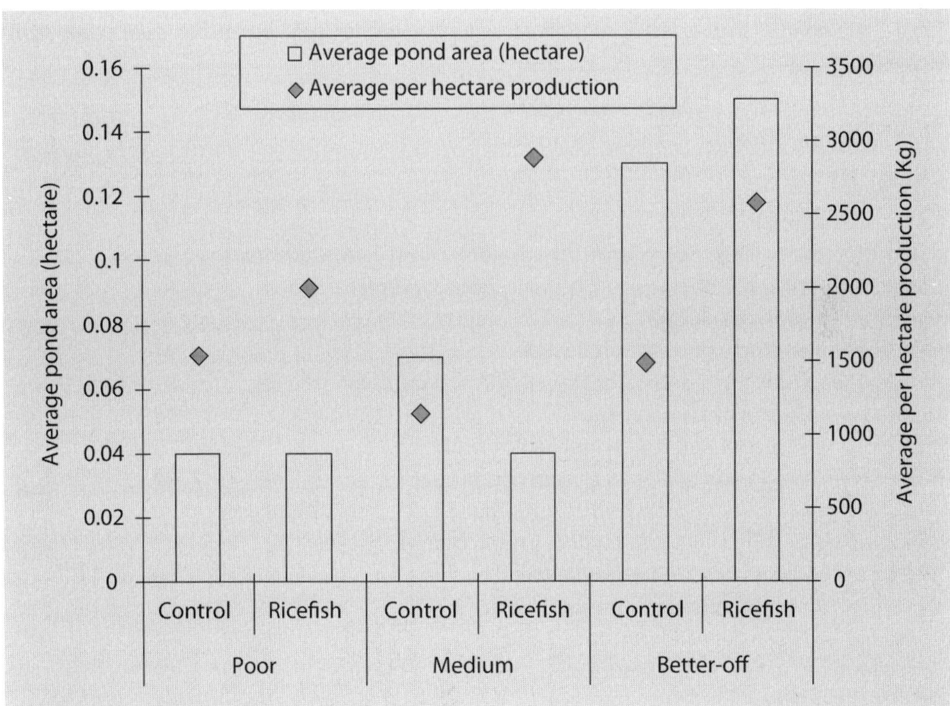

Figure 2.8. Average pond size and production of farming households under different wellbeing categories (Haque et al.*, 2006).*

of people in many of the world's developing countries because of their geographical position on the globe (Gill, 1991)

A host of subtle benefits account for the level of adoption, rather than a single dominant cash-related incentive (Table 2.4). The major factors that constrain adoption on a community level are poor access to groundwater or the availability of alternative, profitable enterprises e.g. potato production, during the spring rice crop season (Barman *et al.*, 2004). Attempts to further increase production by deepening rice fields into fishponds typically results in a decline in production; mixed sex tilapia seed production over a longer time cycle also results in more erratic quality of seed as the size of the seed is a less reliable indicator of its age. The promotion of rice-field based tilapia seed production among established food fish farmers and conventional nurseries has also given erratic results, partly because such operators often tried to use ponds rather than rice fields and returns were not competitive with production of carp fry.

Lack of quality tilapia broodfish in areas of introduction was the major reason for low productivity among adopters (Barman *et al.*, 2004) and is also likely to have implications

Table 2.4. Complex socio-cultural reasons for adoption of rice field based fish seed production technologies.

Reasons	Adopting households (%)
No use of chemicals	97
Fish consumption during *Boro** harvest-lean season of fish (nature and market)	93
Compatible with available household labour especially women, children	87
Consumption of fish by children	87
Restocking own pond to increase production	87
Women like to cook fish regularly for household consumption	83
No negative effect on rice production	83
Increases abundance of non-stocked fish in rice-fields	83
Farmers affinity to this technology, easier than other agricultural activities	80
Availability of rice-field fish after *Boro** reduces need for distress sales of rice	70
Window of availability of on-farm produced fish increased	70
Rice-plot becomes the centre for development	70
Compatible for consumption by extended family	67
Consumption of fish during low income month	67
Gift seed and food fish to relatives and neighbours	60
Effective use of rice-plot	57
Strategic use of pump water	53
Easy to catch fish during wet season	50

* *Boro* = irrigated spring rice crop.

for sustaining genetic quality. Lack of land or poor suitability of the land available is a major disincentive to adoption by rural households.

But adoption followed by later rejection is more often linked to changes in land tenure as land owners understand the relatively high returns of fingerling production and seek to change the terms of agreement (Figure 2.9). Sustainability of seed production in such systems is thus linked to broader aspects of power relations within rural areas.

Promoting decentralised approaches

The issue of making high quality germplasm available to rural communities has challenged promoters of both traditional and modern varieties within crop and livestock sectors as well as aquaculture (e.g. Tripp, 2001). A focus on developing and then maintaining centralised, often Government funded, seed multiplication facilities has often led to private sector approaches being ignored. This is partly because of continuing technocentric, top-down views but also linked to concerns that seed produced by farmers for farmers might be inherently poorer quality (Fleisher *et al.*, 2002). This has resulted in a continued shortage of high quality Nile tilapia seed in northern Vietnam as official promotion and endorsement of large-scale centralised production of mono-sex seed has proved inadequate and approaches to producing quality mixed sex seed have been ignored. This is despite successful piloting of commune-level conditioning of broodfish prior to distribution to individual households for breeding in shallow irrigated rice fields and indications of very positive production and returns on the mixed sex Nile tilapia production (Nguyen *et al.*, 1997). Higher level endorsement of decentralised approaches can therefore be critical to any larger scale uptake as initial efforts to promote the practice over significant rural areas requires significant resources and, crucially, effective

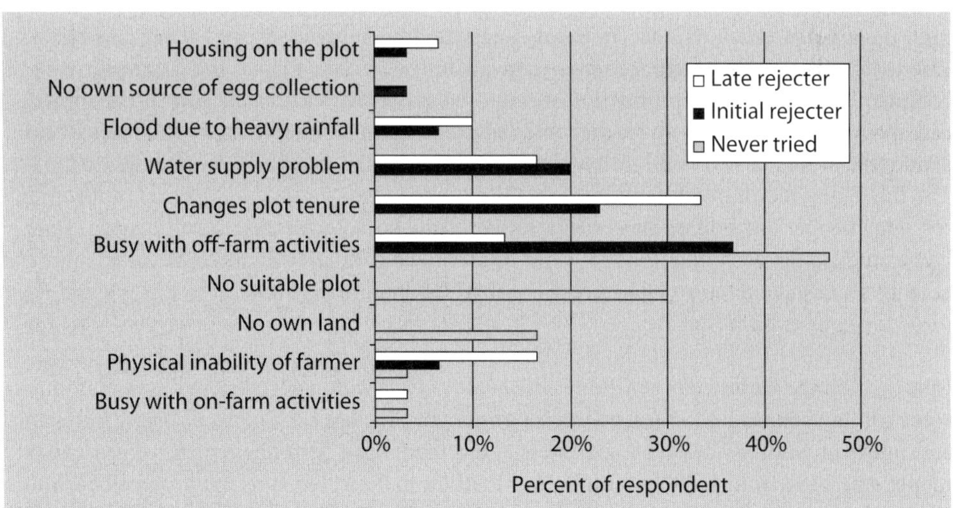

Figure 2.9 Reasons for which respondents (%) did not adopt fish seed production in the rice field based systems (Haque et al.*, 2006).*

public-private partnerships that build on established networks. Mechanisms for promoting and then supporting the spread of decentralised seed networks require further analysis but research to date in Bangladesh investigating the cost effectiveness of using different grass roots promoters suggest many lack capacity for any effective role. The potential of linking larger organisations with significant local capacity to support aquaculture and small established fish seed businesses with networks of rice farmers is currently under consideration.

Discussion

The potential of decentralised approaches to seed production is contrary to the prevailing orthodoxy regarding aquaculture development in several countries of Asia that focuses on centralised production and distribution of improved strains. The rapid spread of seed production among even poorer households producing irrigated spring rice with resultant impacts on livelihoods through seed production, local trading and more productive food fish production has resulted from minimal initial extension effort. Households practicing seed production directly or benefiting indirectly quickly spread beyond the initial extension focus of poorer farming households to landless fish seed traders and better off landowners with greater production capacity. Even when centralised supplies of seed and distribution networks are well developed, such as in Northwest Bangladesh (Lewis *et al.*, 1993; AIT/NFEP, 2002) decentralised seed production can have a major impact on the productivity of rural food fish production. Attitudes to production of 'quality' seed are difficult to change once established among policy makers. In particular perceptions that mixed sex fish necessarily equate with poor quality and that development of tilapia as a major cultured species requires introduction of mono-sex, are inaccurate (Little and Hulata, 2000). These perceptions can constrain the acceptance and benefits possible from promotion of decentralised strategies.

Decentralised tilapia seed production produces high quality seed during the season of high demand in much of Asia. In areas dominated by supplies of seed from conventional carp hatcheries and nurseries improved availability of Nile tilapia and common carp is complementary to current production and trading networks. Technically simple-the approach can spread rapidly farmer-to-farmer once established in a suitable area. The roles of non-producers, particularly fry traders have been important to the speed and effectiveness of uptake. The low entry requirement (investment, land, knowledge) make the approach appropriate for poor people but households over a range of the socio-economic strata benefit. Poorer households, often without access to water bodies to restock seed themselves, often value the activity for cash generation relatively more than the better off who tend to restock and raise more for household subsistence.

Introduction and maintenance of the genetic quality of fish within decentralised systems requires some continued linkages to the networks of promoters. The level and costs of such maintenance and future upgrading have yet to be assessed but there have been few reports of poor quality tilapia emerging from such systems so far. Institutional mechanisms for promotion require grass roots capacity – and the central and sustained role of rural entrepreneurs is critical.

Take home points

1. Promoting decentralised seed production and supply can benefit poor rural people inadequately served by conventional approaches.
2. Access to irrigated rice fields, even through sharecropping or leasing, provides opportunities for seed production by the poor but marketing of the seed locally has broader impacts on poorer people who gain as traders and other intermediaries.
3. Large-sized, high quality tilapia seed available at the critical time for rain-fed fish culture enhances the productivity and returns from seasonal fish culture.
4. Sufficient numbers of high quality broodfish available locally through promoters with capacity to support poor households with no fish culture experience is critical to successful adoption.
5. Building awareness among policy makers of the potential for decentralised seed approaches and their contribution to poverty-focused aquaculture development is important.

References

ADB, 2004. Small-scale freshwater rural aquaculture development: Case studies in Bangladesh, Philippines and Thailand. Supplementary Appendix. AIT/DOF, 2001. 155 pp.

AIT/DOF, 2000. Fish seed quality in Northeast Thailand. State of the System report. Aquaculture Outreach Programme, AIT, Bangkok. 31 pp.

AIT/NFEP, 2001. Fish seed quality in Asia: Fish seed quality in Northwest Bangladesh, State of the system report. Asian Institute of Technology, Bangkok, Thailand.

Barman, B.K. and D.C. Little, 2006. Nile tilapia seed production in irrigated spring rice-fields in Northwest Bangladesh-an approach appropriate for poorer farmers? Aquaculture, 261, 72-79.

Barman, B.K., D.C. Little and P. Edwards, 2002. Small-scale fish culture in Northwest Bangladesh - a participatory appraisal focussing on the role of tilapias. In: Edwards, P., Little, D.C. and Demaine, H. (Eds.). Rural Aquaculture. AIT, Bangkok. CAB International, pp. 227-244.

Barman, B.K., D.C. Little, and J. Janssen, 2004. Tilapia and common carp fingerling production in rice fields of farmers in Northwest Bangladesh: Adoption and dissemination strategies. Working paper 10. Institute of Aquaculture, University of Stirling, UK. 43 pp.

Belton, B. 2006. An assessment of the sustainability of tilapia culture in central Thailand. MSc Dissertation, University of Stirling.UK. 118 pp.

Bhujel, R.C., A. Yakupitiyage, W.A. Turner and D.C. Little, 2001. Selection of a commercial feed for Nile tilapia (*Oreochromis niloticus*) broodfish breeding in a hapa-in-pond system. Aquaculture 194, 303-314.

Dan, N.C. and Little, D.C. 2000. The culture performance of mono-sex and mixed sex new season and overwintered fry in three strains of Nile tilapia *Oreochromis niloticus* in northern Vietnam. Aquaculture 184, 221-231.

Dey, M., 2000. The impact of genetically improved farmed Nile tilapia in Asia. Aquaculture Economics and Management 4, 107-124.

Edwards, P., H. Demaine, S. Komolmarl, D.C. Little, N.L. Innes-Taylor, D. Turongruang, A. Yakupitiyage, T.J. Warren, 1991. Towards the improvement of fish culture by small-scale farmers in Northeast Thailand. J. Asian Farm. Syst. Assoc.1, 287-302.

Faruque, G., D.C. Little, J.A. Young and M.M. Dey, 2006. Impacts of the production and marketing of freshwater aquaculture products on stakeholders in Bangladesh. Paper presented in a seminar on Aquaculture Production and Marketing 2020, WorldFish Center, Dhaka, Bangladesh, 20 March 2006.

Fleisher, G, H. Waibel and G. Water-Eghols, 2002. Transforming top-down agricultural extension to a participatory system: A study of cost and prospective benefits in Egypt. Public Administration and Development 22, 309-322.

Gill, G.J, 1991. Seasonality and Agriculture in the Developing World: a Problem for the Poor and Powerless. Cambridge University Press Gregory, R. and K. Kamp, 1999. Aquaculture extension in Bangladesh experience from Northwest Fisheries Extension Project (NFEP) 1989-92. FMS and NFEP Bangladesh.

Haque, M.M, B.K. Barman, M.A. Wahab and D.C. Little, 2006. Fish seed production in irrigated rice-fields: Livelihood impacts and factors affecting adoption and rejection process at household level. Paper given International Seminar on Aquaculture Production and Marketing for 2020, WorldFish Center, Dhaka, Bangladesh, 20 March 2006.

Kundu, N, B.K. Barman, M. Rahman and D.C. Little, 2006. Paper given International Seminar on Aquaculture Production and Marketing for 2020, WorldFish Center, Dhaka, Bangladesh, 20 March 2006.

Lewis, D.J., R.G. Gregory and G.D, Wood, 1993. Indigenising extension: farmers, fish-seed traders and poverty focused aquaculture in Bangladesh. Development and Policy Review 11, 185-194.

Little, D.C. and P. Edwards, 2004. Impact of nutrition and season on pond culture performance of mono-sex and mixed-sex Nile tilapia (*Oreochromis niloticus*). Aquaculture 232, 279-292.

Little, D.C. and G. Hulata, 2000. Strategies for tilapia seed production. In: Beverage, M.C.M. and D.J. MacAndrew (eds.) Tilapias: Biology and Exploitation, Kluwer Academic Publishers. Fish and Fisheries Series 25, Dordrecht, pp. 267-326.

Little, D.C., R.C. Bhujel and T.A. Pham, 2003. Advanced nursing of mixed-sex and mono-sex tilapia (*Oreochromis niloticus*) fry, and its impact on subsequent growth in fertilised ponds. Aquaculture 221, 265-276.

Nguyen, C. D., T.M. Thien, D.V. Trung, 1997. Tilapia breeding in ricefields in Vietnam. NAGA 20, 23-25.

Tripp, R., 2001. Seed Provision and Agricultural Development, Oxford, James Currey.

World Bank, 2006 Aquaculture: Changing the Phase of the Waters-Meeting the Promise and Challenge of Sustainable Aquaculture, Report No. 36622-GLB, Agriculture and Rural Development, The World Bank, Washington DC, USA.

Breeding program requirements in developing countries may differ depending on circumstances

Raul W. Ponzoni, Nguyen Hong Nguyen and Hooi Ling Khaw
The WorldFish Center, Jln. Batu Maung, 11960 Batu Maung, Penang, Malaysia

Abstract

Aquaculture in developing countries is largely based on unimproved fish strains. There is ample evidence indicating the potential of genetic improvement programs and a range of selection methods may be used. Examples of the application of mass, cohort, within family, and combined between-within family are given. The methods are discussed in terms of their effectiveness and suitability. It is concluded that in principle most of the methods can work well, provided the selection program is started with a population with a broad genetic base and that during its conduct a balance is struck between selection intensity and containment of inbreeding. Programs that entail the collection of very detailed information are more suitable for national research organisations, whereas simpler ones may have greater appeal to commercial operators.

Keywords: aquaculture, genetic improvement, selection methods

Introduction

Aquaculture is likely to play a major and ever increasing role in meeting human needs for protein. In terrestrial animal and plant species genetic improvement programs have made a substantial contribution to productivity and viability. By contrast, most aquaculture stocks in current use in developing countries are genetically similar or inferior to wild, undomesticated stocks. Hence, there is ample justification for the planning, design and implementation of genetic improvement programs for aquatic animal species. A range of methods of varying complexity is available for selection purposes, but their suitability for different circumstances is not always clear. In this paper we briefly present the main selection methods that have been used or advocated, and discuss their virtues and shortcomings. When possible, we make reference to practical examples of their application.

Selection Approaches

General

We present the different selection approaches in increasing order of complexity, beginning with the simplest one. In each case, we refer to specific requirements that may constitute a limitation for their implementation in developing countries. Note that we assume that there is genetic variation for the trait(s) of interest in the population undergoing selection and that it does not suffer from problems (e.g. bottlenecks, inbreeding) created by earlier genetic mismanagement. Such problems could undermine the effectiveness of any selection program (e.g. Teichert-Coddington and Smitherman, 1988; Huang and Liao, 1990).

Individual or mass selection

The terms 'individual selection' and 'mass selection' are often used interchangeably, and they refer to selection solely based on the individual's phenotype. It has been a common strategy with fish because of its simplicity. It does not require individual identification or the maintenance of pedigree records, hence it may be considered the least costly method. In principle, it can produce rapid improvement if the heritability of the trait(s) under selection is high. Under those circumstances, however, there is risk of inbreeding due to inadvertent selection of progeny from few parents producing the best offspring, especially if progeny groups are large. For growth rate and morphological traits (easily assessed, expressed in both sexes) it can be quite suitable. By contrast, individual selection is not suitable for situations in which the estimation of breeding values requires slaughter of the animals (e.g. carcase and flesh quality traits) or challenge of some sort (e.g. selection for salinity tolerance or for disease resistance).

Hulata *et al.* (1986) carried out two generations of mass selection for growth rate with Nile Tilapia (*Oreochromis niloticus*) and observed no improvement over the original base population. They attributed the lack of response to selection to a number of possible factors, including inbreeding and genetic drift. They concluded that mass selection was not a promising method unless measures could be taken to control inbreeding. Rezk (2004) reported selection responses in harvest weight of 35 and 14 percent in *O. niloticus* and *O. aureus* respectively, after four generations of selection. There were no clear reasons for the difference in response between both species since the selection protocol applied in either case was the same (140 individuals of each sex were selected each generation). WorldFish (unpublished) records indicate that the experience with Silver Barb (*Barbonymus gonionotus*) in Bangladesh and Thailand and Common Carp (*Cyprinus carpio*) in Vietnam has been of satisfactory response to selection in early generations up to the fourth or fifth, declining sharply thereafter.

Gjerde *et al.* (1996) and Villanueva *et al.* (1996) provide procedures to determine the optimum design of mass selection programs in fish with constrained levels of inbreeding. The latter authors combine expected rates of genetic progress (ΔG) with expected rates of inbreeding (ΔF) in a linear objective function ($\Phi = \Delta G - \lambda \Delta F$). Maximization of the function Φ for appropriate values of λ gives the optimum number of sires and dams when specific constraints on the level of inbreeding are imposed. For $\lambda=0$ no restriction on inbreeding is imposed. Using this approach the authors show that for a population size of 300 and no restrictions on inbreeding the optimum design is eight sires and 16 dams, and inbreeding increases at a rate of 3.9 percent per generation. When inbreeding is restricted to 0.25 percent then the optimum design is 85 sires and 85 dams, and the rate of genetic gain is halved due to the inbreeding constraint. These design considerations have to be carefully applied in species such as Tilapia. Fessehaye *et al.* (2006) found that in *O. niloticus* under mass spawning in large hapas there was a large variance in male reproductive success, with one third of the males siring more than 70 percent of the offspring. This lead to a rate of inbreeding of approximately twice of that predicted from the effective population size. Overall, the evidence suggests that simple, unstructured, mass selection will result in problems unless the number of parents is large, and even so, chance could have a negative effect. Some form of structuring to control the parental contribution to the next generation appears necessary. If controlled pair matings can be carried out, the results of Bentsen and Olesen (2002) can be used to formulate the design of the breeding program. These authors investigated the effect of number of parents selected and of number of progeny tested per pair for a range of population sizes and heritability values. They show that inbreeding rates can be kept as low as one percent per generation if a minimum of 50 pairs is mated and the number of progeny tested from each pair is standardised to 30 to 50 progeny. Note that although not requiring individual identification of the fish, the schemes suggested by these authors entail the conduct of pair matings, initial maintenance of the progeny of such pair matings in separate enclosures, and controlled contribution of each full sib family to the next generation at the time the fish are assigned to communal rearing. The guidelines provided by Bentsen and Olesen (2002) can be very valuable if they can be put into practice. However, we have found that in some developing countries the conduct of a large number of pair matings, the subsequent containment of the full sib groups, and the sampling of a standard number of progeny to contribute to the next generation were tasks beyond the available resources, and we had to change the strategy to that described in the following section.

Selection within cohorts and exchange of breeders

Eknath (1991) reports the genetic deterioration taking place in hatcheries in India due to poor brood stock management. Among other measures to remedy the situation he suggested that brood stock could be arbitrarily divided into several groups. Mating could then be performed between individuals from the different groups on a rotational basis to avoid inbreeding. In this section we develop that notion further, based on the mating design used by McPhee *et al.* (2004) for weight selection in redclaw crayfish (*Cherax quadricarinatus*). These authors divided the population into cohorts, namely, groups sampled from a previously established foundation population. A selection line was created, consisting of 20 cohorts, where each cohort had 15 female and 10 male foundation parents. A control line of eight cohorts of the same size was also established. One hundred individuals were measured per cohort. Offspring of cohorts were hatched and grown in separate pens within a pond. At harvest time individuals of the heaviest weight in each cohort were chosen as parents of the next generation in the selection line, whereas individuals of average weight were chosen in the control line. In either case, selection was based on the difference between the harvest weight of an individual and its cohort mean. This within cohort selection aims to eliminate the environmental effect of cohorts on growth differences among individuals. The same number of individuals was selected from each cohort. Animals selected in one cohort were mated with those selected in another one to avoid mating related animals. After four generations of selection harvest weight in the selection line was 1.25 times greater than in the control line. In this mating design the importance of pen effects can be judged by computing the ratio (e) of the between pen variance component and the total variance (sum of between and within pen variance components). Although the exact number of parents contributing to the next generation is not known the rate of inbreeding can be calculated for the worst case, that is, that only one pair per cohort left offspring. Generally, inbreeding at generation t, would be calculated as:

$$F_t = 1 - (1 - \Delta F)^t \text{ where } \Delta F = 1/(8f) + 1/(8m)$$

where f and m are the number of females and males leaving progeny in each generation. With only one pair leaving progeny per cohort the equation becomes:

$$\Delta F = 1/(8c) + 1/(8c) = 1/4c$$

where c is the number of cohorts. We may then write:

$$F_t = 1 - (1 - 1/4c)^t$$

and we can use this equation to predict the maximum inbreeding after t generations. By designing the selection program in such a way that even in the case that only one pair from each cohort produced progeny the inbreeding rate was not excessive, then we would be able to ensure that we would not run into problems due to inbreeding. With regards to the exchange of breeders between cohorts, this could be achieved by shifting the males born in one cohort to another one in a pattern as described by Nomura and Yonezawa (1996), following for instance Cockerham's cyclical mating system (Cockerham, 1970). If we have information about the likely number of parents leaving offspring in a cohort (e.g. Fessehaye *et al.*, 2006) then we can further refine the design. In practice, we have found that, in contrast to single pair matings, selection within cohorts with exchange of breeders between cohorts following a prescribed pattern is a feasible design even with limited resources. Field personnel feel comfortable with it, and will thus rigorously adhere to the instructions provided.

Within family selection

The method requires identification of the families. This may be achieved by maintaining them in separate tanks, cages, hapas or any other means of containment, without necessarily tagging the fish. The criterion of selection is the deviation of each individual from the mean of the family to which it belongs. Within family selection is especially advantageous when there is a large component of environmental variance common to members of the same family. Full sib groups reared in unreplicated hapas or any other form of containment fall into this category (e.g. see estimates in Ponzoni *et al.*, 2005). Under such circumstances selection between families would be misleading from a genetic viewpoint because of the confounding between genetic merit and common environmental effects. The method can make very effective use of facilities. If replacements are chosen so that every family contributes the same number of individuals to the next generation (e.g. choose one female and one male from each family) the effective population size twice the actual (Falconer and Mackay, 1996). However, not all the additive genetic variance is available for selection, only a fraction r will be available, where r is the coefficient of relationship among the family relatives in question (i.e. 0.5 and 0.25 for full and half sibs, respectively). The heritability of within family deviations is consequently reduced:

$$h^2_w = [(1 - r)/(1 - t)]\, h^2$$

where t is the intraclass correlation. For example, if for a particular trait the heritability in the population is 0.2 and the families are full sib groups, then $h^2_w = 0.11$. The lower within family heritability can be compensated for by the high within family selection intensity that can be applied without increasing the rate of inbreeding. The selection intensity within families will be limited only by the number of individuals tested per family. The number of families involved in the program will determine the lower limit of inbreeding, which can easily be controlled by applying a rotational mating system such as that earlier suggested for selection within cohorts.

The use of within family selection was recommended for South East Asian countries by Uraiwan and Doyle (1986). It was successfully applied in the selection program that resulted in an improved Tilapia strain developed in the Philippines by the Freshwater Aquaculture Center (FAC) of Central Luzon State University. The strain is known by a variety of names, FAC-selected, FaST, and IDRC strain (in recognition of the support received from the International Development Research Centre of Canada). The selection program and the strain's performance have been described by Abella *et al.* (1990), Bolivar *et al.* (1994), Bolivar and Newkirk (2000, 2002) and Camacho *et al.* (2001). The selection line started from a base population combining four strains of Tilapia, namely, Israel, Singapore, Taiwan and a 'FAC' strain available at the time. Nineteen full sib groups were established, and the basis of selection was body weight at 16 weeks of age. The heaviest male from a given family was mated to the two heaviest females of another family to avoid inbreeding. After 12 generations of selection the genetic gain in body weight has been estimated at 12.4 percent per generation. All the selection program was conducted (from spawning to selection) in outdoor concrete tanks, but the strain also performed well in hapas and ponds. Camacho *et al.* (2001) comment that within family selection was easy to manage, and that taking care of inbreeding by means of a rotational mating posed no difficulties. The method reduces the need for tagging large numbers of individuals. They estimate that the implementation of a selection method that entailed the individual identification of large numbers of fish and a period of communal rearing would be more expensive and difficult to implement. Note that Bolivar and Newkirk (2000) and Ridha (2004, 2006) compared FAC-selected with GIFT (Genetically Improved Farmed Tilapia), the latter strain resulting from a program combining individual and family selection (dealt with in a later section of this paper), and found that there were no differences in growth rate between both strains, but in some of the experiments (Bolivar and Newkirk, 2000; Ridha, 2004) GIFT had greater (23 percent) survival rate. However, both strains were significantly more productive than other, unimproved, strains.

Combined selection

We use the term 'combined selection' in a broad sense, meaning selection that is based on individual information as well as on information coming from relatives (e.g. full and half sibs, progeny). In this case all of the additive genetic variance is available for selection and the use of information from relatives increases the accuracy of the estimation of breeding values. Furthermore, relatives' records can be used to estimate breeding values for traits that require slaughter of the animals (i.e. carcase and flesh quality traits) or that entail a risky challenge (i.e. disease resistance, tolerance to some environmental component). This is not possible with the other methods (e.g. mass selection or within family selection). Earlier work with fish (e.g. Eknath *et al.*, 1998) used selection index theory to combine individual, full sib and half sib information. A selection index can be very useful in combining such information, but the approach has limitations that have been overcome with BLUP procedures (see Van Vleck, 1993 for a detailed account). Best Linear Unbiased Prediction (BLUP) procedures rely on mixed model methodology for the estimation of individuals' genetic merit. In the case of aquaculture, all systematic effects (e.g. batch, sex, production environment, age variation) associated with traits of interest can be accounted for in the model fitted to the data. The maternal and common environmental effects due to separate rearing of full-sib families before tagging can also be separated effectively from the additive genetic variance. One particular advantage of BLUP procedures is that genetic gain can be estimated from the mean of the estimated breeding values in each year or generation of selection provided there are genetic links. There are now readily available computer programs (e.g. PEST, Groeneveld, 1990; ASReml, Gilmour *et al.*, 2002) that can estimate breeding values using full pedigree information. A drawback of BLUP selection is that if truncation selection on estimated breeding values is practised, it also results in higher levels of inbreeding than individual selection, especially for lowly heritable traits. This is because BLUP uses family information, leading to co-selection of relatives. Hence, optimal genetic contribution theory should be used with the aid of software such as EVA (Berg, 2004) or TGRM (Kinghorn, 2000) in order to maximise genetic gain while controlling the rate of inbreeding.

In order to use combined selection, identifiable families have to be produced. This is usually achieved by pair mating or by external fertilisation of ova and sperm of known parental origin. The progeny of the different families must be marked in some way so that they can be communally stocked, and tested for genetic evaluation purposes. Preferably, all the individuals to be tested should have a unique identification, and the pedigree will take care of all family relations. Aquatic animals are generally very small at spawning. They are kept in their family (usually full sib) groups until they are large enough to be tagged. This often results in an appreciable common environmental effect in traits such as growth rate. Tagging itself is an issue. The most commonly used tags with fish are Floy Tags® and PIT (passive integrated transponder) tags, the latter being about five times more expensive than the former, but far superior in terms of retention rate. With Floy Tags we have lost up to 60 percent of the tags in some groups due fragility of the thread that holds the plastic tag. In the hands of unskilled staff the combined effects of poor reproductive rates, large common environmental effects, and high tag losses can negate the theoretical virtues of the approach. The potential advantages of combined selection cannot be disputed, but before embarking upon it, one must be sure that the physical and human resources at hand are adequate for the tasks that will have to be undertaken.

Three documented examples of the successful application of combined selection to the improvement of fish in developing countries will be cited here (in all cases growth rate was the main focus of selection): (1) The GIFT project in Philippines, which reported genetic gains of 12 to 17 percent per generation in Nile Tilapia, over five generations (Eknath *et al.*, 1998); (2) The Jayanti Rohu (*Labeo rohita*) selective breeding project in India, which reported a genetic gain of 17 percent per generation over five generations (Reddy *et al.*, 1999; K.D. Mahapatra, personal communication); and (3) The selection project of a Malawian indigenous Tilapia,

Oreochromis shiranus, where the accumulated gain over two generations was 13 percent (Maluwa, 2005). GIFT and Jayanti Rohu have been tested extensively on farm and proven to outperform other strains used by farmers. We earlier noted that the GIFT and FaST strains have very similar growth performance, but GIFT has shown greater survival rate, possibly due to the broader genetic basis in the population originally assembled and to the greater effective population size relative to FaST. Because the program with *O. shiranus* is still in its early stages, the strain has not yet been tested extensively on farm.

These three programs (GIFT, Jayanti Rohu, *O. shiranus*) have a number of features in common: (1) They all started with the assembly of a base population drawn from different sources in order to capture genetic variation; (2) Controlled matings of identified females to identified males were conducted and complete pedigrees were maintained; (3) Full sib groups were kept together until tagging; (4) Approximately, 50 to 200 fish per full sib group were tagged and destined to communal rearing in a range of production environments in order to estimate genotype by environment interactions. In the case of GIFT and Jayanti Rohu a selection index combining individual, full sib and half sib information was used to rank individuals on genetic merit, whereas BLUP breeding values were estimated in *O. shiranus*. BLUP procedures are also used in the selection of GIFT in the population that was transferred to Malaysia (Ponzoni *et al.*, 2005). The sound design coupled with rigorous conduct and analysis accounts for the gains achieved in these programs. Furthermore, data sets of this nature, developed over a number of generations, provide great research opportunities in the area of estimation of phenotypic and genetic parameters, as well as of environmental effects and genotype by environment interactions. As a by-product of the genetic improvement program, opportunities for local staff capacity building are created around it. If captured, these opportunities can result in the training of staff to a level that enables them to independently plan and conduct genetic improvement programs. Note that the amount of information that can be extracted from a pedigreed population is much greater than from a non-pedigreed one. This general model, packaged in what has been called 'GIFT Technology' has been advocated by the WorldFish Center for implementation in several developing countries (WorldFish Center, 2004).

Concluding remarks

In this paper we focused on selection approaches that in our perception can be managed with the resources that are available in developing countries. We did not comment on more sophisticated ones such as the 'walk-back' (Doyle and Herbinger, 1994; Sonesson, 2005) or the 'PROSPER' (Chevassus *et al.*, 2004) methods. Whereas we recognise their potential value, we believe that due to the resources they require they would be difficult to implement in the developing countries where aquaculture development is most needed (e.g. Sub Saharan Africa).

A rigorous comparison of different methods based on published evidence is not possible. In practice, the outcome of a program will be affected by many factors other than the selection method itself. When planning a new program in a developing country, a way of approaching the problem could be to begin thinking about and outlining the most complete one (i.e. full pedigrees and BLUP estimates), and to simplify it gradually until it becomes feasible with the available resources, working backwards through the methods we presented. The final decision before implementation will be a matter of judgment. For instance, a national program developed with a NAR would most likely have a larger component of capacity building than one for a commercial operator. In any case, one should ensure that the program is manageable within the limits of the available physical and human resources, and that it has a high probability of continuing beyond project duration. Note, for instance, that mass, within cohort and within family selection cannot be used to select for traits that cannot be recorded in the live breeding candidates. Hence, application of these methods may render the improvement program unsustainable in the long term. Starting with a population with ample genetic variation is a trademark of successful fish genetic improvement programs. Although this in itself is not a sufficient condition for success, it is indeed a necessary condition. Sophisticated designs and genetic evaluation procedures are no substitute or remedy for a genetically deteriorated base population. The failure of some attempts to achieve genetic improvement with aquatic animals may have been due more to weaknesses in the base population than to the selection method utilised. Irrespective of the method of choice, continued genetic improvement will hinge upon the adequate balance between high selection intensity and the maintenance of low inbreeding rate.

References

Abella, T.A., M.S. Palada and G.F. Newkirk, 1990. Within family selection for growth rate with rotation mating in *Oreochromis niloticus*. Proc. 2nd Asian Fisheries Forum, pp. 515-518.

Bentsen, H.B. and I. Olesen, 2002. Designing aquaculture mass selection programs to avoid high inbreeding rates. Aquaculture 204, 349-359.

Berg, P., 2004. User's Guide EVA version 1.5. Danish Institute of Agriculture Science, Denmark. 20 pp.

Bolivar, R.B., Z.P. Bartolome and G.F. Newkirk, 1994. Response to within-family selection for growth in Nile Tilapia (*Oreochromis niloticus L.*). Proc. 3rd Asian Fisheries Forum, pp. 548-551.

Bolivar, R.B. and G.F. Newkirk, 2000. Response to selection for body weight of Nile Tilapia (*Oreochromis* niloticus) in different culture environments. Proc. 5th ISTA I, pp. 12-23.

Bolivar, R.B. and G.F. Newkirk, 2002. Response to within family selection for body weight in Nile tilapia (*Oreochromis niloticus*) using a single-trait animal model. Aquaculture 204, 371-381.

Camacho, A.S., T. Abella and M.M. Tayamen, 2001. Fish Genetics research and development in the Philippines. ICLARM Conf. Proc. 64, 71-76.

Chevassus, B., E. Quillet, F. Krieg, M. Hollebecq, M. Mambrini, A. Faure, L. Labbe, J. Hiseux, M. Vandeputte, 2004. Enhanced individual selection for selecting fast growing fish: the "PROSPER" method, with application on brown trout (*Salmo trutta fario*). Genet. Sel. Evol. 36, 643-661.

Cockerham, C.C., 1970. Avoidance and rate of inbreeding. In: K. Kojima (ed.), Mathematical Topics in Population Genetics, Springer Verlag, New York, pp. 104-127.

Doyle, R.W. and C. Herbinger, 1994. The use of DNA fingerprinting for high-intensity, within-family selection in fish breeding. Proc. 5th WCGALP 19, 364-371.

Eknath, A.E., 1991. NAGA. The ICLARM Quarterly 738, 13-14.

Eknath, A.E., M.M. Dey, M. Rye, B. Gjerde, T.A. Abella, R. Sevilleja, M.M. Tayamen, R.A. Reyes and H.B. Bentsen, 1998. Selective breeding of Nile Tilapia for Asia. Proc. 6th WCGALP 27, 89-96.

Falconer, D.S. and T.F.C. Mackay, 1996. Quantitative Genetics. Longman Group Ltd., Harlow, UK, 464 pp.

Fessehaye, Y., Z. El-Bialy, M.A. Rezk, R. Croojimans, H. Bovenhuis and H. Komen, 2006. Mating systems and male reproductive success in Nile tilapia (*Oreochromis niloticus*) in breeding hapas: A microsatellite analysis. Aquaculture 256, 148-158.
Gilmour, A.R., B.J. Gogel, B.R. Cullis, S.J. Welham and R. Thompson, 2002. ASReml User Guide Release 1.0. VSN International Ltd, Hemel Hampstead, HP1 1ES, UK.
Gjerde, B., H.M. Gjoen and B. Villanueva, 1996. Optimum designs for fish breeding programmes with constrained inbreeding Mass selection for a normally distributed trait. Livest. Prod. Sci. 47, 59-72.
Groeneveld, E., 1990. PEST User's Manual. Dept. Anim. Sci., Univ. of Illinois, USA.
Huang, C.M. and I.C. Liao, 1990. Response to mass selection for growth rate in *Oreochromis niloticus*, Aquaculture 85, 199-205.
Hulata, G., G.W. Wohlfarth and A. Halevy, 1986. Mass selection for growth rate in the Nile tilapia (*Oreochromis niloticus*). Aquaculture 57, 177-184.
Kinghorn, B.P., 2000. The tactical approach to implementing breeding programs. In: B.P. Kinghorn, J.H.J. Van der Werf and M. Ryan (eds.), Animal breeding – use of new technologies. The Post Graduate Foundation in Veterinarian Science of the University of Sydney, Australia, pp. 291-308.
Maluwa, A.O.H., 2005. PhD Thesis. Norwegian University of Life Sciences, Norway.
McPhee, C.P., C.M. Jones and S.A. Shanks, 2004. Selection for increased weight at 9 months in redclaw crayfish (*Cherax quadricarinatus*). Aquaculture 237, 131-140.
Nomura, T. and K. Yonezawa, 1996. A Comparison of Four Systems of Group Mating for Avoiding Inbreeding.Genet. Sel. Evol. 28, 141-159.
Ponzoni, R.W., A. Hamzah, S. Tan, N. Kamaruzzaman, 2005. Genetic parameters and response to selection for live weight in the GIFT strain of Nile Tilapia (*Oreochromis niloticus*). Aquaculture 247, 203-210.
Reddy, P.V.G.K., B. Gjerde, B., K.D. Mahapatra, R.K. Jana, J.N. Saha, M. Rye and P.K. Meher, 1999. Selective breeding Procedures for Asian Carps. CIFA, Bhubaneswar, India.
Rezk, M., 2004. Genetic improvement of tilapia, mass selection approach. In: Overview of WorldFish Center's Work on Genetic Improvement 1999-2003, WorldFish Center Report, Penang, Malaysia, pp. 48-52.
Ridha, M.T., 2004. Preliminary study on growth, feed conversion and production in non-improved and improved strains of the Nile Tilapia. Proc. 6th ISTA I, 60-72.
Ridha, M.T., 2006. Comparative study of growth performance of three strains of Nile tilapia, *Oreochromis niloticus*, L. at two stocking densities. Aquaculture Research 37, 172-179.
Sonesson, A.K., 2005. A combination of walk-back and optimum contribution selection in fish: a simulation study. Genet. Sel. Evol. 37, 587-599.
Teichert-Coddington, D.R. and R.O. Smitherman, 1988. Lack of response by *Tilapia nilotica* to mass selection for rapid early growth. Transactions American Fisheries Society 117, 297-300.
Uraiwan, S. and R.W. Doyle, 1986. Replicate variance and the choice of selection procedures for tilapia (*Oreochromis niloticus*) stock improvement in Thailand. Aquaculture 57, 93-98.
Van Vleck, L.D., 1993. Selection Index and Introduction to Mixed Model Methods. CRC Press Inc., Florida, USA. 481 pp.
Villanueva, B., J.A. Woolliams and B. Gjerde, 1996. Optimum design for breeding programmes under mass selection with an application in fish breeding. Anim. Sci. 63, 563-576.
WorldFish Center, 2004. GIFT Technology Manual: An aid to Tilapia selective breeding. WorldFish Center, Penang, Malaysia.

Chapter 3

Managing nutrients in fishponds

Nutrient balances in ponds

Marc C.J. Verdegem
Animal Sciences Group, Aquaculture and Fisheries Institute, Wageningen University, P.O. Box 338, 6700 AH Wageningen, The Netherlands; marc.verdegem@wur.nl

Abstract

Globally the vast majority of aquaculture production of fish and crustaceans is from ponds. In spite of the overwhelming importance of ponds as an aquaculture production unit our basic understanding of the pond ecosystem is still limited. Future increases in fish and crustaceans availability will to a large extent, depend on our ability to raise pond production. As land and water resources are becoming increasingly scarce, opportunities to bring new pond areas in production are limited and increases in global pond production will mainly have to be realised by raising pond productivity. The challenge is to do this while mitigating the environmental impacts from pond culture. Possible routes to follow include (1) improving the conversion of primary and microbial production into fish and shrimp production, (2) integrating pond technology with recirculation technology and (3) developing integrated aquaculture-agriculture systems. Irrespective of the route chosen, a better understanding of the underlying processes explaining nutrient dynamics in ponds will be instrumental.

Keywords: pond, fish, shrimp, nutrients, water-use

Introduction

The goal of this manuscript is to sketch out why research towards better understanding of nutrient balances in ponds is important. These goals are set considering some of the major challenges for pond based production in the future.

Importance of pond aquaculture

An overview of the global production in 2003 of the principal cultured fish and crustacean species are given in Table 3.1a and 3.1b. Twenty-one species provide 23 million metric ton per year. From the list, only Atlantic salmon (1.1 million MT) and rainbow trout (0.5 million MT) are not primarily raised in ponds. Generalising, one can say that culture of fish and shrimp mainly takes place in ponds.

Table 3.1 also reviews the principal species used in polyculture in Asia. Traditionally, different species are stocked together in ponds with the goal to better utilise the natural food produced in the pond (Figure 3.1). The majority of species listed are stocked in polyculture ponds.

*Table 3.1a. Most important aquaculture species ranked according to global production; species commonly used in Asian polyculture marked with * (FISHSTAT+; www.fao.org). Column on polyculture indicates if the species is mainly cultured in ponds.*

English name	Polyculture	Scientific name	Metric ton
Silver carp	*	*Hypophthalmichthys molitrix*	3,828,248
Grass carp	*	*Ctenopharyngodon idella*	3,682,994
Common carp	*	*Cyprinus carpio*	3,239,712
Bighead carp	*	*Aristichthys nobilis*	1,928,622
Crucian carp	*	*Carassius carassius*	1,794,167
Nile tilapia	*	*Oreochromis niloticus*	1,367,679
Atlantic salmon		*Salmo salar*	1,131,241
Whiteleg shrimp	*	*Litopenaeus vannamei*	723,858
Rohu	*	*Labeo rohita*	713,267
Giant tiger prawn	*	*Penaeus monodon*	666,071
Catla	*	*Catla catla*	566,051
Milkfish	*	*Chanos chanos*	552,043
White amur bream	*	*Parabramis pekinensis*	524,927
Mrigal carp	*	*Cirrhinus mrigala*	514,662
Rainbow trout		*Oncorhynchus mykiss*	490,652
Chinese river crab	*	*Eriocheir sinensis*	368,050
Channel catfish	*	*Ictalurus punctatus*	301,192
Black carp	*	*Mylopharyngodon piceus*	270,279
Japanese eel	*	*Aguilla japonica*	222,469
Fleshy prawn	*	*Fenneropenaeus chinensis*	195,385
Giant river prawn	*	*Macrobrachium rosenbergii*	180,221
Total			23,261,790

Are ponds efficient production systems?

Although ponds are important culture systems, they are not very efficient systems from a nutrient use point of view. In semi-intensive ponds, 5-25% N, 20% organic carbon (OC) and 5-18% P of the nutrient input is recovered in harvested fish or shrimp. In intensive ponds, the recovery is higher with 11-27% N, 26-65% OC and 20-32% P of the nutrient input being recovered. Most of the nutrients end up in the sediment. In semi-intensive ponds, of the input 66-70% N, 38-46% OC and 35-86% P accumulate in the sediment (Avnimelech and Lacher, 1979; Sinha *et al.*, 1980; Boyd, 1985; Edwards, 1993; Acosta-Nassar *et al.*, 1994; Green and Boyd, 1995). Highest recovery rates in harvested fish were reported in super-intensive indoor recirculation systems, with 50% N and 53% P recovery (Schneider *et al.*, 2005).

In tropical, well-fertilised ponds, primary production fluctuates between 4 and 8 g C m^{-2} d^{-1}. Assuming that 50% of organic matter (OM) is carbon, this means that 29,200–58,400 kg

Table 3.1b. List of additional fish species commonly used in Asian polyculture.

English name	Scientific name
Blue tilapia	*Oreochromis aureus*
Calbasu	*Labeo calbasu*
Fathead minnow	*Pimephales promelas*
Giant gourami	*Osphronemus goramy*
Gonia	*Labeo gonius*
Grey mullet	*Mugil cephalus*
Kissing gourami	*Helostoma temminckii*
Mozambique tilapia	*Oreochromis mossambicus*
Mud carp	*Cirrhinus molitorella*
Shire tilapia	*Oreochromis shiranus shiranus*
Silver barb	*Puntius gonionotus*
Silver sharkminnow (Nilem)	*Osteochilus hasseltii*
Straightfin barb	*Barbus paludinosus*
Tambaqui	*Colossoma macropomum*
Threespot barb	*Barbus trimaculatus*
Wu-Chang fish	*Megalobrama amblycephala*

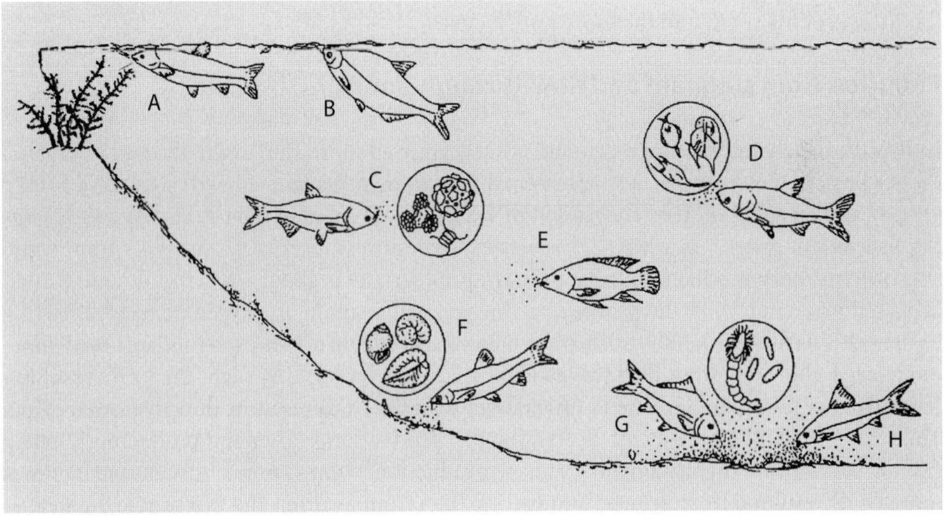

Figure 3.1. Chinese polyculture pond. Each species has it own preferences. Grass carp (A) and Wuchang bream (B) prefer grasses and macrophytes; Silver carp (C) eats phytoplankton; Bighead carp (D) focuses on zooplankton; tilapia (E) eats plankton and detritus from the water column and sediment; Black carp (F) prefers snails; and common carp (G) and mud carp (H) search the sediment for small benthic animals and organic matter (taken from Azim and Wahab, 2005).

organic dry matter is produced per ha per year. In ponds, with an annual fish production of 1000–4000 kg ha^{-1} yr^{-1}, this represents a dry matter production of 250–1000 kg ha^{-1} yr^{-1}. In consequence, only 0.85–1.7% of the primary production is harvested as fish biomass.

Carbon, nitrogen and phosphorus are important building blocks of organic matter. Various stable (non-radio active) isotopes of each element exist, and the isotope ratio can be measured in plankton, detritus, macrophytes, benthic organisms and periphyton, feed and fish. By comparing isotope ratios of the different food types and the fish, it is possible to make a rough estimate of the contribution of the various food types to fish production. Stable isotope studies indicate that, even in supplementary fed ponds, 50-80% of the fish or shrimp production is based on natural food items (Schroeder, 1983; Anderson *et al.*, 1987). In consequence, about half of the feed administered to ponds is nothing more than an expensive organic fertiliser, indicating that there are substantial opportunities to improve feeding strategies in ponds.

Hence, when supplemental diets are applied to ponds, then only 0.4–0.8% of the primary production is retained in fish production. Although a 45-50% protein diet will yield faster growth than a 30% protein diet, tilapia pond farmers for example prefer to use the low protein diet, because nutrients not used by the fish, contribute to the production of natural food organisms (algae, periphyton, zooplankton, and detritus) which in turn complement the fish diet. This is partially confirmed by observations of production in ponds where fish are confined to a small area of the pond in cages or pens. The growth will be 10-30% less than the growth of the same biomass of fish swimming free in a similarly sized pond (Schwedler *et al.*, 1989). A possible partial explanation is that free-swimming fish have better access to natural food than confined fish and therefore grow better.

Pollution from stagnant and flow-through ponds

Indirectly, the chemical oxygen demand (COD) gives an indication of the amount of organic matter present. For example, a supplemental 30% protein tilapia feed with a COD of 1400 g O_2 kg^{-1} and an apparent feed conversion of 1.5 will result in an oxygen demand of 1155 g O_2 (kg fish weight gain)$^{-1}$, of which 2/3 is waste (Verdegem *et al.*, 1999). Nutrients from waste stimulate primary production. If the primary productivity is about 4 g C m^{-2} d^{-1} and it is all broken down aerobically, the primary production will result in an oxygen demand of 104 kg O_2 ha^{-1} d^{-1}. Assuming 3,500 kg fish are produced directly on a 30% protein tilapia feed, then, on average, the COD from feed wastes will be 7 kg O_2 ha^{-1} d^{-1}. As such, the COD resulting from primary production will be 15 times larger than the COD resulting directly from feeding. This ratio of 15 will increase with decreasing fish production, e.g. a pond producing 2,000 kg fish ha^{-1} yr^{-1} with a slightly smaller primary production (3.5 g C m^{-2} d^{-1}), will have 21 times more COD resulting from primary production than from feeding. This is a simple theoretical calculation, and the pollution from ponds is highly influenced by the water exchange rate and degree of water mixing in the water column. These factors must also be taking into consideration when evaluating the pollution from ponds.

The self-aeration capacity of ponds limits the production capacity. Farmers can increase production by replacing organically-enriched water (high oxygen demand) with clean water (low oxygen demand). Supposing a tilapia farm producing 30 metric tons ha^{-1} yr^{-1} replaces 15% of the water volume daily. In spite of high water replacement rates the primary production remains important in such ponds. Nutrient enriched bottom water with a COD close to 100 mg O_2 l^{-1} will be mostly replaced (Boyd, 1982). On an annual basis, 54,750 kg O_2 ha^{-1} will be discharged, compared to a waste production due to feeding of 23,100 kg O_2. The resulting dissolved oxygen demand of the discharge is 2.4 times higher than the oxygen demand resulting from feeding. When draining a stagnant pond at harvesting the resulting oxygen demand will be 1/3 of the feed related oxygen demand. As the oxygen demand exerted on receiving surface waters is so high, many governments set quality standards for aquaculture effluents, which farmers can meet applying simple technologies. For example, passing pond effluents through stabilisation or waste settlement ponds will effectively reduce the oxygen demand exerted by aquaculture effluents.

Do we understand the basic ecology of ponds?

Ponds are complex production eco-systems. Although there is an enormous amount of information available on ecological processes in aquatic ecosystems, some very basic skills necessary for good system management are still insufficiently developed:

- It is nearly impossible to make a balanced nutrient budget for basic nutrients like N or P. Even in closed recirculation systems, it is normal to find 15-25% of the budget unexplained. What is worse, the budget can be under or over-estimated (unpublished data). This raises serious questions about, either our understanding of the N and P cycle (do we measure all the right things?) or the correctness of the analytical methods used (do we measure correctly?).
- Practical models to predict pond aquaculture production with a user interface useful to farmers are still lacking. The most commonly used decision support software is POND© (pdacrsp.oregonstate.edu/datatools/datatools_txt.html). Most other models were developed as research tools and focus on one aspect of the ecosystems, e.g. N-flow (Jiménez-Montealegre *et al.*, 2002) or for integrated agriculture-aquaculture systems (Jamu and Piedrahita, 2002a,b).
- The contribution of natural food to production is still insufficiently understood. In shrimp culture, the addition of carbohydrate had a profound effect on the heterotrophic bacteria count in both water column and sediment (Hari *et al.*, 2004, 2006). Bacterial flocks or biofilms are rich sources of protein, fatty acids, vitamins and mineral that apparently contribute to the shrimp diet (Tacon *et al.*, 2002). Indirect estimates can be made of the contribution from bacterial flocks or biofilms to shrimp production, however the mechanisms of food selection and uptake need further research in order to optimise pond nutrition practices.

If the goal is to fully control production in ponds, then more research into basic ecological processes in ponds will be necessary.

Future of pond production?

By 2050, fish consumption is predicted to increase from the present 13-16 kg to 35 kg caput^{-1} yr^{-1} (Wijkstrom, 2003). The latter will not be possible if relying on present pond culture technology, simply because there is not enough water available to do so. Today, the bulk of inland aquaculture production still takes place in extensive ponds with an average production of less than 2,000 kg ha^{-1} yr^{-1}. Counting water losses due to drainage at harvest, and evaporation and seepage losses, 45 m^3 of water are consumed per kg fish produced (Verdegem *et al.*, 2006). Raising aquaculture production in ponds to meet future consumer demands in 2,000 kg ha^{-1} yr^{-1} ponds means an additional water consumption of ± 400 m^3 caput^{-1} yr^{-1}. This amount is more than 3 times the residential per capita water consumption in developed countries and is similar to the total renewable per capita water resources in many Middle Eastern and North African countries (Rosegrant *et al.*, 2002). However, the real increase in water consumption would be much higher, as population growth still needs to be considered.

Water use in aquatic production systems can be drastically reduced through recirculation technology, such as applied in the Netherlands, for various species, including African catfish (*Clarias gariepinus*), European eel (*Anguilla anguilla*) and turbot (*Psetta maxima*). The realised feed conversions in catfish, eel and turbot recirculating farms are 0.85, 1.3 and 1.1, respectively (Verdegem *et al.*, 2006). Water use in these systems is expressed per kg feed because it describes the effectiveness of the water treatment unit of the farm. Low values are realised, especially in systems with a denitrification unit. Water exchange ranges between 0.1 m^3 per kg feed for catfish and eel, and 0.8 m^3 per kg feed for turbot (Eding and Kamstra, 2002). This reduces water use per kg fish produced to the range of 0.085-0.880 m^3.

Hence, in order to meet future demands, the challenge is to increase pond production intensity, and to do this in a sustainable way. Possible routes to follow include:

- Improving the conversion of primary and microbial production into fish and shrimp production, and to upgrade the use of natural food production in ponds.
- Integrating pond technology with recirculation technology, while intensifying production.
- Developing integrated aquaculture-agriculture systems in which a higher percentage of the input nutrients end up in harvestable products, while reducing the percentage of lost or unused nutrients in the production cycle.

For each route, a better understanding of the underlying processes explaining nutrient dynamics in ponds will be instrumental.

References

Acosta-Nassar, M.V., J.M. Morell and J.E. Corredor, 1994. The nitrogen budget of a tropical semi-intensive freshwater fish culture pond. Journal of the World Aquaculture Society 25, 261-270.

Anderson K.R., P.L. Parker and A. Lawrence, 1987. A $^{13}C/^{12}C$ tracer study on the utilization of presented feed by a commercially important shrimp *Penaeus vannamei* in a pond growout system. Journal of the World Aquaculture Society 18: 148-155.

Avnimelech, Y and M. Lacher, 1979. A tentative nutrient balance for intensive fish ponds. Bamidgeh 31: 3-8.

Azim, M.E. and M.A. Wahab, 2005. Periphyton based pond polyculture. In: Azim, M.E., M.C.J. Verdegem, A.A. van Dam and M.C.M Beveridge (Eds.), Periphyton: ecology, exploitation and management. CAB International, UK, pp. 207-222.

Boyd, C.E. 1982. Water quality management for pond fish culture. Develoments in Aquaculture and Fisheries Science, 9. Elsevier Science Publishers, The Netherlands, 318 pp.

Boyd, C.E., 1985. Chemical budgets for channel catfish ponds. Transactions of the American Fisheries Society 114, 291-298.

Eding, E. and A. Kamstra, 2002. Netherlands farms tune recirculation systems to production of varied species. Global Aquaculture Advocate 5, 52-54.

Edwards, P., 1993. Environmental issues in integrated agriculture-aquaculture and wastewater-fed fish culture systems. Environment and aquaculture in developing countries. ICLARM conference proceedings 31, 139-170.

Green, B.W. and C.E. Boyd, 1995. Chemical budgets for organically fertilized fish ponds in the dry tropics. Journal of the World Aquaculture Society 26, 284-296.

Hari, B., M.B. Kurup, J.T. Varghese, J.W. Schrama and M.C.J. Verdegem, 2004. Effects of carbohydrate addition on production in extensive shrimp culture systems. Aquaculture 241, 179-194.

Hari, B., M.B. Kurup, J.T. Varghese, J.W. Schrama and M.C.J. Verdegem, 2006. The effect of carbohydrate addition on water quality and the nitrogen budget in extensive shrimp culture systems. Aquaculture 252, 248-263.

Jamu, D.M. and R.H. Piedrahita, 2002a. An organic matter and nitrogen dynamics model for the ecological analysis of integrated aquaculture/agriculture systems: I. Model development and calibration. Environmental Modelling and Software 17, 571-582.

Jamu, D.M. and R.H. Piedrahita, 2002b. An organic matter and nitrogen dynamics model for the ecological analysis of integrated aquaculture/agriculture systems: II. Model evaluation and application. Environmental Modelling and Software 17, 583-592.

Jiménez-Montealegre, R., M.C.J. Verdegem, A. van Dam and J.A.J. Verreth, 2002. Conceptualization and validation of a dynamic model for the simulation of nitrogen transformations and fluxes in fish ponds. Ecological Modelling 147, 123-152.

Rosegrant, M.W., X. Cai and S.A. Cline, 2002. World water and food to 2025: Dealing with Scarcity. Washington DC: International Food Policy Research Institute.

Schneider, O., V. Sereti, E.H. Eding and J. Verreth, 2005. Analysis of nutrient flows in integrated intensive aquaculture systems. Aquacultural Engineering 32, 379-401.

Schroeder, G.L., 1983. The role of natural foods in tilapia growth: a study based on stable isotope analysis. In: L. Fishelson and Z. Yaron (Eds.), International Symposium on Tilapia in Aquaculture, Nazareth, Israel 8-13 May 1983, Israel, pp. 313-322.

Schwedler, T.E., J.R. Tomasso and J.A. Collier, 1989. Production characteristics and size variability of channel catfish reared in cages and open ponds. Journal of the World Aquaculture Society 20, 158-161.

Sinha, V.R.P., H.A. Khan, D.P. Chakraborty and M.V. Gupta, 1980. Preliminary observations on nitrogen balance of some ponds under composite fish culture. Aquacultura Hungarica (Szarvas) II, 105-116.

Tacon, A.G.J., J.J. Cody, L.D. Conquest, S. Divakaran, I.P. Forster and O.E. Decamp, 2002. Effect of culture system on the nutrition and growth performance of Pacific white shrimp *Litopenaeus vannamei* (Boone) fed different diets. Aquaculture Nutrition 8, 121-137.

Verdegem, M.C.J., E.H. Eding, J.M. van Rooij and J.A.J. Verreth, 1999. Comparison of effluents from pond and recirculating production systems using formulated diets. World Aquaculture 30, 28-36.

Verdegem, M.C.J., R.H. Bosma and J.A.J. Verreth, 2006. Reducing water use for animal production through aquaculture. International Journal of Water Resources Development 22, 101-113.

Wijkstrom, U.N., 2003. Short and long term prospects for consumption of fish. Veterinary Res. Comm. 27, Suppl. 1, 461-468.

Multi-species fishpond and nutrient balance

Mustafizur M. Rahman and Marc C.J. Verdegem
Aquaculture and Fisheries Group, Wageningen University, P.O. Box 338, 6700 AH Wageningen, The Netherlands; mustafizur.rahman@wur.nl

Abstract

The efficiency of pond aquaculture largely depends on species combination, stocking density and ratio, and the quantity and quality of nutrient inputs. In semi-intensive polyculture, fish production is mainly based on natural productivity, which greatly depends on nutrients availability in the water column. The addition of benthivorous fish enhances the release of nutrients from the sediment, stimulating phytoplankton productivity and accelerating the flux of nutrients to higher trophic levels. Density of benthivorous fish is very important. Optimum density of benthivorous and planktivorous fish in polyculture can improve synergistic effects, which increase nutrient retention efficiency in fish and decrease nutrient losses. Besides decreasing nutrient loss this type of polyculture maximises resource utilisation. This type of aquaculture is very appropriate in Asia, where investments in waste treatment are minimal or non-existent.

Keywords: nutrients, polyculture, carp, synergism, behaviour

Introduction

Multi-species aquaculture (polyculture) is based on the assumption that each species stocked has a feeding niche that does not completely overlap with the feeding niche of the other species present. Therefore, by stocking multiple species, a more complete use is made of the space and natural food available in the pond, maximising resource utilisation. In some cases, one species enhances food availability, and hence production of another species, a phenomenon that is referred to as 'synergism' (Swingle, 1966; Hepher *et al.*, 1989). When synergism occurs, nutrient use efficiency is increased while environmental impacts from fish farming will be reduced. The number of species combinations that can be made in polyculture is enormous. In addition, numerous biotic and abiotic factors also influence farming success. Therefore, in experiments designed to understand synergism, environmental conditions should be standardised as much as possible, and the number of species stocked should be limited.

Polyculture is commonly applied in semi-intensive ponds in which the flow of nutrients through the food web depends to a large extent on nutrient availability in the water column. In undisturbed water bodies, diffusion across the sediment-water interface is the main mechanism by which nutrients become available in the water column (Wetzel, 1975; Avnimelech and Kochva, 1999). In aquaculture ponds, the fluxes of nutrients between sediment and water column are enhanced by fish-driven re-suspension, which quickly becomes more important than diffusion (Ritvo *et al.*, 2004; Tarvainen *et al.*, 2005). Fish-driven re-suspension is particularly important in low-input ponds, where fish production depends almost entirely on natural food availability. Sometimes farmers know that resuspension of nutrients has a positive effect on production; e.g. in Bangladesh harrowing the pond bottom to re-suspend nutrients is a traditional method to enhance pond production (Beveridge *et al.*, 1994) The objectives of this paper are (1) to review how interactions among fish species affect nutrient concentrations, nutrient accumulation, natural food availability and total fish production, and (2) to analyse the underlying mechanisms explaining synergism for growth and production.

Synergistic effects in multi-species pond

Due to synergistic effects, the growth and production of some fish species may be higher in polyculture than monoculture (Table 3.2). Silver carp *Hypophthalmichthys molitrix* grown in polyculture with common carp *Cyprinus carpio* attains a higher harvest weight than when grown in monoculture (Yashouv, 1971; Hepher *et al.*, 1989). Rohu *Labeo rohita* and total pond production increase by factors of 1.4 and 2, respectively, in the presence of common carp (Rahman *et al.*, 2006). Synergistic effects among fish species can be explained mainly on the basis of two interrelated processes: (1) increase in natural food availability and (2) the improvement of environmental conditions in the pond. Both processes enhance growth and production of fish. The mechanism through which the stocking of an additional species contributes to increased food availability and improved environmental condition largely depends on its specific place in the pond food web (Milstein, 1992). Synergistic effects are

Table 3.2. Reported yield in different mono and polyculture systems. All ponds were fertilised. Some ponds received in addition artificial feed (referred to as "Feed"), some ponds received substrate for periphyton development (referred to as "Periphyton").

Nutrient source	Culture systems	Yield ($t\ ha^{-1}\ yr^{-1}$)*	Reference
-	Monoculture (silver carp)	2.1	Hepher *et al.*, 1989
-	Polyculture (silver carp, common carp, hybrid tilapia, grey mullet & grass carp)	3.9	Hepher *et al.*, 1989
-	Monoculture (silver carp)	0.24	Hepher *et al.*, 1989
-	Polyculture (silver carp, common carp, tilapia, mullets & grass carp)	2.0	Hepher *et al.*, 1989
Periphyton	Monoculture (rohu)	2.4	Azim *et al.*, 2001
Periphyton	Monoculture (catla)	1.4	Azim *et al.*, 2001
Periphyton	Polyculture (rohu & catla)	3.1	Azim *et al.*, 2001
Feed	Monoculture (rohu)	5.4	Rahman *et al.*, 2006
Feed	Polyculture (rohu & common carp)	10.2-11.8	Rahman *et al.*, 2006
-	Monoculture (rohu)	3.9	Rahman *et al.*, 2006
-	Polyculture (rohu and common carp)	5.9-7.0	Rahman *et al.*, 2006

*Reported values were extrapolated to annual production. Seasonal differences in growth and production were not considered.

more pronounced between benthivorous and herbivorous species. Benthivorous fish have very large effects on the pond ecosystem. Most benthivores, when feeding on benthic material increase oxygen availability in the sediment, accelerating the mineralisation and re-suspension of nutrients, which in turn has a large impact on the abiotic and biotic quality of the overlying water column (Northcote, 1988). Common carp is a very well known benthivore that has a larger impact on pond productivity than most other benthivorous species. For example, Wahab *et al.* (2002) achieved a 60% higher yield in rohu with common carp as bottom feeder compared to mrigal *Cirrhinus cirrhosus*. In an another study, Parkos III *et al.* (2003) observed larger increases in phytoplankton biomass and in total phosphorus availability in the presence of common carp than in the presence of channel catfish *Ictalurus punctatus*.

At higher densities, common carp may influence production negatively through: (1) over-grazing of zooplankton and benthic macro-invertebrates to an extent that recovery is not possible, (2) increases in pond soil redox potential values to >200 mV through increased oxygen supply, which may in turn cause the precipitation of soluble phosphorus (PO_4-P) as inorganic particles (e.g. with iron as iron (III) phosphate) (Holdren and Armstrong, 1980; Bostrom *et al.*, 1988; Boyd, 1995) and (3) an increase in turbidity due to re-suspended particles, which reduce light penetration and hence photosynthesis. However, considering both positive and negative effects on production, stocking densities of benthivorous fishes are important. Rahman *et al.* (2006) observed that total fish production was significantly lower in rohu ponds (1.5 rohu m^{-2}) where 1 common carp m^{-2} was stocked compared to ponds with 0.5 common carp m^{-2}. A possible explanation might be the inter-species competition for food with increasing fish density. Forester and Lawrence (1978) found that high density of common carp decreased standing crop of bluegill *Lepomis macrochirus* (Rafinesque) through food competition, which resulted in the bluegill ingesting their own eggs.

Another important factor is nutrient input as feed or fertiliser. The type of nutrients input may regulate the synergistic effects among benthivorous fishes (Figure 3.2). For example, higher nutrient inputs may increase the synergistic effects of benthivorous fishes on fish production. Rahman *et al.* (2006) observed the highest total fish yield of 11,800 kg $ha^{-1}yr^{-1}$ in high input ponds at 15,000 rohu plus 5,000 common carp ha^{-1}. This was 1,800 kg higher than the previous highest production record found in the literature for south Asia (10,100 kg $ha^{-1}yr^{-1}$; Mathew *et al.*, 1988). Obviously, there is an optimum level of nutrients input. More research is needed in this aspect.

Natural foods and its utilisation in multi-species pond

Selection of suitable species combinations to optimise production and food utilisation in ponds is very important in order to optimise nutrient use efficiency (Milstein *et al.* 1988). In fertilised ponds, fishes mainly feed on phytoplankton, zooplankton, benthic macro-invertebrates and detritus. The amounts of these natural foods in lakes and ponds (Table 3.3) are influenced by management factors, such as fish species combinations in polyculture, fish stocking density and ratio, and nutrient input quality and quantity (Milstein, 1993; Diana *et al.*, 1997). Fish feeding habits have an important influence on natural food availability, both directly by consumption and indirectly through influencing the food web and nutrient availability. For instance, bottom

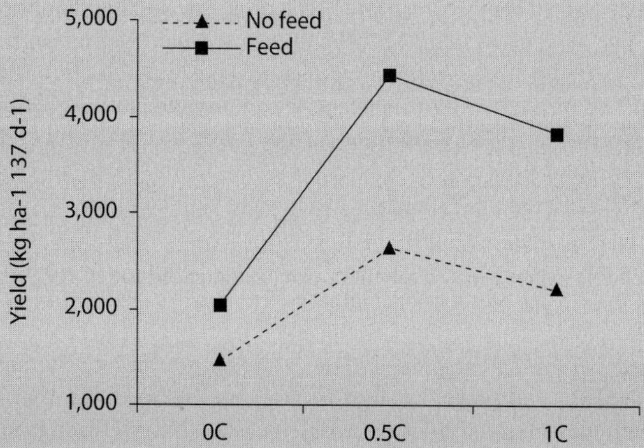

Figure 3.2. Interaction effects of common carp density and artificial feed on the total fish yield. 0C, 0.5C and 1C indicate treatments without common carp, with 0.5 common carp and 1 common carp m^{-2}, respectively (Rahman et al.*, 2006).*

Table 3.3. Reported biomass of natural foods in fishponds. (FM = fresh matter; DM = dry matter; int. = integrated).

Food types	Biomass	Unit	Pond system	References
Phytoplankton	44-65	gFM m^{-3}	Pig-fish int.	Yang, 1994
Phytoplankton	29-36	gDM m^{-2}	Int. fish	Takamura *et al.*, 1995
Phytoplankton	1-71	gFM m^{-3}	Shrimp	McIntosh *et al.*, 2001
Phytoplankton	121	gDM m^{-2}	Fish	Tang, 1970
Phytoplankton	265-352	mm^3FM m^{-3}	Fish	Rahman *et al.*, 2006
Zooplankton	3-5	gDM m^{-2}	Int. fish	Takamura *et al.*, 1995
Zooplankton	10-13	gFM m^{-3}	Pig-fish int.	Yang, 1994
Zooplankton	10	gDM m^{-2}	Fish	Tang, 1970
Zooplankton	2	gDM m^{-2}	Eutrophic	Iwakuma *et al.*, 1989
Zooplankton	42-67	mm^3FM m^{-3}	Fish	Rahman *et al.*, 2006
Zooplankton	4-10	gDM m^{-3}	Nursery	Molodtsova-Zaikina, 1977
Benthos	9-36	gDM m^{-2}	Nursery	Molodtsova-Zaikina, 1977
Benthos	15-70	gDM m^{-2}	Fish	Wade & Stirling, 1999
Macroinvertibrate	4	gDM m^{-2}	Fish	Oerlti, 1993
Macroinvertibrate	3.2-6.2	cm^3FM m^{-3}	Fish	Rahman *et al.*, 2006
Detritus	130	gDM m^{-2}	Pig-fish int.	Yang, 1994
Detritus	83	gDM m^{-2}	Pig-fish int.	Yang, 1994

feeders, such as common carp, searching for benthic macro-invertebrates re-suspend sediments, thereby influencing nutrient availability in the water column, which in turn affects photosynthesis and subsequently phytoplankton and zooplankton production. Rahman *et al.* (2006) found increases in phytoplankton (8-33%) and zooplankton (36-60%) biomass in the presence of benthivorous common carp. However, without planktivorous fish, nutrients accumulated in the plankton will not be transferred into fish biomass. The faecal pellets produced by planktivorous species make part of the plankton production available to benthivorous species like common carp or even omnivorous species like tilapias. Therefore, it makes sense to combine benthivorous or planktivorous fish, as each species benefits through increased food availability.

Fish density also plays an important role in food availability and its utilisation. If fish density is too great grazing pressure could be so high as to prevent recovery. In this case fish production will be negatively affected. For example, higher densities of phytoplanktivorous fishes reduce phytoplankton, which in turn adversely affects the production of zooplankton and the growth of zooplanktivorous fishes. With increasing stocking density competition increases, fish shift to less preferable foods as their preferred food items become depleted (Milstein, 1992). Rohu will eat more phytoplankton when zooplankton availability is less and will shift from phytoplankton to zooplankton when zooplankton is more abundant in the presence of common carp. Rohu ingested smaller quantities of phytoplankton and zooplankton when the abundance of these food items was reduces under high density of common carp (Rahman *et al.*, 2006).

Fish behaviour in multi-species pond

Social interaction between/among species may influence fish grazing, swimming and resting behaviours. For example, when three species of sunfish, the bluegill (*Lepomis macrochirus*), green sunfish (*L. cyanellus*) and the pumpkinseed (*L. gibbosus*), are stocked separately in ponds, each species preferred to forage mostly on invertebrates in the vegetation, but when they are stocked together, the diet of the green sunfish showed no change, the bluegill concentrated partially on prey from the water column, especially zooplankton, and the pumpkinseed largely fed on prey from the sediment (Werner and Hall, 1976, 1977, 1979). So, fishes behave differently when stocked in different species combinations.

In the presence of common carp, grazing, swimming and social behaviour of rohu were found to change. When rohu and common carp were cultured together, rohu spent 47-52% of the time in the company of common carp, increased their active time by 51-62% and decreased intra-specific schooling time by 55-71%. The net result is that rohu spent relatively more time grazing near the bottom instead of in the water column where the density of zooplankton is less (Rahman *et al.*, personal communication). Hence, in the presence of common carp, rohu ate more zooplankton, which resulted in better growth (Rahman *et al.*, 2006). Behaviour of common carp also changed with density. In treatments where common carp densities were highest, common carp spent relatively more time swimming in the water column at the expense of time grazing near the bottom. Spending more time swimming in the water column and less time grazing near the bottom might be the cause of reductions in individual growth rates of common carp at high densities. More research is needed to understand the underlying mechanisms of the observed changes.

Nutrients balance in multi-species pond

Earthen ponds receiving artificial feeds are the most widely used type of aquaculture production systems. The principal fate of artificial feed in ponds is supplying organic nutrients driving primary production and bacterial decomposition. Stable isotope studies have indicated that most fish production in ponds is based on natural food items, even in pellet fed ponds where 50-80% of fish and shrimp production is based on natural food (Schroeder, 1983; Anderson *et al.*, 1987). Therefore, pond aquaculture can benefit from improved strategies for natural food production and utilisation. This will also lead to less pollution from pond aquaculture. Even in ponds receiving protein-rich pellets, it was shown that only 11-35% of the supplied N and 13-36% of P was retained as fish biomass (Table 3.4). A large fraction of the unused N and P accumulated in the system affecting water quality in the overlaying water column. Most of these nutrients accumulated in the sediment where in absolute terms the quantities are 100-1000 times higher than in the water column (Biro, 1995). Re-suspension can transfer these nutrients back into the water column. However, re-suspension not only affects water quality

Table 3.4. Reported accumulation efficiency (%) of input nutrients by fish, phytoplankton, water and sediment.

	Accumulation (%)			
Types	N	P	System	References
Ictalurus punctatus	26.8	30.1	Earthen pond	Boyd, 1985
Oreochromis niloticus	18-21	16-18	Earthen pond	Green and Boyd, 1995
Hybrid tilapia	17.5	-	Earthen pond	Acosta-Nassar *et al.*, 1994
Sparus aurata	26	21	Marine pond	Neori and Krom, 1991
Sparus aurata	29	36	Marine pond	Krom *et al.*, 1985
Sparus aurata	21	26	Marine pond	Krom and Neori, 1989
Oncorhynchus mykiss	18.9	13.2	Circular tank	Foy and Rosell, 1991
Carp	11	32	Earthen pond	Avnimelech and Lacher, 1979
Penaeus monodon	24	13	Shrimp pond	Briggs and Funge-Smith, 1994
Penaeus monodon	22	-	Shrimp farm	Jackson *et al*, 2003
Phytoplankton	0.74	-	Earthen pond	Acosta-Nassar *et al.*, 1994
Particulate matter	46	50	Marine pond	Krom and Neori, 1989
Water	10.4	-	Earthen pond	Acosta-Nassar *et al.*, 1994
Water	14	21	Marine pond	Neori and Krom, 1991
Water	13	22	Marine pond	Neori and Krom, 1991
Detritus	10	17	Marine pond	Krom and Neori, 1989
Sediment	70	35-40	Earthen pond	Green and Boyd, 1995
Sediment	31	84	Shrimp pond	Briggs and Funge-Smith, 1994
Sediment	11	15	Marine pond	Neori and Krom, 1991
Sediment	14	-	Shrimp farm	Jackson *et al.*, 2003
Loss to atmosphere	0.92	-	Earthen pond	Acosta-Nassar *et al.*, 1994

but also the fraction of nutrient input accumulated in benthos, zooplankton, phytoplankton and culture animals. Rapid mineralisation and utilisation of nutrients not retained by culture organisms will improve water quality and natural food availability, improving the nutrient conversion ratios.

Nitrogen and phosphorus retention efficiency in fish biomass have been shown to almost double in polyculture with rohu and common carp, resulting in more nutrients passing through the pond food web and less nutrients accumulating in the sediments (Rahman *et al.*, personal communication). In polyculture, nitrogen and phosphorus accumulation in the sediment was reduced by 17-22% and 34-36%, respectively (Figure 3.3). It remains to be seen if a similar effect would be observed with species with more flexible feeding habits like tilapia compared to rohu. Also, it remains interesting for future research to compare the effects of man-made bottom disturbance through harrowing with the effects of fish driven re-suspension on nutrient use efficiency and production.

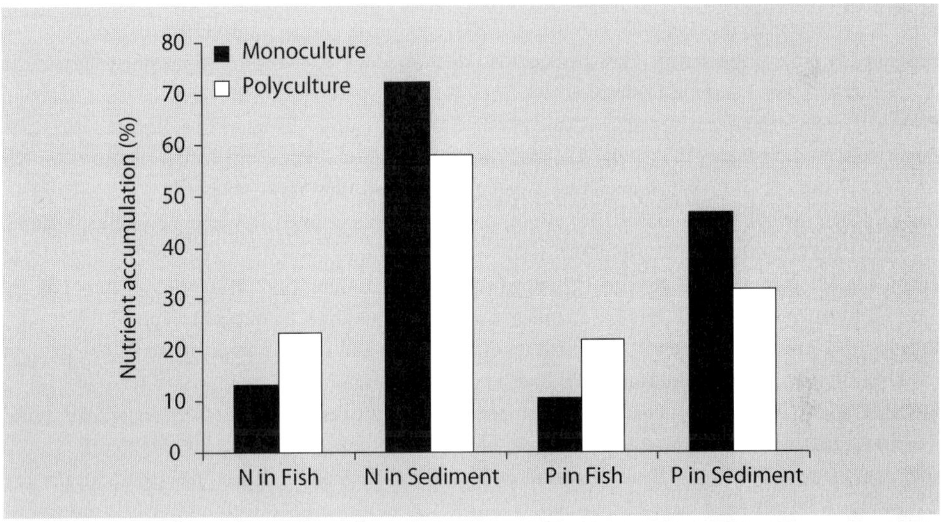

Figure 3.3. Comparison between mono (rohu) and polyculture (rohu plus common carp) on N and P accumulation (%) in fish and sediment. Higher total nutrients input and total fish densities were used in polyculture than monoculture. N and P values of polyculture are the mean of two polyculture system (rohu plus 0.5 and 1 common carp m^{-2}) (reproduced from Rahman et al., personal communication).

Conclusion

Inclusion of a benthivorous species at appropriate ratios and densities considerably contributes to increased production and greater nutrient efficiency in polyculture systems. As a result, also pollution from aquaculture per kg fish produced is reduced. These findings are particularly important in Asia, where investments in waste treatment are minimal or non-existent.

References

Acosta-Nassar, M.V., J.M. Morell and J.E. Corredor, 1994. The nutrient budget of a Tropical Semi-Intensive Freshwater Fish Culture Pond. J. World Aquac. Soc. 25, 261-270.

Anderson, P.K., P.L. Parker and A. Lawrence, 1987. A $^{13}C/^{12}C$ tracer study of the utilization of presented feed by a commercially important shrimp *Penaeus vannamei* in a pond grow out system. J. World Aqua. Soc. 18, 148-155.

Avnimelech, Y. and M. Kochva, 1999. Sedimentation and resuspension in earthen fish ponds. J. World Aqua. Soc. 30, 401-409.

Avnimelech, Y. and M. Lacher, 1979. A tentative nutrient balance for intensive fish ponds. Bamidgeh 31, 3-8.

Azim, M.E., M.A. Wahab, A.A. van Dam, M.C.M. Beveridge, E.A. Huisman and M.C.J. Verdegem, 2001. Optimization of stocking ratios of two Indian major carps, rohu *Labeo rohita* Ham. and catla *Catla catla* Ham. in a periphyton-based aquaculture system. Aquaculture 203, 33-49.

Beveridge, M.C.M., M.A. Wahab and S. Dewan, 1994. Effects of daily harrowing on pond soil and water nutrient levels and on rohu fingerling production. Progressive Fish-Culturist 56, 282-87.

Biro, P., 1995. Management of pond ecosystems and trophic webs. Aquaculture 129, 373-386.

Bostrom, B., J.M. Andersen, S. Fleischer and M. Jansson, 1988. Exchange of phosphorus across the sediment-water interface. Hydrobiol. 170, 229-244.

Boyd, C.E., 1985. Chemical budget for channel catfish ponds. Trans. American Fish. Soc. 114, 291-298.

Boyd, C.E., 1995. Exchange of Dissolved Substances between Soil and Water. In: Boyd, C.E. (Ed.), Bottom Soils, Sediment and Pond Aquaculture. Chapman and Hall, New York, 348 pp.

Briggs, M.R.P. and S.J. Funge-Smith, 1994. A nutrient budget of some intensive shrimp ponds in Thailand. Aquacult. Fish. Manage. 25, 789-811.

Diana, J.S., J.P. Szyper, T.R. Batterson, C.E. Boyd and R.H. Piedrahita, 1997. Water quality in ponds. In: H.S. Egna and C.E. Boyd (Eds.), Dynamics of pond aquaculture. CRC press LLC, pp. 53-72.

Forester, T.S. and J.M. Lawrence, 1978. Effects of grass carp and carp on populations of bluegill and largemouth bass in ponds. Trans. Am. Fish. Soc. 107, 172-175.

Foy, R.H. and R. Rosell, 1991. Loadings of nitrogen and Phosphorus from a Northern Ireland fish farm. Aquaculture 96,17-30.

Green, B.W. and C.E. Boyd, 1995. Chemical budgets for organically-fertilized fish ponds in the dry tropics. J. World Aqua. Soc. 26, 284-296.

Hepher, B., A. Milstein, H. Leventer and B. Teltsch, 1989. The effect of fish density and species combination on growth and utilization of natural food in ponds. Aquacult. Fish. Manage. 20, 59-71.

Holdren, G.C., Jr. and D.E. Armstrong, 1980. Factors affecting phosphorus release from intact sediment cores. Environ. Sci. Technol. 14, 79-86.

Iwakuma, T., K. Shibata and T. Hanazato, 1989. Production ecology of Phytoplankton and Zooplankton in a eutrophic pond dominated by *Chaoborous flavicans* (diptera: Chaoboridae). Ecol. Res. 4, 31-53.

Jackson, C., N. Preston, P.J. Thompson and M. Burford, 2003. Nutrient budget and effluent nitrogen components at an intensive shrimp farm. Aquaculture 218, 397-411.

Krom, M.D. and A. Neori, 1989. A total nitrogen budget for the experimental intensive fishpond with circularly moving seawater. Aquaculture 83, 345-358.

Krom, M.D., C. Porter and H. Gordin, 1985. Nutrient budget of marine fish pond in Eilat, Israel. Aquaculture 51, 65-80.

Mathew, P.M., B.K. Singh and D.P. Chakrabarty, 1988. A case study stocking of fry in six species composite fish culture. J. Maharashtra Agric. Univ.13, 79-82.
McIntosh, D., C. Collin, C. McKeon and C. Huntsinger, 2001. Phytoplankton biomass and community progression in an inland, low-salinity shrimp pond. Aquaculture, Book of Abstracts, World Aquaculture Society, 422 pp.
Milstein, A., 1992. Ecological aspects of fish species interactions in polyculture ponds. Hydrobiol. 231, 177-186.
Milstein, A., 1993. Water quality and freshwater fish culture intensification: the Israeli example. Aquacult. Fish. Manage. 24, 715-724.
Milstein, A., B. Hepher and B. Teltch, 1988. The effect of fish species combination in fish ponds on plankton composition. Aquacult. Fish. Manage. 19, 127-137.
Molodtsova-Zaikina, A.I., 1977. Increasing the efficiency of rearing young sturgeons in the 1st cycle. Mosk Rybn-Khoz 2, 11-15.
Neori, A. and M.D. Krom, 1991. Nitrogen and Phosphorus budgets in an intensive marine fishpond: the importance of microplankton. In: C.B. Cowey and C.Y Cho (Eds.), Proceedings of the First International Symposium on Nutritional Strategies in Management of Aquaculture Waste. University of Guelph, Ontario, Canada, pp. 187-205.
Northcote, T.G., 1988. Fish in the structure and function of freshwater ecosystems: a "top-down" view. Can. J. Fish. Aquat. Sci. 45, 361-379.
Oertli, B., 1993. Leaf litter processing and energy flow through macroinvertebrate in a woodland pond (Switzerland). Oecologia 96, 466-477.
Parkos III, J.J., V.J. Santucci Jr. and D.H. Wahl, 2003. Effects of adult common carp (*Cyprinus carpio*) on multiple trophic levels in shallow mesocosms. Can. J. Fish. Aquat. Sci. 60, 182-192.
Rahman, M.M., M.C.J. Verdegem, L.A.J. Nagelkerke, M.A. Wahab, A. Milstein and J.A.J. Verreth, 2006. Growth, production and food preference of rohu *Labeo rohita* (H.) in monoculture and in polyculture with common carp *Cyprinus carpio* (L.) under fed and non-fed ponds. Aquaculture (in press).
Ritvo, G., M. Kochba and Y. Avnimelech, 2004. The effects of common carp bioturbation on fishpond bottom soil. Aquaculture 242, 345-356.
Schroeder, G.L., 1983. The role of natural foods in tilapia growth: a study base don stable isotope analysis. In: N. Fishelson and Z. Yaron (Eds.), International Symposium on Tilapia in Aquaculture. Nazareth, Israel, pp. 313-322.
Swingle, H.S., 1966. Biological means of increasing productivity in ponds. Proceedings World Symposium Warm-water Pondfish Culture, FAO Fish. Rept. 44, 243-257.
Takamura, N., X.-B. Zhu, H.-Q. Yang, X.-Y. Jiang, J.-L. Li and Z.-P. Mei, 1995. Characteristics of plankton communities in Chinese integrated fish ponds: effects of excessive grazing by planktivorous carp on plankton communities. Hydrobiol. 315, 211-225.
Tang, Y.A., 1970. Evaluation of Balance between Fishes and Available Fish Foods in Multispecies Fish Culture Ponds in Taiwan. Trans. Amer. Fish Soc. 4, 708-718.
Tarvainen, M., A.-M. Ventela, H. Helminen and J. Sarvala, 2005. Nutrient release and resuspension generated by ruffe (*Gymnocephalus cernuus*) and chironomids. Freshwat. Biol. 50, 447-458.
Wade, J.W. and H.P. Stirling, 1999. Fertilization of earth ponds 3: Effects on benthic macroinvertibrates and ecological interrelationship with aquatic macrophyte. J. Aquat. Sci.14, 19-27.
Wahab, M.A., M.M. Rahman and A. Milstein, 2002. The effect of common carp, *Cyprinus carpio* (L.) and mrigal, *Cirrhinus mrigala* (Hamilton) as bottom feeders in major Indian carp polycultures. Aquacult. Res. 33, 547-556.

Werner, E.E. and D.J. Hall, 1976. Niche shifts in sunfishes: experimental evidence and significance. Science 191, 404-406.

Werner, E.E. and D.J. Hall, 1977. Competition and habitat shift in two sunfishes (*Centrarchidae*). Ecology 58, 869-876.

Werner, E.E. and D.J. Hall, 1979. Foraging efficiency and habitat switching in competing sunfish. Ecology 60, 256-264.

Wetzel, R.G., 1975. Limnology. Saunders College Publishing, Philadelphia, Pennsylvania, USA, 743 pp.

Yang, H., 1994. Studies on the ecological-economic benefits of "pig-grass-fish" farming model. The selected paper of basic integrated fish farming theory and fishery bioeconomic research, pp. 71-88.

Yashouv, A., 1971. Interaction between the common carp (*Cyprinus carpio*) and the silver carp (*Hypopthhalmichthys molitrix*) in fish ponds. Bamidgeh 23, 85-92.

Nutrient accumulation in Tilapia pond sediment and its agricultural uses

Amararatne Yakupitiyage, S.L. Ranamukhaarachchi, Yang Yi and Rahman Mizanur
Aquaculture and aquatic Resources Management field of Study, School of Environment, Resources and Development, Asian Institute of Technology (AIT), Bangkok, Thailand

Abstract

Six experiments were conducted at the Aquaculture research facility of the Asian Institute of Technology, Pathumthani, Thailand, focusing on nutrient accumulation in fishpond sediment and its potential agricultural uses. It was found that 63-72 kg m^{-2} sediment accumulated during a single production cycle of 180 to 400 days. The upper 10 cm layer of pond sediment contained around 2.1 g N kg^{-1} and close to 190 mg P kg^{-1}. The fertiliser values of sediment of a 1 ha pond were equivalent to about 3.0 t urea and 0.7 t triple super phosphate. Accumulation of both P and N ceases after 5-6 months of fish culture, but sediment K-content continuous to increase during the trial. Further studies are required to determine the most cost effective way of removing pond sediments at the appropriate time (after 5–6 months of fish culture) to maximise nutrient recovery and minimise nutrient losses. Pond sediment improved moisture content, pH and aggregate stability of alluvial acid sulphate soil in Central Thailand, and the yield of morning glory and chilli pepper. Depending on the availability and the purpose of application, i.e. fertiliser supplementation or soil conditioner, farmers may use either 25% or 50% of pond sediment mixed with farm soils. Supplementary use of locally available N and K fertilisers and manures is also recommended. Further research would also help establish the soil conditioning effects of pond sediments on marginal sandy soils using leguminous crops. Marginal sandy soils are frequent in areas were the poor quality resources are reflected in the poverty of the agricultural population.

Keywords: fishpond, sediment, nutrients, fertiliser, vegetables

Introduction

A large amount of the nutrients applied to fishponds remains unused, although a large fraction accumulates as organic matter (OM) and bottom sludge (Boyd, 1995). Through re-suspension, chemical transformations and microbial breakdown of OM, nutrients in sediment are partially released to pond water (Funge-Smith and Briggs, 1994). This is a continuous process, which affects primary productivity and subsequent fish yield (Jamu and Piedrahita, 2002). Scientific information regarding quantification of nutrients and OM accumulation and microbial breakdown over time in sediment of various pond aquaculture systems is scarce.

Pond sediments represent a potential source of environmental pollution if not managed judiciously. To avoid or reduce environmental pollution, the utilisation of nutrients and OM of pond sediment for economic activities (such as crop production, nursery culture of plants, horticultural activities, etc.) can be considered as a viable option. In fact, some farmers reuse the nutrients accumulated in pond sediments to supplement fertiliser requirements of crops. Scientific information regarding nutrient quality, potential fertiliser value and efficiency of sediment use for crop production are scant. Therefore our on-station research experiments focused on nutrient accumulation in fishpond sediment and its potential agricultural uses through a series of six experiments.

Methodology

The on-station component of the research focused on the use of tilapia pond sediment using Nile tilapia (*Oreochromis niloticus*), which is grown widely in Asia. A total of 6 pond- and plot-based trials were conceived to quantify physical and chemical properties of tilapia pond sediment and to assess it agricultural potential using two vegetables, morning glory (*Ipomoea reptans L.*) and chilli pepper (*Capsicum annuum longum*), which were common to all 3 study areas (Bangladesh, North East Thailand and Vietnam).

Sedimentation and nutrient build up in tilapia ponds under different input systems

Organic matter and nutrient accumulation were quantified in a factorial experiment with three types of pond inputs [*viz.* inorganic fertiliser (IF), IF plus chicken manure (IF+CM), and IF plus commercial fish feed (IF+Feed)] and two culture period (*viz.* 180 and 400 days) with three replicates in earthen fishponds stocked with Nile tilapia at 2 fish m^{-2}. Fertilisation rate of IF and IF+CM treated ponds were 4 kg N and 1 kg P ha^{-1} d^{-1}. These doses were administered using either IF alone or with chicken manure, the latter being provided at a rate of 20 kg ha^{-1} d^{-1} (dry weight) on a weekly basis. In addition to IF, pelleted feed (30% protein) was applied to IF+Feed ponds daily at rates of 5%, 3% and 1% body weight of fish for 1-2, 3-4 and 5-13 months, respectively.

Fertiliser requirements of morning glory

Four levels of N (0.1, 0.2, 0.3 and 0.4 g N pot^{-1}), two levels of P (0.05 and 0.10 g P pot^{-1}) and two levels of K (0.07 and 0.14 g K pot^{-1}) were tested in a factorial experiment in pots to determine the nutrient requirements for morning glory. Plant tissues were analysed at the harvesting stage and used to calculate fertiliser requirements following soil tests and fertiliser use efficiencies of commonly used fertilisers. Fertiliser requirements were determined following soil tests, and nutrient recoveries by the crop from specific fertilisers using nutrient use efficiencies.

Fertiliser value of tilapia pond sediment for morning glory production

The potential of sediment as a fertiliser supplement for morning glory production was assessed first using 100% air dried sediment in pots in a lath house in a 3 x 3 factorial experiment with three levels of N (0.38, 0.55 and 0.98 g N pot^{-1}) and three levels of P (0, 0.05 and 0.10 g P pot^{-1}).

Value of tilapia pond sediment in terms of phosphorus fertilisation and soil conditioner

Two pot experiments were conducted to quantify the benefits and potential of tilapia pond sediment for P supplementation, improving soil aggregation, moisture content, soil pH and yield of morning glory. Sediment was collected from the top 10 cm of the pond bottom after harvest of fish in six months-old tilapia ponds fertilised with inorganic fertilisers and pelleted feed. Sediment analysis indicated that P was very high, while N was not adequate to meet crop requirement. Five treatments composed of different ratios of sediment: topsoil [0:100 (S_0), 25:75 (S_{25}), 50:50 (S_{50}), 75:25 (S_{75}) and 100:0 (S_{100})] were tested for dry matter (DM) production and nutrient uptake by morning glory, and soil aggregates stability, pH, moisture and organic matter (OM) content of soils.

Economic returns of morning glory using pond sediment as a fertiliser supplement

Sediments collected from the three pond treatments (IF; IF+CM; IF+Feed) of the first experiment were used (Table 3.5). Morning glory was grown in triplicate in 1 m^2 plots on pond dykes. The economic returns of morning glory cultured on the dyke using pond sediment were estimated.

Nutritive value of tilapia pond water and sludge for chilli pepper cultivation

A plot trial was carried out to assess the fertiliser value of pond water and sediments from tilapia pond culture for chilli pepper cultivation. Four 200 m^2 earthen ponds were stocked with tilapia at a rate of 2 fish m^{-2}. Ponds were fertilised with urea and triple super phosphate (TSP) at rates of 28 kg N and 7 kg P ha^{-1} $week^{-1}$. Twenty-four chilli pepper plots (2.0 m x 0.6 m) were prepared on pond dykes, and six treatments: (1) control with no sludge and fertilisers (S_0); (2) application of sludge one time a week ($S_{1/wk}$); (3) two twice a week ($S_{2/wk}$); (4) once every two weeks ($S_{1/2wk}$); (5) application of sludge + half of the recommended dose of phosphorous (P) + recommended dose of N and K ($S_{1/2wk}$ + $N_1P_{0.5}K_1$); and (6) recommended doses of N, P and K ($N_1P_1K_1$)] were randomly assigned to plots. There were four replicates per treatment. The sludge (162 g DM) and water (36 litres) from tilapia ponds were collected into a tank using a submersible pump and applied to individual chilli pepper plots according to the above treatments. Plots were irrigated daily, except during rainy days.

Table 3.5. Properties of sediment and farm soil.

Potting media	pH	OM	T-N	Av. N	TP	Av. P	K	Silt
		$g\ kg^{-1}$			$mg\ kg^{-1}$			%
Sediment	6.7	25.4	1.61	0.10	115	85	50	25
Farm soil	5.1	20.0	1.13	0.10	35	20	40	18

Results

Sedimentation and nutrient accumulation in ponds under different input systems

Total fish biomass and net extrapolated yield were significantly higher in feed applied ponds. The effect of culture period on net extrapolated yield was insignificant. Pond inputs, culture periods and their interaction had significant effects on the rates of sedimentation and OM accumulation in the sediment. Sedimentation was 860, 1340 and 2170 g m^{-2} day^{-1} in IF, IF+CM and IF+Feed ponds, respectively, while the values for OM were 17 and 54 g m^{-2} day^{-1}, respectively. Concentrations of OM, N and K in sediments were significantly higher in IF+Feed ponds, while the treatment effects on concentrations of P, and the micronutrients B, Cu, Fe, Mn and Zn were not statistically significant. The N and P contents of sediments were medium and very high, respectively, while depending on the micronutrient, concentrations ranged between high and very high. Greater accumulation of micronutrients obviously increased the potential of sediment as a crop fertiliser. Amount of sediment of top 10 cm bottom layer was significantly higher in IF ponds due to significantly higher bulk density as well as clay content. The quantities of sediments contained in a tilapia pond with an area of 1 ha to a depth of 10 cm was 720, 667 and 627 t in IF, IF+CM and IF+Feed treated ponds, respectively. The sediments contained 1.97, 1.95 and 2.18 g N kg^{-1} and 190.7, 182.0 and 192.5 mg P kg^{-1}, in IF, IF+CM and IF+Feed treated ponds, respectively. Fertiliser values of the top 10 cm of sediment in a 1 ha pond were equivalent to about 3.0 t urea and 0.7 t triple super phosphate.

Sedimentation of suspended and insoluble particles and organic matter were considerably higher in fed ponds and ponds fertilised with organic manures. Experimental results revealed that sediments became saturated with both N and P after 5 to 6 months of fish culture, while the accumulation of K continued throughout the duration of the trials (Figure 3.4). Most of the N and P added after 6 months was unaccounted for and presumably lost through seepage, de-nitrification and de-mineralisation.

Fertiliser requirements of morning glory

Plant tissues were analysed at harvesting and it was found that the N, P and K removed by morning glory were 0.63, 0.12 and 0.48 g pot^{-1}, respectively, with an optimum total dry matter (DM) 69 g pot^{-1}. The values correspond to 9.2 g N, 1.7 g P and 7.0 g K kg DM^{-1} of morning glory and to 356 kg N, 68 kg P and 269 kg K ha^{-1} of a morning glory crop.

Fertiliser value of tilapia pond sediment for morning glory production

Nitrogen application of up to 0.55 g pot^{-1} significantly increased DM production, and P and K uptake of morning glory. The effects of P application on DM production, and N, P and K uptake were not statistically significant, which confirmed that all the P requirements of morning glory were satisfied by the sediment. Sediment also provided 37% N and 24% K nutrient requirements for morning glory. Supplementation of N and K by inorganic fertilisers at a rate of 0.55 and 0.48 g pot^{-1}, respectively, provided a yield of morning glory of 91 g DM pot^{-1}, which is equivalent to 51 t ha^{-1} $crop^{-1}$. Sediment alone supported the production of 56

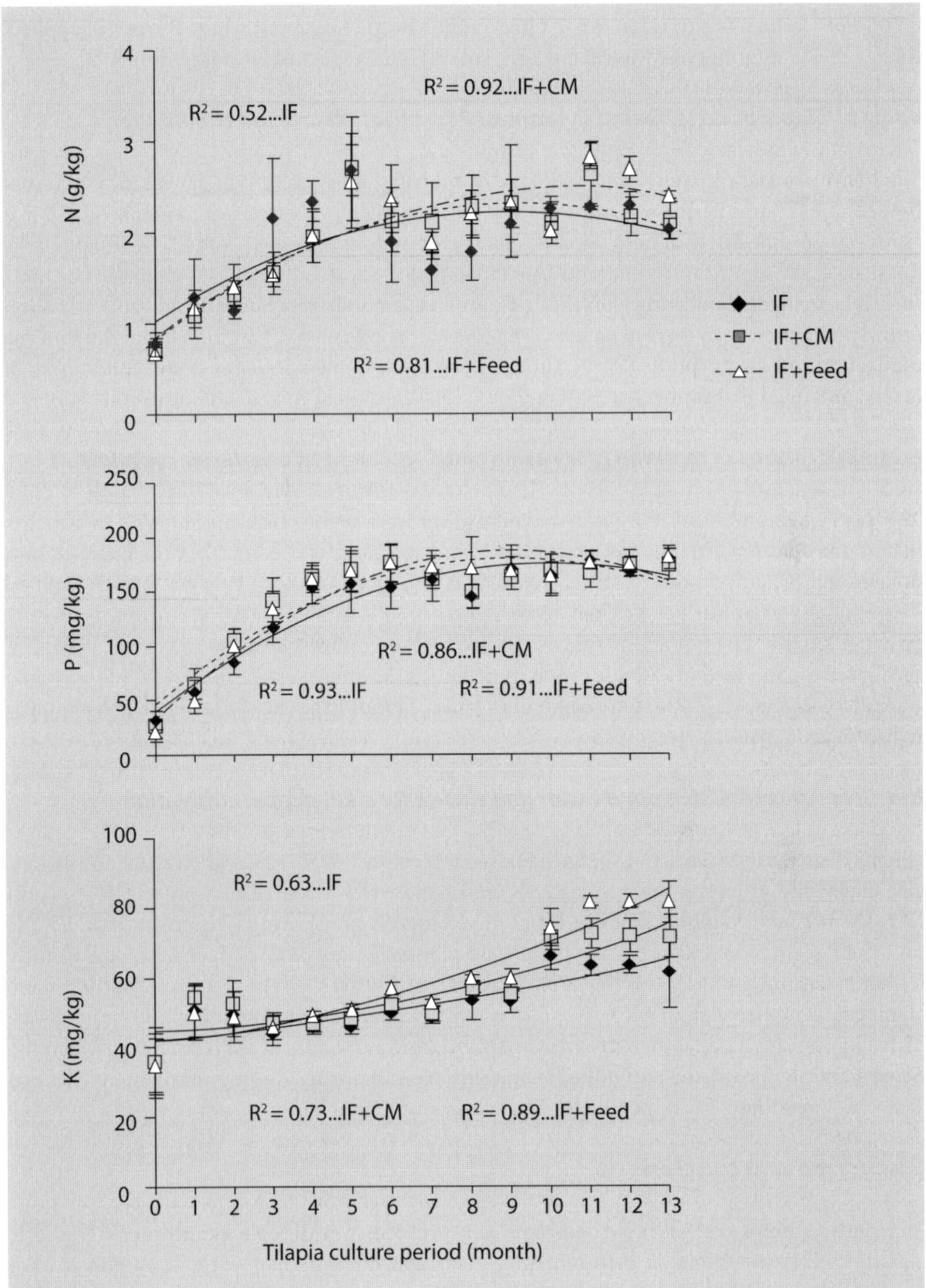

Figure 3.4. The accumulation of N, P and K in the sediment of tilapia ponds during the culture period in three treatments (IF = inorganic fertiliser; IF+CM = IF plus chicken manure; IF+Feed = IF plus commercial fish feed).

g DM pot^{-1}, which is equivalent to 32 t ha^{-1} $crop^{-1}$. In the treatment where P was not applied, the residual P after harvest was 60 mg kg^{-1}, still high in respect of crop requirements.

Value of tilapia pond sediment in terms of P fertilisation and soil conditioner

Total DM production and N and P uptake of morning glory were not significantly different between treatments S_0 and S_{25}, but were significantly different between treatments S_0 and S_{50}. However, the differences in fresh yields of edible parts among S_0, S_{25}, S_{50} and S_{75} treatments were not statistically significantly different. This shows that the application of 25% sediment can meet the P requirement of morning glory. Soil pH, water stable soil aggregates and moisture retention linearly increased with increasing levels of sediment application. Therefore, depending on the availability and the purpose of application, i.e. fertiliser supplementation or soil conditioner, it is recommended that farmers use either 25% or 50% sediment mixed with farm soils.

Economic returns of morning glory using pond sediment as a fertiliser supplement

The fresh yields and gross margin of morning glory were significantly higher in all treatments than in the control (T0). The fresh yields were not significantly different, but the gross margin differed significantly among treatments S_0NPK, $S_{60}NP_0K$ and $S_{120}NP_0K$, while S_0NPK resulted in the highest economic benefit (Table 3.6). However, 30% supplementation of pond sediments provides slightly low but acceptable economic returns. Labour cost for sediment collection accounted for 50 – 60% of variable production costs if the official wage rate of AIT international standard was applied. Active participation of family labour in aquaculture would substantially reduce labour costs.

Nutritive value of tilapia pond water and sludge for chilli pepper cultivation

Chemical analysis showed that sludge contained 67 mg kg^{-1} NH_4^+-N, and NO_3^--N, 97 mg kg^{-1} available P, and 41 mg kg^{-1} exchangeable K, and pond water contained 2.8 mg l^{-1} NH_4^+-N and NO_3^--N, 0.2 mg l^{-1} soluble reactive P, 0.5 mg l^{-1} total P, and 1.6 mg l^{-1} exchangeable K. The pod yield, plant biomass (excluding pods), and plant height of chilli pepper were significantly higher in treatments $S_{1/2wk} + N_1P_{0.5}K_1$ (T5) and $N_1P_1K_1$ (T6) than the rest, and also there were significantly higher removal of N, P and K by chilli pepper pods and leaves in both treatments. There was no significant difference in growth and yield responses of chilli pepper between the control and sludge treated plots. The fresh yields of chilli pepper were approximately 17.0 and 16.3 t ha^{-1} in T5 and T6, respectively.

Discussion and conclusions

The study established that pond sediment can provide considerable quantities of N and K and the full requirement of P to morning glory and chilli pepper without compromising fish growth and yields. The three treatments, either inorganic fertiliser alone or plus chicken manure or plus commercial fish feed, contributed similarly to N and P enrichment in sediment and pond water.

Table 3.6. The partial budget analysis of Morning glory on experimental plots in Baht (between brackets the gross return transposed in Baht m^{-2}).

Treatments**	Gross income	Variable cost *	Gross return	
T0. S_0	28.1 ±7.4	19.6	8.4 ±7.4 [a]	(10)
T1. $S_0 N_{21.5}P_{5.7}K_{27.1}$	110.7 ±12.4	23.5	87.2 ±12.1 [f]	(100)
T2. S_{60}	76.8 ±3.8	39.6	37.2 ±3.8 [bc]	(43)
T3. $S_{60}N_0P_0K_{27.1}$	83.2 ±5.4	42.4	40.7 ±5.4 [c]	(47)
T4. $S_{60}N_{21.5}P_{5.7}K_{27.1}$	112.7 ±7.0	42.8	70.0 ±7.0 [e]	(80)
T5. S_{120}	87.1 ±3.2	59.6	27.5 ±3.2 [b]	(32)
T6. $S_{120}N_{21.5}P_{5.7}K_{27.1}$	94.9 ±4.4	62.3	32.6 ±4.4 [bc]	(37)
T7. $S_{120}N_{21.5}P_{5.7}K_{27.1}$	118.1 ±3.0	62.6	55.5 ±3.0 [d]	(64)

*Labour cost was 20 and 40 Baht for applying 60 or 120 kg of sediment respectively.
**S=sediment. The quantities (kg) indicated in subscript were mixed to 10 cm of topsoil; 60 and 120 kg of sediments correspond to 30 and 60% of the 10cm top soil.
NPK in g $plot^{-1}$; N as urea, P as triple super phosphate and K as muriat of potassium.
Means of each parameter in the same column having dissimilar superscript(s) differ significantly ($P<0.05$).

All P requirement and partial N and K requirements of crops can be satisfied by nutrient trapped in pond sediments; the fertiliser values of sediment of 1 ha pond area were equivalent to about 3.0 t urea and 0.7 t triple super phosphate. Pond sediment improved moisture content, pH and aggregate stability of alluvial acid sulphate soil from Central Thailand. Depending on the availability and the purpose of application, i.e. fertiliser supplementation or soil conditioner, farmers may use either 25% or 50% sediment mixed with farm soils. Supplementary use of locally available fertilisers and manures is recommended. Remaining residual levels of P after harvesting indicate that mixing sediment with agricultural soils would help make use of additional P. Further research using the available P for culturing consecutive leguminous N-producing crops to get insight in the depletion of P from the pond sediments is recommended. It is expected that improvements in plant health in areas where sandy soils were predominant would help reduce the incidence of pests and diseases and hence reduce pesticide use in agriculture.

Sediment nutrient content does not increase after 6 month probably because of seepage and leaching. Mechanising the removal of nutrient saturated bottom sediments would be beneficial to farmers as this would optimise nutrient reuse and reduce environmental pollution. Labour cost of manual recovery are too high and available technologies for mechanical abstraction from Viet Nam and China should be tested. On-farm research on the dissemination of appropriate technologies for cost-effective sediment removal is warranted.

In principle, crops grown using pond sediment are safe for human consumption. However, micronutrient contents in pond sediments were quite high, which warrant further research to identify their sources and occurrences, as do their effects on crops, fish and the environment under different pond culture management practices.

References

Boyd, C.E. 1995. Bottom soils, sediment and pond aquaculture. Chapman and Hall, New York. 348 pp.

Funge-Smith, S.J. and Briggs, M.R.P. 1994. The origins and fate of solids and suspended solids in intensive marine shrimp ponds in Thailand. Summary report from Sonykhla Region of Thailand. Institute of Aquaculture. Stirling, UK. 20 pp.

Jamu, D.M. and R.H. Piedrahita, 2002. An organic matter and nitrogen dynamics model for the ecological analysis of integrated aquaculture/agriculture systems II. Model evaluation and application. Environmental Modelling and Software 17: 586-592.

Using the fishpond as an improved nutrient trap for recycling nutrients on farm in North East Thailand and Mymensingh, Bangladesh

Kabir M. Shamsul[1], M. Abdul Wahab[1], Chittra Arjinkit[3], Danai Turongrouang[3], Roel H. Bosma[2] and Marc C.J. Verdegem[2]
[1] Bangladesh Agriculture University, Dep. Fisheries Management, Mymensingh, Bangladesh
[2] Aquaculture and Fisheries Group, Wageningen University, P.O. Box 338, 6700 AH Wageningen, The Netherlands; marc.verdegem@wur.nl
[3] Sisaket College of Agriculture & Technology, 91/14 Kusikum road, Sisaket, Thailand

Abstract

Various experiments were conducted in Bangladesh and Northeast Thailand to evaluate the use of fishponds as nutrient traps in integrated farming systems. All ponds accumulated nutrients, but the amount accumulated was influenced by the types of inputs, the species stocked and densities used. Results suggest, that the higher the percentage of input nutrients retained in fish production, the better the cost : income ratio. The latter is very important, especially for poor farmers in the prevailing cash constrained local economies. Pond farming systems used by poor farmers yielded a higher cost : income ratio than systems applied by better-off farmers due to the polyculture and to the use of the pond sediment on the crops at the dyke. However, more nutrients accumulated in the sediment of better-off farmers, due to the additional application of feed. This higher pond sediment nutrient content makes it a better fertiliser for dyke-based crop production. How to reduce the labour demand for the use of pond sediments for crop production is an important topic for further research.

Keywords: fishpond, dyke, nutrients, sediment, profit

Introduction

Because ponds function as traps that accumulate runoff nutrients and recycle waste from the household, surrounding homestead, embankments and livestock, an efficient use of these nutrients might contribute to the improvement of livelihoods. Several studies indicate that farms with more links between the various sub-enterprises use nutrients more efficiently and improve the livelihoods of the farm households (Edwards *et al.*, 1988). Integrating an aquaculture component into a farming system may result in sustainable pond-dyke systems producing a range of products (Lightfoot *et al.*, 1996; Ruddle and Zhong, 1988), but may also lead to environmental pollution (Prein, 2002). In Bangladesh and Thailand several on-farm experiments were conducted to identify and improve methods for on-farm recycling of nutrients. In Bangladesh detailed monitoring of the nutrient dynamics of the pond-dyke system provided a comparative baseline, followed by one year of experiments focussing on the value of the pond mud and the choice of the fish species for efficient nutrient use. In

Thailand the fieldwork focussed on trials of low-input methods to increase fish production and recycling of nutrients. The issues for the trials were identified through Participatory Community Appraisals (PCA) and State-of-the-System (SoS) workshops.

Methodology

In a series of two to four workshops the farmers discussed the results of the SoS, raised researchable issues for their community related to fish production, received information and technical know-how and assisted in the design of field experiments. The on-farm trials were also accompanied by monthly workshops for monitoring and evaluation. Each time the collected data were discussed with the farmers, who had the opportunity to discuss options to solve problems relating to the farming techniques used. E.g. in Thailand, after fish nursing for 4 weeks the feeding ration was adjusted and after 8 weeks, the fingerlings were distributed and techniques for fish rearing were transferred to the farmers.

Data were regularly registered in databases. For statistical analysis one-way and repeated measures ANOVA and correlation/regression were used relying on the statistical packages of SPSS (version 10) and MSTAT. After the experiments the farmers, extension workers and research team gathered to discuss the effectiveness of the intervention to different categories of farmers and the constraints faced during the experiments and general problems related to pond-dyke systems.

Comparison between low input and high input pond-dyke systems in Bangladesh

From July 2002 to February 2003, the inputs and outputs of ponds and dykes were recorded by three low-input (LIS) and three high-input (HIS) farmers, which were randomly selected in the area of Mymensingh in Bangladesh. This was continued from June 2003 to May 2004 for two rich and two poor farmers. Input-output samples were collected and nitrogen content was determined in the laboratory by micro Kjeldahl method (AOAC, 1980). For conversion into input-output of kg nitrogen, the %N value determined in the laboratory analyses and the available secondary data (Gopalan *et al.*, 1993) were used. Fish, livestock (cattle and goat in homestead), poultry (chicken and duck in homestead), vegetables and rice (seedling and rice) were considered as different components in the farming system.

Pond production under different management strategies

In Bangladesh an on-farm trial was conducted in 12 earthen farmers' ponds of a 400-2431 m^2 area in Noshirpur of Muktagacha Upazila under Mymensingh District from July 2004 to March 2005. There were three treatments with four replications each. Farmers under treatment-1 (T-1) raised common carp in a polyculture (silver carp, catla, rohu, mrigal, gonia, and/or thai sarpputi) in which feed and fertilisers were rarely applied; under treatment-2 (T-2), 10% mono-sex GIFT(tilapia) was stocked along with carp species, where feeding and fertilisation were similar to treatment-1, and under treatment-3 (T-3), 10% monosex GIFT (tilapia) was stocked along with carp species, where complementary feed and fertiliser were applied.

In 9 villages, belonging to 6 communities of North East Thailand, 38 men and 5 women farmers participated in the trials with sex-reversed tilapia. The on-farm experiment had 2 phases: nursing and fattening. The fish was sampled for measuring length and weight before the monitoring workshops. For the fattening three treatments were applied (Table 3.7). The nutrient inputs available in the control (S-1) were termites, chicken manure, cow dung, compost, rice bran, vegetable waste and household waste. The goal of the second treatment (S-2) was to show that fish can grow using fertilisation with manure from cattle, pig or chicken and urea, and of third treatment (S-3) to demonstrate that supplementing with a home compound feed was commercially viable. The recommended components for the home compound feed were: rice bran, fish-meal and vegetables. If urea, or money to buy it, were not available farmers used a NPK fertiliser or manure. The final weight and the probable cash income was calculated estimating the average weight of a sample of 50 fishes and assuming an estimated survival rate of 80% and a market price of 30 baht/kg. The partial budget considered the fry and rearing costs and the real or opportunity cost of all feed components, including manures and wastes.

Evaluation of pond mud as fertiliser for growing vegetables

To explore the potential of pond sludge on vegetable production an on-farm experiment was conducted in Noshirpur village of Muktagacha Upazila close to Mymensingh in Bangladesh during the period of December 2004 to February 2005. Pond mud, collected from the three different previously tested ponds systems (T-1, T-2 and T-3) was used to grow red amaranth in soils with three different mixing rates of the 10-cm top soil with pond mud (M-1: 33% pond mud; M-2: 66% pond mud; M-3: 100% pond mud). The experiment was laid out in the split-plot design with 9 treatments and 3 replications; each of the plot area was 2.5 *1.2 m^2.

Results

In Bangladesh two major groups of fish culturists were identified in Mymensingh area. The subsistence group belonging to the less well-off people raise mostly various carps species in a polyculture with on-farm waste; they grow fish for family consumption with the intention of selling the excess production. The well-off farmers are mostly involved in monoculture of Thai catfish (exotic fish) (*Pangasius sutchi*), and apply regularly commercial feed. It was concluded that for the improvements in the nutrient management different strategies for high and low input systems might be identified after a detailed monitoring. In Thailand the intensive use

Table 3.7. The three treatments of on-farm fish fattening trials in North East Thailand.

Treatment	S-1	S-2	S-3
Fish.m^{-2}	2	4	6
Feeding method	Traditional with on-farm available wastes	Green water using fertiliser	Green water using fertiliser, supplemented with homemade feed

of the available ponds was limited due to their location far from the homesteads and the low survival rate of stocked fingerlings.

Nutrient flow in whole farm households in Bangladesh

In Bangladesh, the biomass input and output from the pond were higher in HIS than that of LIS, while the biomass output from the dyke was higher in LIS than that of HIS (Table 3.8). The monitoring showed a dramatic difference in nutrient flows and nutrient budget between rich and poor farms (Figure 3.5). Both categories of farmers could have 5 farm components (fish, crop, vegetable, livestock and poultry), but the size of the components was larger among the well-off compared to poor. Considering all the facts and figures, it may be concluded that the LIS system was more profitable in comparison to HIS. The poor farmers may follow LIS giving

Table 3.8. Input-output status (mean value ± S.E.) in low and high input systems.

Parameters	Unit*1000	Low input	High input	
Pond input	kg ha^{-1}	9.2 ± 3.0	72.8 ± 5.2	***
Pond input cash	Taka ha^{-1}	18.5 ± 5.8	641.2 ± 58.5	***
Pond input non-cash	Taka ha^{-1}	15.8 ± 9.4	459.6 ± 48.0	***
Pond input total cash	Taka ha^{-1}	34.2 ± 3.7	1,100.9 ± 52.9	***
Pond input	kg N ha^{-1}	0.7 ± 0.02	2.0 ± 0.08	***
Dyke input	kg ha^{-1}1	6.6 ± 0.5	7.1 ± 0.2	
Dyke input cash	Taka ha^{-1}	3.1 ± 1.2	3.5 ± 1.0	
Dyke input non-cash	Taka ha^{-1}	3.9 ± 0.7	3.6 ± 0.07	
Dyke input total cash	Taka ha^{-1}	7.1 ± 1.8	7.1 ± 1.1	
Pond output	kg ha^{-1}	1.5 ± 0.2	3.2 ± 0.6	***
Pond output cash	Taka ha^{-1}	48.9 ± 4.9	1.303.8 ± 64.8	**
Pond output non-cash	Taka ha^{-1}	19.6 ± 11.5	13.7 ± 6.1	
Pond output	kg N ha^{-1}	0.036 ± 0.004	0.7 ± 0.03	***
Dyke output	kg ha^{-1}	7.8 ± 2.4	7.0 ± 4.2	
Dyke output cash	Taka ha^{-1}	37.5 ± 18.1	54.4 ± 30.3	
Dyke output non-cash	Taka ha^{-1}	33.3 ±11.1	10.5 ± 6.1	
Pond gross income	Taka ha^{-1}	68.5 ± 16.8	1,317.5 ± 65.9	***
Pond net income	Taka ha^{-1}	34.4 ± 18.5	216.6 ± 48.9	*
Pond cost-net income	ratio	1.1 ± 0.6	0.20 ± 0.0	
Dyke gross income	Taka ha^{-1}	70.8 ± 10.7	65,0 ± 31.9	
Dyke net income	Taka ha^{-1}	63.8 ± 12.3	57.8 ± 3.9	
Dyke cost-net income	ratio	11.2 ± 4.3	7.4 ± 2.9	

1 Euro = 69 Taka; the subscript indicates whether the value for the two systems in different at 0.05 (*), at 0.01 (**) or at 0.001 (***) confidence value.

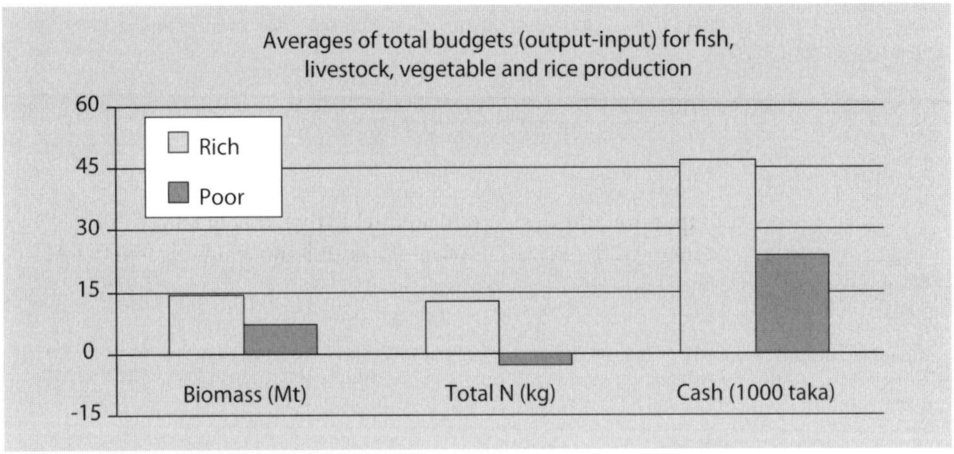

Figure 3.5. Summarised results of the monitoring of four farms in Mymensingh.

special attention to increase per hectare pond production through proper use of resources. The rich farmers may follow HIS with special attention to the best utilisation of the productive pond water and mud to increase per hectare production through utilisation of pond dykes.

Performance of pond production under different management strategies

In Bangladesh, existing carp polyculture (T-1) can be improved by stocking 10% Tilapia and applying feed-fertiliser (T-3)-(Table 3.9), mainly due to increased yield of carp (Figure 3.6a and 3.6b). Adding Tilapia without supplementary feed (T-2) resulted in higher efficiency of nutrient use in the pond. The nutrient N budget of the feed-fertiliser treatment showed an oversupply, and the use of pond sediment as fertiliser was tested.

The farmers in North East Thailand recognised the importance of nursing the fingerlings before stocking the ponds; perhaps the quantity of feed should have been adapted more often (Arjinkit, 2006). At present many still use hapas for nursing and they are teaching the technology to neighbours, friends and relatives. Though the estimated gross margin from fattening the fish was negative for some farmers (Table 3.10), all farmers perceived a benefit. Reasons for this discrepancy are the facts that an opportunity cost for on-farm wastes was included and the fertiliser value of the pond water and sediment was not considered in the gross margin analysis. A complete farm budget might show a more positive picture. Due to the large variation between replicate farms none of the observed differences was statistically significant. The distance between the pond and the farm did not affect the gross margin, but whether the farmer came to his pond every day improved the gross margin (r^2=0.4). The use of urea was positively correlated with the weighted gross margin (r^2=0.89).

Close to 40% of the 36 farmers preferred S-2 because it saved labour time, expenditures were low, the fish growth was quit well and the green water could also be used for irrigation. To decrease cost while maintaining or improving fish growth 25% of the farmers proposed to

Table 3.9. The mean values if the input-output status in different fish management systems in Mymensingh, Bangladesh.

		Different management systems		
Parameters	Units	T-1	T-2	T-3
Fish species		Carp	Carp & Tilapia	Carp & Tilapia
Feed input	kg	1,500	1,640	7,465
Fertiliser	kg	325	880	1,055
Input	kg ha^{-1}	2,212 ^{b}	3,009 ^{b}	8,702 ^{a}
Input	kg N ha^{-1}	43 ^{b}	65 ^{b}	314 ^{a}
Cash input	Taka ha^{-1}	18,125 ^{b}	18,951 ^{b}	65,122 ^{a}
Non-cash input	Taka ha^{-1}	7,556 ^{a}	8,446 ^{a}	14,098 ^{a}
Total input	Taka ha^{-1}	25,681 ^{b}	27,396 ^{b}	79,221 ^{a}
Output	kg ha^{-1}	1,841 ^{b}	2,083 ^{bc}	5,312 ^{a}
Output	kg N ha^{-1}	42 ^{b}	49 ^{b}	123 ^{a}
Cash output	Taka ha^{-1}	83,639 ^{b}	89,700 ^{b}	248,830 ^{a}
Non-cash output	Taka ha^{-1}	10,934	15,253	17,239
Total output	Taka ha^{-1}	94,573 ^{b}	104,952 ^{b}	266,069 ^{a}

1 Euro = 69 Taka.
^{a,b}Means of each parameter in the same row having dissimilar superscript(s) differ significantly ($P<0.05$).

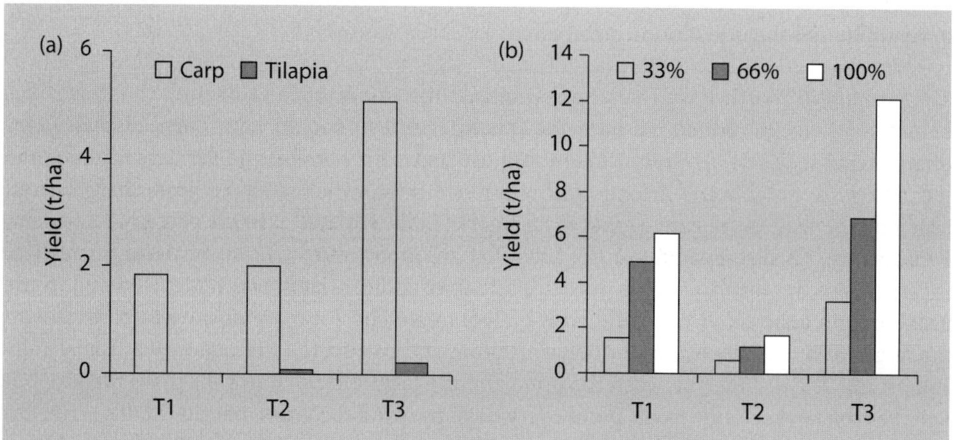

Figure 3.6. Yield of fish in different systems (a) and of red amaranth with different levels of pond sediment in the top soil (b). T1 = Carp; T2= Carp + 10%Tilapia; T3= T2 + feed-fertilisers.

Table 3.10. The number of farmers (n) and some statistical parameters for the estimated average body weight (g), the weighted cost and the weighted gross margin (Baht/fish stocked) of the fattening trial in Thailand.

	Average body weight			Weighted cost *			Weighted gross margin *		
Method	S-1	S-2	S-3	S-1	S-2	S-3	S-1	S-2	S-3
n	7	14	8	7	14	8	7	14	8
Mean	72	95	114	1.28	1.24	1.79	0.45	1.04	0.95
SEM	17	11	15	0.17	0.09	0.24	0.48	0.29	0.48
Min	19	29	63	0.71	0.77	0.88	-1.05	-0.75	-1.76
Max	134	156	162	1.97	2.15	3.28	2.51	2.97	2.07

*The total cost and the gross margin were weighted by the number of fish stocked.
SEM = Standard Error of Mean; 1 Euro = 48 Bath.

lower fish densities for S-3 to 4 fish m^{-2}. The farmers found that while working alone it was more difficult to improve their aquatic system than other components of their farm. Working in small groups and sharing knowledge increased their learning.

Evaluation of pond mud as nutrients for growing vegetable

Yield of red amaranth was significantly higher in the plots, where pond mud from the high input system was used. The mud from the treatments M-2, M-3 and T-3 improved significantly the organic matter and nitrogen content of the soil, while pond mud from M-2 resulted in lower values (Table 3.11). Some fish species can have a large impact on the quality of pond sediment as fertiliser. Adding 10% tilapia to the pond fish population, keeping all other inputs similar, reduced the production of red amaranth by more than 50% (treatments T-1 and T-2 in Table 3.11). Application of home-made feeds, improves the quality of pond sediments for crop production. The higher the quantity of pond sediment used, the higher the production of red amaranth. The highest production was achieved using 100% pond sediment in the 10 cm topsoil. Removing pond sediment is labour intensive. However, in case of seasonal ponds, crops could be produced in the ponds during the dry period. This aspect needs further investigation. Total system nutrient budgets could not be estimated due to labour constraints necessary for detailed monitoring. The highest yield (12.13 t ha^{-1}) was obtained by the interaction of system 3 and mixing 3 and the lowest (0.57 t ha^{-1}) by the interaction of system 2 and mixing 1.

Conclusion

In low cash economies, cost-benefit ratios are very important. The farming systems used by poor farmers show a better cost-benefit ratio than the systems applied by richer farmers, both in Bangladesh and Northeast Thailand. Giving careful consideration to available on-farm inputs and pond production strategies, experiments in Thailand showed that it is possible to further improve the cost-benefit ratio.

Table 3.11. Initial and final soil characteristics of pond-dyke plots for different fish culture systems (see Table 3.9) and mixing rates.

Parameters	Fish culture systems			Sediment mixing levels		
	T-1	T-2	T-3	M-1	M-2	M-3
Initial organic matter (%)	3.846 ^{b}	2.293 ^{c}	5.095 ^{a}	2.827 ^{c}	3.691 ^{b}	4.716 ^{a}
Final organic matter (%)	2.826 ^{b}	1.629 ^{c}	3.715 ^{a}	2.104 ^{c}	2.659 ^{b}	3.408 ^{a}
Initial total nitrogen (%)	0.106 ^{b}	0.057 ^{c}	0.138 ^{a}	0.069 ^{c}	0.106 ^{b}	0.126 ^{a}
Final total nitrogen (%)	0.090 ^{b}	0.052 ^{c}	0.126 ^{a}	0.063 ^{b}	0.094 ^{ab}	0.110 ^{a}

^{a,b,c}Means of each parameter in the same row having dissimilar superscript(s) differ significantly ($p<0.05$).

Pond sediments are a valuable fertiliser for red amaranth production, and pond management has a large impact on the quality of pond sediments for crop production. More research is needed to identify strategies to reduce labour necessary to use pond sediments for crop production on pond-dykes.

References

AOAC (Association of Official Analytical Chemists), 1980. Official Methods of Analysis. 12th ed. Washington DC, USA. 1050 pp.

Arjinkit, C, R.H. Bosma, D. Turongrouang, D.C. Little, and M.C.J. Verdegem, 2006. On-farm trials of improved nursing and fattening sex-reversed gift tilapia in Sisaket province, Northeast Thailand. In: WAS, 2006. Book of abstracts. Poster presentation, Firenze May 5 to 10.

Gopalan, C., B.V.R. Sastri and S.C. Balasubramanian, 1993. Nutritive value of Indian foods. National Institute of Nutrition, Indian Council of Medical Research, Hyderabad-500 007, India, pp. 47-59.

Edwards, P., R.S.V. Pullin and J.A. Gartner. 1988. Research and education for development of integrated crop-livestock-fish farming systems in the tropics. ICLARM Studies and Reviews, Manila, Philippines, 16: 53 pp.

Lightfoot, C., M. Prein, J.K. Ofori, 1996. The potential impact of integrated agriculture-aquaculture systems on sustainable farming, In: Prein, M., J.K. Ofori. and C. Lightfoot (eds.). Research for the future development of aquaculture in Ghana. ICLARM Conf. Proc. 42, 51-56.

Prein, M., 2002. Integration of aquaculture into crop-animal systems in Asia. Agricultural Systems 71, 127-146.

Ruddle, K. and G.F. Zhong. 1988. Integrated Agriculture-Aquaculture in South China: The Dike-pond System of the Zhujiang Delta. Cambridge University Press, Cambridge, 173 pp.

Chapter 4

Managing nutrients in IAA systems

From nutrient balances towards soil organic matter dynamics

Jetse Stoorvogel
Laboratory of Soil Science and Geology, Wageningen University, P.O. Box 37, 6700 AA Wageningen, The Netherlands; jetse.stoorvogel@wur.nl

Abstract

The loss of soil fertility is a slow but important form of land degradation that puts the future productivity of many agricultural systems at risk. This paper looks back at the use of soil nutrient balances as an indicator for soil fertility loss. Since the early 1990s, soil fertility losses have been quantified by balancing the net inputs of soil nutrients against the outputs of the reserves of soil nutrients. Though these nutrient balances have been useful in quantifying the extent of the problem, they have been less successful in finding solutions. Nutrient balances do not consider the interactions between various nutrient flows. It is therefore difficult to use them to evaluate potential interventions. At the same time, it is difficult to predict the long-term changes in soil fertility from a current rate of change. In this paper, two alternative approaches to deal with these problems are proposed, for further development in the future. Both take the calculations in the Nutmon toolbox for nutrient monitoring as a starting point. One, called Dyn-Nutmon, includes some of the key interactions necessary to evaluate alternative management practices. The other approach, SOM-Nutmon, deals with the dynamics of the soil organic matter pool. Both systems are similar in complexity to the calculations in Nutmon, in order to keep data requirements within the limits of typical data availability. In this paper, a Nutmon dataset from Embu district in Kenya is used to illustrate the various approaches. The variation in cropping conditions makes the dataset eminently suitable to illustrate and discuss the pros and the cons of the two systems.

Keywords: soil, fertility, Nutmon, nitrogen, Kenya

Introduction

Nutrients are exported or lost from all agricultural systems. This is not a problem as long as they are replaced by fertilisation or natural regeneration. In many African farming systems, however, lost nutrients are not properly replaced, with the result that soil fertility declines. In the past, the main way to restore soil fertility was by fallowing; this was effective when population pressure and yields were low. Now, however, African farmers need to apply external inputs to maintain soil fertility, but most do not use them adequately. Agricultural researchers have been struggling with the soil fertility issue for a long time. Programmes recommending fertiliser use have been successful in determining optimal fertilisation rates (e.g. FURP, 1987; RCC-Embu, 1998) but have tended to focus on mineral fertilisers and the short-term effects in terms of crop production. In many places, adoption of the recommendations has been poor. Numerous reasons have been put forward for the low adoption rates, including credit constraints, lack of social acceptance and land tenure (Feder *et al.*,1985; Salasya *et al.*, 1998; Franzel, 1999; Ouma *et al.*, 2002; Koning and Smaling, 2005). It is also increasingly being realised that the application of mineral fertilisers is in itself not a long-term solution for maintaining soil fertility. Though mineral fertiliser boosts yields, especially in marginal areas, it also increases leaching and denitrification. Relatively little of the mineral fertiliser results in an improvement of soil fertility. Additional measures are required to affect the soil nutrient balance positively; the POND program suggests that fishponds might function as a nutrient trap. In addition, the increased production creates opportunities to leave fields fallow or introduce green manuring and agro-forestry systems. These more integrated approaches (e.g. Ransom *et al.*, 1993; Giller *et al.*, 2006; Roose and Barthès, 2001) are often referred to as integrated nutrient management (Gruhn *et al.*, 2000). Although many of these concepts are well known, it has proved difficult to provide empirical evidence of their efficacy. Long-term field experiments are rare and agricultural statistics are typically confounded by variability in soil and weather conditions and in agricultural practices. This makes it very difficult to isolate the impact on soil fertility.

How can the changes in soil fertility be quantified if they are so difficult to measure? A study commissioned by FAO in the late 1980s introduced the concept of soil nutrient balances (Stoorvogel and Smaling, 1990; Stoorvogel *et al.*, 1993): an accounting exercise considering five different inputs of nutrients (mineral fertiliser, organic fertiliser, wet and dry deposition, nitrogen fixation, and sedimentation) and five different outputs of nutrients (crop products, crop residues, leaching, gaseous losses, and erosion). At regional level, the inputs and outputs were estimated from agricultural statistics, regional soil surveys and climate data in combination with estimates produced by statistical models (e.g. of nutrient leaching and soil erosion). Later, the methodology was refined to create the Nutmon decision-support model for application at farm level: this included a monitoring framework to gather data on the various nutrient flows (Van den Bosch *et al.*, 2001).

Whether or not fishponds may function as a nutrient trap will be evaluated with a soil nutrient balance analyses using the Nutmon model in the next paper of this chapter (Muendo *et al.*, 2007). Soil nutrient balances have proven to be a successful approach and have been applied in many case studies (de Jager *et al.*, 2001; Priess *et al.*, 2001; Haileslassie *et al.*, 2006, Roy *et al., 2003; FAO, 2005*). The key objectives of Nutmon are to determine current rates of change in soil fertility and to identify the main processes driving the soil nutrient balance. The latest version of Nutmon helps analyse the current nutrient status and as such provide a diagnosis of the problem, but it does not have the capability to evaluate either the long-term solution or the short-term interventions that may reverse the degradation (Vanlauwe and Giller, 2006). Recognising this, various authors have begun to look at more dynamic approaches for determining the appropriate strategies (e.g. Rowe *et al.*, 2006; Schlecht and Hiernaux, 2004, Tittonell *et al.*, 2006, Giller *et al.*, 2006). This paper looks at what a model must have if it is to be used for a more integrated long-term analysis in order to develop solutions using the existing Nutmon datasets and conceptual framework. Two prototype models have been developed that deal with some of the feedback mechanisms and the organic matter dynamics. They are illustrated using a Nutmon dataset from Embu district in Kenya.

The study area and Nutmon survey

The decline of soil fertility has been well studied in many research projects (Roy *et al.*, 2003). In east Africa, especially in Kenya, various Nutmon studies have been carried out (Van den Bosch, 1998b, de Jager *et al.*, 2001). The present paper is based on one such study carried out in Embu district on the slopes of Mount Kenya. Embu district exhibits the typical agro-ecological profile of the windward side of Mount Kenya: from the hot, dry, lower zones in the Tana River Basin to the cold, wet upper zones higher up the mountain. The elevation ranges from 760 to 2,070 metres above sea level and average rainfall increases with altitude: from 640 to 2,000 mm year^{-1}. During previous studies five agro-ecological zones (AEZ) were identified in the district (Figure 4.1). Land use in the area varies from livestock and annual crops in the lower zones to coffee, tea and dairy in the higher parts (Table 4.1). Geologically, there is a distinct difference between AEZ 1-4 comprising basic and ultra-basic igneous rocks (basalts) and low-lying AEZ 5 comprising basement system rocks (predominantly gneisses).

Some years ago, a participatory rural appraisal (PRA) was conducted in the district in order to identify the different farming systems and the major constraints to farming in the different AEZ. The findings, augmented with secondary data, were used to select three farms in each AEZ for on-farm monitoring of nutrient stocks and flows. In total, therefore, 15 farms were monitored. Over a full year, farmers were visited monthly and interviewed about nutrient management on the farm. For details on the questionnaires and survey, see De Jager *et al.*

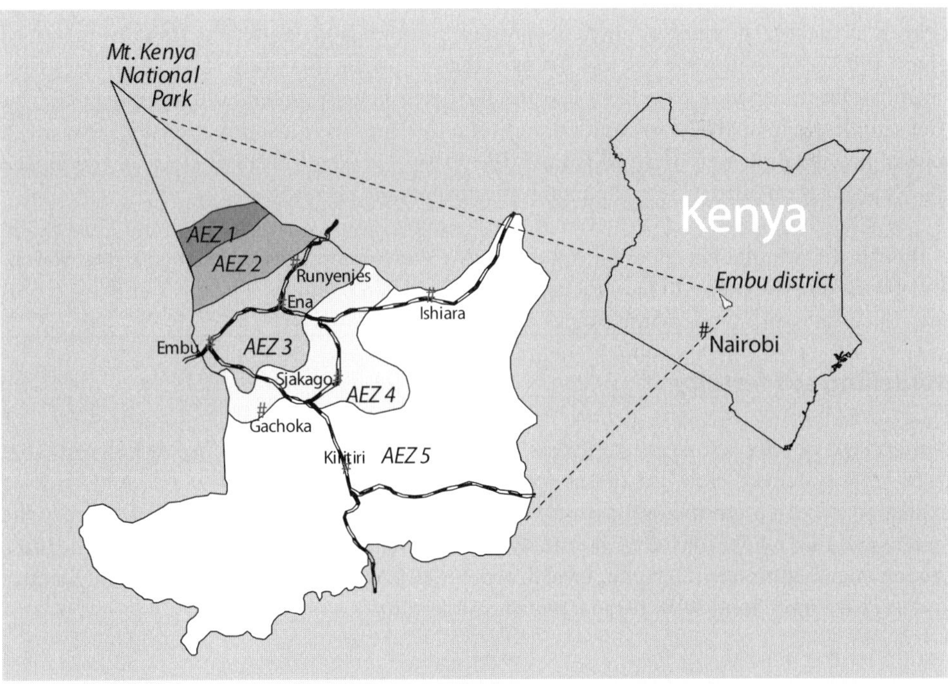

Figure 4.1. Location of the agro-ecological zones in Embu district (Sombroek et al., *1982).*

Table 4.1. General characteristics of the different agro-ecological zones.

Characteristic	AEZ 1	AEZ 2	AEZ 3	AEZ 4	AEZ 5
Altitude (m.a.s.l.)	1,770	1,590	1,320	980	830
Temperature (°C)	16.8	18.2	20.2	21.4	22.6
Rainfall (mm/yr)	1,750	1400	1,200	900	800
Main soil types	andosol	nitosol	nitosol	nitosol	arenosol
	nitosol			cambisol	
Main land use	tea	tea	coffee	tobacco	livestock
	dairy	coffee	maize	food crops	shifting
					cultivation
		dairy			

m.a.s.l. = metres above sea level.

(1998[a]) and Van den Bosch *et al.* (2001). The survey included several major categories of information: inputs in fields and livestock, outputs from fields and livestock, time that animals were confined, use of manure and household waste, growth of herd, inputs and outputs to and from the food stock, family labour and off-farm income. Crop products and residues were sampled on all farms. Meat, milk and the most common external inputs were sampled on farms and in local markets. Samples were analysed for total N, P and K. The soil nutrient stock was defined as the total amount of nutrients present in the organic matter fraction, the adsorbed phase and in soil solution in the top 30 cm of the soil profile. Between 5 and 25 samples were taken per farm, depending on farm size and the heterogeneity of soils and cropping patterns. The samples were analysed for total N and total C. Clay fractions, bulk densities and mean annual precipitation were derived from FURP (1987). The soil nutrient stock in the topsoil (0–30 cm) was calculated from the total soil nutrient concentrations.

Although the original Nutmon dataset includes 15 farms and data on nitrogen, phosphorus and potassium, the present study focuses on the nitrogen balances for only 5 of these farms, each one representing one of the 5 agro-ecological zones.

Modelling soil fertility

Nutmon was developed to quantify the soil nutrient balance and identify the key nutrient flows in farming systems. Below, the methodology it uses to quantify the soil nutrient balances is explained. Next, a methodology is proposed that considers some of the key interactions in the system and thus allows for the *ex ante* analysis of alternative management practices. The third section takes Nutmon to the higher level in order to deal with the soil organic matter dynamics and to give a more long-term perspective on soil fertility.

Nutmon

Nutmon has been developed as a decision-support model for monitoring soil nutrient balances. It allows nutrient flows on the farm to be monitored, enabling estimates to be made of the corresponding soil nutrient balances. As mentioned in the introduction, the balances are based on an accounting exercise in which the net balance equals the sum of the inputs minus the sum of the nutrient outputs. Table 4.2 provides an overview of the various inputs and outputs that are considered in the system. In Nutmon, the farming system is subdivided into a number of compartments including:

- the primary production units (PPUs) which are the fields with the various crops or grassland;
- the secondary production units (SPUs) which are the various groups of animals on the farm;
- the household (HH) defined as the actual household with all the people that spend a significant amount of time on the farm;
- the redistribution units (RUs), which are locations within the farm where nutrients are collected or accumulate and from where nutrients are redistributed over the farm (including e.g. garbage heaps and places where animals are confined); and
- the stock of crop products and fertiliser on the farm.

During monitoring, various management-related nutrient flows are recorded (IN 1, IN 2, OUT 1, and OUT 2). Other flows are not recorded, so have to be estimated at a later stage, using simple transfer functions or models. It should be noted that these functions and models may also require specific data-gathering for calibration and validation. It is also important to realise that various models to estimate these flows are statistical models estimated on the basis of available experimental data from Africa. Nutmon will carry out the calculations for the various compartments and the sum of the various compartments will yield the farm balance. Details on the Nutmon methodology are provided in the extensive literature on the Nutmon model (e.g. Smaling and Fresco, 1993, de Jager *et al.*, 1998b). Nutmon has proven to be a valuable tool for evaluating farming systems under all kinds of agro-ecological conditions (e.g. Van den Bosch

Table 4.2. Nutrient inputs and outputs.

Nutrient inputs		Nutrient outputs	
IN 1	Mineral fertiliser	OUT 1	Crop product
IN 2	Organic fertiliser[1]	OUT 2	Crop residues[2]
IN 3	Atmospheric deposition	OUT 3	Leaching
IN 4	Nitrogen fixation	OUT 4	Denitrification
IN 5	Sedimentation	OUT 5	Erosion
		OUT 6	Human excreta

[1]Including feed, organic fertiliser, off-farm grazing, and imported food.
[2]Including crop residues but also manure from the SPU and grass through grazing.

et al., 1998b, Gachimbi *et al.*, 2002). It allows researchers to determine where major losses of nutrients take place and improves their understanding of the farming system.

A dynamic nitrogen balance model (Dyn-Nutmon)

Datasets collected with the Nutmon model typically provide a complete overview of farm management and all corresponding nutrient flows. However, Nutmon cannot be used to evaluate alternative management practices: for example, it does not include the effect of fertiliser on production. Adding just one small module to the model enables a more dynamic analysis to be done This section presents a dynamic nutrient balance model called Dyn-Nutmon, based on the concepts of the basic Nutmon model. For a more dynamic analysis of the farming system, various feedbacks need to be incorporated. Figure 4.2 gives a general overview of the model and the various nutrient flows that Dyn-Nutmon considers. The most important feedback is probably the effect of changes in management on the production. The key changes to the Nutmon calculation are listed below.

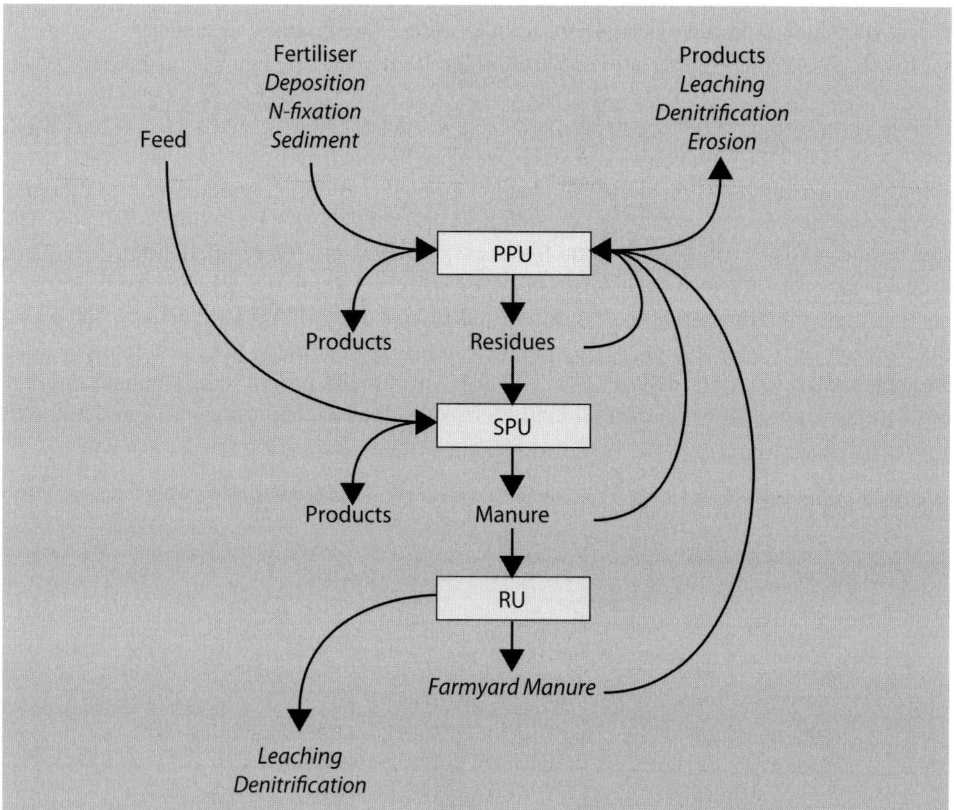

Figure 4.2. Main soil nutrient flows between the PPU, SPU and RU as modelled in Nutmon. Nutrients in bold are measured during monitoring; nutrients in italics are estimated using transfer functions.

Typically, numerous different PPUs, SPUs and RUs are observed on a farm. Each unit has its intrinsic properties. It is impracticable to deal with each individual unit, especially in the complex African farming systems with, for example, intercropping, mixed cropping, fallow systems, and perennial crops. Therefore, for Dyn-Nutmon the various PPUs were aggregated into a single overarching PPU with a single set of joint inputs and outputs. In addition, weighted averages were calculated for the various land characteristics (e.g. slope, clay) and crop properties (e.g. dry matter and nitrogen concentration). Similarly, the SPUs and RUs were aggregated. Dyn-Nutmon only includes the household by means of a fixed contribution of household waste (as observed in the survey). Other household inputs and outputs are ignored. As a result, the farming system is described by three compartments representing the PPU, SPU, and RU.

The Nutmon analysis yields inputs to the PPU and outputs in terms of e.g. crop products, fodder, and grass. The relationship between the various inputs and outputs is not simulated. This is probably one of the main impediments to the evaluation of interventions. During the aggregation process for the PPU, a number of basic properties describing the system are described: the current productivity, the harvest index, average nitrogen concentration of inputs and outputs, and the level of inputs. While assuming constant nutrient concentrations, harvest index and the contribution of leguminous crops, Dyn-Nutmon simulates the crop production. In this case, it is simulated as an exponential relationship between nutrient availability (N_{avail}) and nutrient uptake (N_{uptake}).

$$N_{uptake} = c_0 + c_1 \cdot e^{-c_2 \cdot N_{avail}} \qquad \text{(Eq. 1)}$$

The three constants (c_0, c_1, and c_2) are estimated for each AEZ using observed available nitrogen and nitrogen uptake, maximum nitrogen uptake, and nitrogen use efficiency for the current systems. Nitrogen availability (kg ha^{-1} yr^{-1}) is calculated as

$$N_{avail} = 1000 \cdot S_{depth} \cdot BD \cdot N_{soil} \cdot N_{min} + N_{sup} \qquad \text{(Eq. 2)}$$

where:

S_{depth} = the effective soil depth in m;
BD = the bulk density in kg dm^{-3};
N_{soil} = the soil nitrogen content in %;
N_{min} = estimated mineralisation rate in %; and
N_{sup} = the nitrogen supplied through the various inputs in kg ha^{-1} yr^{-1}.

All losses in the PPU are defined as in the standard Nutmon model. In Nutmon, inputs and outputs are monitored in redistribution units, but in Dyn-Nutmon they are calculated on the basis of a fixed efficiency. The current efficiency of the various redistribution units is set at 30%, which corresponds to the rates observed (ratio of nitrogen input to nitrogen output) in various Kenyan Nutmon datasets.

When alternative farming practices are evaluated, the secondary production unit is maintained (in other words: the number of animals is kept constant). During the simulation the production of crop residues will change. In principle, fluctuations in crop residues are compensated for by purchasing animal feed. A surplus of food enters the redistribution units and will leave as part of the farmyard manure. This is often observed in the extensively managed zero-grazing units where crop residues are left on the ground and are mixed with manure. Note that the feeding requirements used in Dyn-Nutmon are the same as those defined in Nutmon.

Crop residues may be incorporated into the soil, part may be fed to the animals, and the remainder may be sold. The percentages are observed in the farm survey and kept constant during the simulation.

A soil organic matter model: SOM-Nutmon

Typically, changes in soil fertility are linked to soil organic matter dynamics. As a result, in order to deal with long-term changes in soil fertility, one has to look at the dynamics of soil organic matter. A major limitation of soil nutrient balance models is that they focus on mineral N (and other elements) although under most conditions there will be no accumulation of mineral N in the soil. As a result, in order to study long-term changes in soil fertility one needs to study the soil organic matter dynamics of the system rather than the nutrient concentrations. Though there are various models for soil organic matter dynamics (Parton *et al.,* 1992; Jenkinson and Coleman, 1994; Tittonell *et al.,* 2006), they are based on the same principles. The main differences are the number of carbon pools that are considered in the soil. It is difficult to parameterise the models because a large number of input parameters is necessary to describe the processes properly and to make the models generic. The data requirements hamper the models' application: so much data is required that many researchers opt to estimate the input parameters and run the models without proper validation. One way to resolve this shortcoming is to pare down some of the models to their core processes. The model described below, SOM-Nutmon (SOM is the acronym for soil organic matter), is a simplified version of Nutmon. Its basic structure is presented in Figure 4.3. Once again, the farm structure has been simplified, to one PPU, one SPU, one RU, and one HH. The definition of the various units is based on the current farm structure as observed during the Nutmon survey.

SOM-Nutmon calculates the organic matter balance of the PPU. The key processes included are the mineralisation of SOM (typically at rates of 1–3%, depending on the temperature and moisture regime) and the composition of the soil organic matter structure. The input of fresh SOM includes the residual root mass in the field, crop residues, manure, and other organic amendments. Each of the amendments has a specific transformation factor for its conversion into SOM, depending on the C:N ratio and mineral nitrogen fertilisation. In addition, there is a general coefficient representing the regional conditions.

In the overall model, crop production is simulated in the same way as in Dyn-Nutmon. Again, the harvest index is maintained as observed in the Nutmon survey, although the harvest indices may change in response to different scenarios (e.g. fodder production). The amounts of manure follow from the SPU and RU definitions, as was done in Dyn-Nutmon.

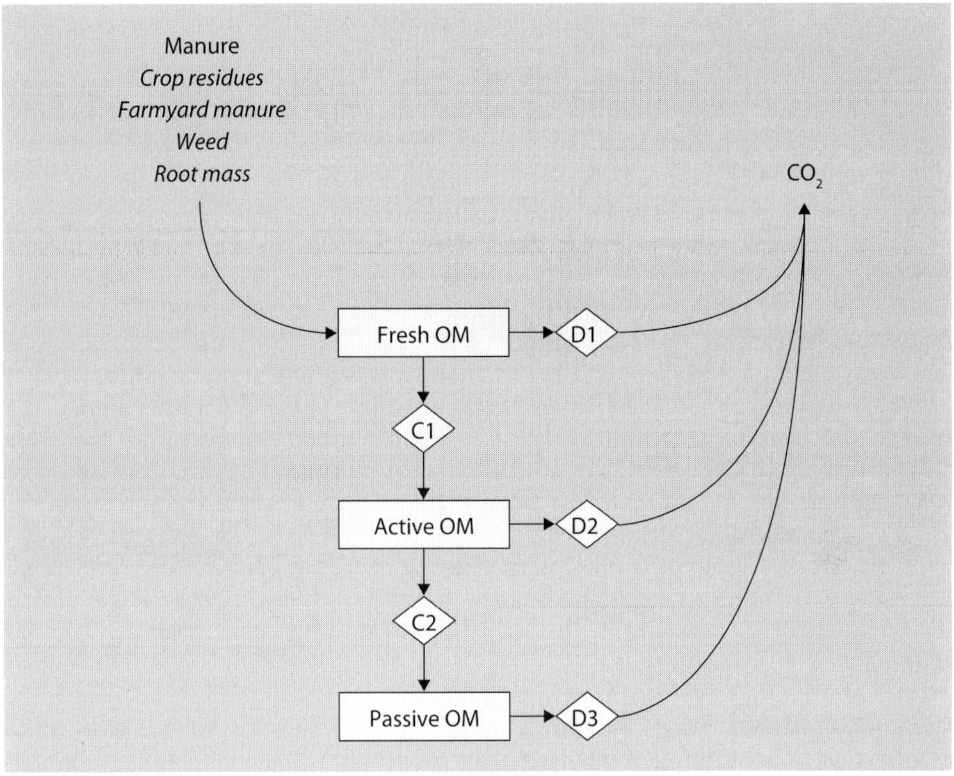

Figure 4.3. Structure of the SOM module in SOM-Nutmon.

The general structure of the model is presented in figure 4.3. Conversion (C1 and C2) and degradation (D1, D3) parameters are estimated on the basis of experimental data (e.g. Wortman *et al.*, 2001; Smaling *et al.*, 1992; Roose and Barthès, 2001; Scheldrick *et al.*, 2003) and model parameters for the Century (Parton *et al.*, 1992) and Rothamsted (Jenkinson and Coleman, 1994) models. It should be noted that depending on the time scale of the simulation, the passive OM pool may be considered inert. In other words, the variables C2 and D3 can be considered to be zero.

Results for Embu province

In this section the three models presented above are used to analyse five farms from Embu province. The results are presented and discussed. Table 4.3 provides the general properties of the 5 farms. The farms at lower altitudes in the province are typically larger and experience drier weather and have poorer soil fertility. The farmers at higher altitudes generally produce cash crops and as a result often apply more mineral fertiliser. The figures for crop production and residue removal are highly variable and depend not only on the production levels but also on the selection of cropping activities on the farm.

Table 4.3. Main properties of the five selected farms from Embu province located in the different agro-ecological zones.

	AEZ 1	AEZ 2	AEZ 3	AEZ 4	AEZ 5
Farm size (ha)	1.2	1.4	3.5	2.1	2.28
Annual rainfall (mm/yr)	2,031	1,938	2,016	1,640	1,197
Average slope (%)	3	25	10	5	25
Slope length (m)	76	35	62	82	33
Nitrogen (%)	0.38	0.24	0.20	0.16	0.07
Bulk density (kg/dm^3)	0.8	1.2	1.5	1.6	1.6
N-fertiliser (kg/ha/yr)	115	79	7	62	0
Crop production (kg/yr)	1,760	670	2,230	2,290	1,960
Residue removal (kg/ha/yr)	6,200	2,070	3,600	5,400	400
Livestock (TLU[1])	3.1	5.5	3.5	3.8	2.9

[1]TLU: tropical livestock units.

Nutmon

Soil nitrogen balances for the five farms are presented in Table 4.4. All five farms show significant losses of soil nitrogen. Although three farms (AEZ1, 2, and 3) apply more nitrogen to the soil through fertilisation, these farms do not show lower depletion rates. Losses through denitrification, leaching and erosion together with crop uptake offset the inputs and determine the soil nitrogen balance. From the Nutmon study in Embu it was concluded that nutrient inputs are essential but that they only mitigate soil nutrient depletion if properly combined with alternative practices to reduce the nitrogen losses (e.g. erosion control, proper fertiliser application techniques to reduce nitrogen leaching, incorporation of organic matter).

Dyn-Nutmon

As the dynamic model is based on the standard Nutmon model, the results of the nutrient balance calculation for the five farms are almost identical to the results presented in Table 4.4. However, with the dynamic model it is possible to evaluate alternative management strategies. A commonly perceived solution to the negative soil nutrient balances in the Kenyan Highlands is to use more mineral fertiliser. This was confirmed by the joint stakeholder meetings at the end of the Nutmon study in Embu but also by the strategy for revitalising agriculture published more recently by the Kenyan government (2004). To illustrate the use of Dyn-Nutmon, the effect of an extra fertiliser application of 50 kg/ha/yr N over and above the current application was evaluated (IN 1 in Table 4.5).

It is commonly argued that the major cause of soil nutrient depletion in Kenya is the low application of mineral fertiliser. Although the Embu data show fertilisation rates above the

Table 4.4. Soil nitrogen balances as calculated by the Nutmon system for the five representative farms in the Embu region (kg N ha^{-1} yr^{-1}).

	AEZ 1	AEZ 2	AEZ 3	AEZ 4	AEZ 5
Inputs					
Mineral fertiliser	115	79	7	62	0
Organic fertiliser/feed	38	40	12	31	29
Atmospheric deposition	6	6	6	6	5
Nitrogen fixation	4	4	2	3	3
Sedimentation	0	0	0	0	0
Total IN	163	128	27	101	36
Outputs					
Crop/animal products	48	14	16	27	29
Crop residues	119	31	21	60	4
Leaching	70	47	32	35	18
Denitrification	26	25	18	15	2
Erosion	11	129	44	8	59
Total OUT	275	246	130	144	112
Depletion rate (kg/ha/yr)	112	117	103	44	75

national averages, the nutrient depletion rates are still considerable. The reasons given for the lack of nutrient inputs include the high price of fertiliser, the lack of credit availability, and risk aversion. Changing fertiliser use impacts all nutrient flows in the system. Increased fertiliser use increases nutrient availability and thus in these nutrient-limited systems it boosts nutrient uptake. However, only part of the added nutrients will be taken up. Another part will be lost, e.g. through leaching and denitrification. The increase in nutrient uptake will result in more crop products and residues. Typically, the products are lost from the system; they may be sold, but if they are consumed by humans, the nutrients end up in latrines. There are several options for the crop residues. If commercial feeds are imported to feed the animals, these will be replaced by the available crop residues. If there is an excess of crop residues or if no commercial feeds are used, the crop residues will be incorporated into the soil. The latter strategy makes more nutrients available for the crop and thus influences the crop production. The model iterates until a new equilibrium is reached. It should be noted that in the current analysis the size of the SPU is kept constant. The resulting nutrient balances for the five farms are presented in Table 4.5.

The results show that although individual nutrient flows do change, the overall soil nutrient balance does not change significantly compared to the extra nutrient input into the system through mineral fertiliser, i.e. an additional nitrogen input of 50 kg/ha reduces nutrient losses by 10 kg/ha. Important effects of an increase in fertiliser use are the increase in crop production and in crop residues. The outcome is that IN 2 remains very similar even though more crop residue is produced. An economic analysis needs to be done to ascertain whether these changes are economically attractive. Such an analysis was beyond the scope of this paper.

Table 4.5. Soil nutrient balances as calculated by the Dyn-Nutmon system for the five representative farms in the Embu region, given an increased fertilisation rate of 50 kg ha^{-1} yr^{-1}.

	AEZ 1	AEZ 2	AEZ 3	AEZ 4	AEZ 5
Inputs					
Mineral fertiliser	165	129	57	112	50
Organic fertiliser/feed	38	40	13	34	29
Atmospheric deposition	6	6	6	6	5
Nitrogen fixation	4	4	2	3	3
Sedimentation	0	0	0	0	0
Total IN	213	178	78	154	87
Outputs					
Crop/animal products	55	21	27	35	51
Crop residues	137	49	35	78	7
Leaching	85	57	42	44	33
Denitrification	32	30	23	18	3
Erosion	11	129	44	8	59
Total OUT	320	285	171	183	153
Depletion rate (kg/ha/yr)	107	107	93	29	66

SOM-Nutmon

Although the soil nutrient balance provides an estimate of the rate of change of soil fertility, it can be questioned whether nutrient balances tell us anything about long-term changes. Below, therefore, using the same Nutmon dataset, it is analysed how the system might evolve over time. The indicator used for these long-term changes is the soil organic matter pool modelled by the SOM-Nutmon model. All the systems were modelled under the same assumptions as the simulations with the Dyn-Nutmon model. The systems are modelled dynamically, i.e. both changes in soil fertility and their feedback on the production are considered. The simulation results for a 20-year period are presented in Table 4.6. The rates for AEZ 5 were significantly smaller than the rates in the other AEZ, mainly due to the smaller reserves of soil organic matter in this AEZ. The results confirm the significant decline in soil fertility that has been identified in the wide range of soil fertility literature. They also indicate that soil organic matter pools are highly dynamic and that long-term predictions will need to consider these dynamics. It is important to realise that these are aggregated results and that there is large within-farm variability in resources allocation, nutrient flows and soil fertility (Tittonell *et al.*, 2005a,b).

Table 4.6. Soil organic matter dynamics over a 20-year period as calculated by the SOM-Nutmon system for the five representative farms in the Embu region with an increased fertilisation rate of 50 kg/ha/yr.

	AEZ 1	AEZ 2	AEZ 3	AEZ 4	AEZ 5
Soil organic matter (t/ha)					
Year 1	91	86	93	77	35
Year 20	56	56	57	49	26
Annual rate of change (t/ha/yr)	-1.8	-1.6	-1.8	-1.4	-0.5

Discussion and conclusions

From the large number of soil nutrient balance studies throughout the African continent it is clear that soil fertility decline is a serious threat. Although the depletion rates are highly variable, most studies present negative soil nutrient balances. Nutmon has proven to be an excellent tool for diagnosing soil nutrient balances and identifying the major soil nutrient fluxes in agricultural systems. However, it is not the only tool available. The other tools that have been developed are just as valuable (Roy *et al.*, 2003). The resource guide for participatory research and action research (Defoer and Budelman, 2000) addresses the soil nutrient balance from a more participatory angle. The Ecopath model (Polovina, 1985; Dalsgaard and Christensen, 1997) is currently commonly applied for integrated agriculture–aquaculture systems. Most of these models are based on resource flow mapping. The model selected must be appropriate for the research questions.

In this paper, two innovative alternatives to the Nutmon model have been proposed. In this particular study, this was deemed preferable to developing a new model or using another existing model. The standardised survey instrument of Nutmon is highly valued and allows for an easy comparison of various case studies. Nutmon does have some limitations, however: (1) dynamic monitoring is time-intensive and requires significant resources, (2) the regression models in Nutmon have been developed on the basis of a limited number of studies from the African continent and are not generic, and (3) Nutmon was developed to estimate the current rates of change. Other shortcomings are given by Færge and Magid (2004). A major advantage of Nutmon is its simple and transparent accounting system with inputs and outputs. As the model does not include the interactions between the various flows, it cannot be used to plan interventions. A simple example illustrates this inherent property of Nutmon. Increasing the use of mineral fertiliser will affect the production of the system. However, the interaction between fertiliser use and crop production is not included in Nutmon but will only be known if farmers with different fertiliser rates were included in the monitoring. In addition, the increased production will lead to increased crop residue production that may result in more manure production.

The static character of Nutmon, providing current rates of change rather than simulating the actual dynamics, has resulted in some misinterpretation of the model results. For example, the results have been extrapolated to the future by multiplying the calculated nutrient balance by a number of years. It is unsound science to extrapolate into the future without dealing with the dynamic character of the system. Declining soil fertility, for example, will result in a decrease in nutrient losses, as nutrient loss (e.g. to production, leaching and erosion) will decrease. Another way Nutmon has been misapplied is in estimating the cost of soil fertility decline by multiplying the soil nutrient balance by the price of mineral fertiliser. The new approaches allow for additional data analysis using existing Nutmon datasets. In addition, the complexity of Nutmon corresponds to the data availability in many studies.

This paper has described two alternatives to the traditional Nutmon system dealing with the *ex ante* evaluation of alternative management practices (Dyn-Nutmon) and with the long-term soil fertility dynamics (SOM-Nutmon). Although they have not been incorporated in the Nutmon model, it is clear from the examples presented above that these approaches open up new perspectives. They allow the surveyed farms to be examined more dynamically and give insight into the actual dynamics of the system. The examples also show that some assumptions made in the past in various Nutmon studies are incorrect. Mineral fertiliser has often been advocated as the solution and in some cases even used as a direct replacement to calculate the cost of nutrient depletion as the product of fertiliser price (expressed in $/kg N) and soil nutrient balance (expressed in δN/ha/yr). Nutrient use efficiencies typically range between 25 and 50%, indicating that a large part of the nutrients, e.g. the soil nitrogen, is lost through leaching, denitrification and erosion. In addition, the nitrogen effectively taken up by the plant will be exported with the crop. In many cases, only the nutrients in crop residues effectively return to the soil. Similarly, this paper has demonstrated that the soil nutrient balance is not the appropriate tool for the evaluation of the temporal dynamics; instead, the soil organic matter dynamics should be examined.

In retrospect this raises an interesting question: does Nutmon calculate realistic nitrogen depletion rates? A simple back-of-the-envelope calculation provides an answer. Generally, the mineralisation rates of soil organic matter are around 2%. Given the soil carbon stocks presented in Table 4.6 and a soil C:N ratio of 10, nitrogen depletions in the different AEZ cannot exceed 150-180 kg N ha^{-1} yr^{-1} for AEZ 1-4 and 70 kg N ha^{-1} yr^{-1} for AEZ 5. Assuming that the mineral N cannot accumulate in the soil, lower nitrogen depletion rates can only be found if significant quantities of organic matter are incorporated into the soil. In the absence of organic matter incorporation, nitrogen depletion rates are probably close to the above numbers

Although Dyn-Nutmon allows for an *ex ante* evaluation of alternative practices, it should be realised that this paper only considers the technical feasibility of these alternative practices. The economic feasibility of alternative management practices and the policy measures needed to make these practices attractive to farmers have been excluded; they merit separate analysis, as illustrated by, for example, Scoones and Toulmin (1998, 1999).

Acknowledgements

This study was carried out in the context of various research projects including (1) the trade-off analysis program (Phase 2) funded the Soil Management Collaborative Research Program of US-AID, (2) the INREF-POND program of Wageningen University, and (3) The Consortium for Agricultural Soils Mitigation of Greenhouse Gases (CASMGS). The author is grateful to the reviewers, editors and Joy Burrough for their constructive comments on content and language.

References

Dalsgaard, J.P.T. and V. Christensen, 1997. Flow modelling with ECOPATH: providing insights on the ecological state of agro-ecosystems. In: P.S. Teng, M.J. Kropff, H.F.M. ten Berge, J.B. Dent, F.B. Lansigan and H.H van Laar (eds.) Applications of Systems Approaches at the Farm and Regional Level. Kluwer Academic Publishers, The Netherlands, pp. 203-212.

De Jager, A., D. Onduru, M.S. van Wijk, J. Vlaming and G.N. Gachini, 2001. Assessing sustainability of low-external-input farm management systems with the nutrient monitoring approach: a case study in Kenya. Agricultural Systems 69, 99-118.

De Jager, A., I. Kariuku, F.M. Matiri, M. Odendo and J.M. Wanyama, 1998a. Monitoring nutrient flows and economic performance in African farming systems (NUTMON): IV. Linking nutrient balances and economic performance in three districts in Kenya. Agriculture, Ecosystems & Environment 71, 81-92.

De Jager, A., S.M. Nandwa and P.F. Okoth, 1998b. Monitoring nutrient flows and economic performance in African farming systems (NUTMON): I. Concepts and methodologies Agric. Ecosystems & Environment 71, 37-48.

Defoer, T. and A. Budelman (Eds.), 2000. Managing soil fertility in the tropics: a resource guide for participatory research and action research. Royal Tropical Research Institute, Amsterdam, the Netherlands.

Færge, J., and J. Magid, 2004. Evaluating NUTMON nutrient balancing in Sub-Saharan Africa. Nutrient Cycling in Agroecosystems 69, 101-110.

FAO, 2005. Scaling Soil Nutrient Balances - Enabling mesoscale approaches for African realities. FAO fertilizer and plant nutrition bulletin 15. FAO, Rome.

Feder, G., E.R. Just and D. Zilberman. 1985. Adoption of agricultural innovations in developing countries: A survey. Economic Development and Cultural Change 33, 255-298.

Franzel, S., 1999. Socioeconomic factors affecting the adoption potential of improved tree fallows in Africa. Agroforestry Systems 47, 305-321.

FURP, 1987. Fertilizer Use Recommendation Project. Final Report. Description of first priority trial sites. Nairobi. National Agricultural Research Laboratories. Nairobi, Kenya.

Gachimbi, L.N., A. de Jager, H. van Keulen, E.G. Thuranira and S.M. Nandwa, 2002. Participatory diagnosis of soil nutrient depletion in semi-arid areas of Kenya. Managing Africa's Soils No. 26. IIED, London, UK.

Giller, K.E., E.C. Rowe, N. de Ridder and H. van Keulen, 2006. Resource use dynamics and interactions in the tropics: scaling up in space and time. Agricultural Systems 88, 8-27.

Gruhn, P., F. Goletti and M. Yudelman, 2000. Integrated Nutrient Management, Soil Fertility, and Sustainable Agriculture: Current Issues and Future Challenges. Food, Agriculture, and the Environment Discussion Paper 32, IFPRI, Washington.

Haileslassie, A., J.A. Priess, E. Veldkamp and J.P. Lesschen, 2006. Smallholders' soil fertility management in the central highlands of Ethiopia: implications for nutrient stocks, balances and sustainability of agroecosystems. Nutrient Cycling in Agroecosystems 75, 135-146.

Jenkinson D.S., and K. Coleman, 1994. Calculating the annual input of organic matter to soil from measurements of total organic carbon and radiocarbon. European Journal of Soil Science, 45, 167-174.

Kenyan Government, 2004. Strategy for Revitalizing Agriculture. Prepared by Ministry of Agriculture and the Ministry of Livestock and Fisheries Development. Nairobi, Kenya.

Koning, N., and E.M.A. Smaling, 2005. Environmental crisis or 'lie of the land'? The debate on soil degradation in Africa. Land use policy 22, 3-11.

Muendo, P.N., J.J. Stoorvogel, M.C.J. Verdegem, E.-N. Gamal and J.A.J. Verreth, 2007. Roles of ponds in integrated aquaculture-agriculture systems. In: A.J. van der Zijpp, J.A.J. Verreth, Le Quang Tri, M.E.F. van Mensvoort, R.H. Bosma and M.C.M. Beveridge (eds.), Fishponds in farming systems. Wageningen Academic Publishers, Wageningen, The Netherlands, pp. 125-133.

Ouma, J.O., F.M. Murithi, W. Mwangi, H. Verkuijl, M. Gethi and H. de Groote, 2002. Adoption of maize seed and fertilizer technologies in Embu District, Kenya. Cimmyt, Kari, Nairobi, Kenya.

Parton, W.J., R. McKeown, V. Kirchner and D. Ojima, 1992. CENTURY Users' Manual, Natural Resource Ecology Laboratory, Colorado State University, Ft. Collins.

Polovina, J.J. 1985. An approach to estimating an ecosystembox model. U.S. Fisheries Bulletin 83, 457–460.

Priess, J.A., G.H.J. de Koning and A. Veldkamp, 2001. Assessment of interactions between land use change and carbon and nutrient fluxes in Ecuador. Agriculture, Ecosystems & Environment 85, 269-279.

Ransom, J.K., J. Ojiem and F.K. Kanampiu, 1993. Nutrient Flux between Maize and Livestock in a Maize-Coffee-Livestock System in Central Kenya. In: Livestock and Sustainable Nutrient Cycling in Mixed Farming Systems of Sub-Saharan Africa, J.M. Powell, S. Fernandez-Rivera, T.O. Williams and C. Renard (eds.). Addis Ababa, Ethiopia: International Livestock Centre for Africa, pp. 411-418.

Roose, E., and B. Barthès, 2001. Organic matter management for soil conservation and productivity restoration in Africa: a contribution from Francophone research. Nutrient cycling in agroecosystems 61, 159-170.

Rowe, E.C., M.T. van Wijk, N. de Ridder and K.E. Giller, 2006. Nutrient allocation strategies across a simplified heterogeneous African smallholder farm. Agriculture, Ecosystems and Environment 116, 60-71.

Roy, R.N., R.V. Misra, J.P. Lesschen and E.M.A. Smaling, 2003. Assessment of soil nutrient balance. Approaches and methodologies. FAO Fertilizer and plant nutrition bulletins 14. FAO, Rome.

RRC-Embu. 1998. Recommendation Guidelines for Crops, Soil fertility, and Livestock Management (up to 1998). Embu, Kenya: RRC-Embu.

Salasya, B.D, W. Mwangi, H. Verkuijl, M.A. Odendo and J.O. Odenya, 1998. An Assessment of the Adoption of Seed and Fertilizer Packages and the Role of Credit in Smallholder Maize Production in Kakamega and Vihiga Districts, Kenya. Mexico, D.F.:CIMMYT and KARI.

Scheldrick, W., J. Keith Syers and J. Lingard, 2003. Contribuition of livestock excreta to nutrient balances. Nutrient cycling in agroecosystems 66, 119-131.

Schlecht, E., and P. Hiernaux, 2004. Beyond adding up inputs and outputs: process assessment and upscaling in modelling nutrient flows. Nutrient Cycling in Agroecosystems 70, 303-319.

Scoones, I., and C. Toulmin, 1998. Soil nutrient balances: what use for policy? Agriculture Ecosystems and Environment 71, 255-267.

Scoones, I., and C., Toulmin, 1999. Policies for soil fertility management in Africa. Report prepared for DFID, IDS and IIED, UK.

Smaling, E.M.A., and L.O. Fresco, 1993. A decision-support model for monitoring nutrient balances under agricultural land use (NUTMON). Geoderma 60, 235-256.

Smaling, E.M.A., S.M. Nandwa, H. Prestele, R. Roetter and F.N. Muchena, 1992. Yield response of maize to fertilizers and manure under different agro-ecological conditions in Kenya. Agriculture, Ecosystems, and Environment 41, 241-252.

Sombroek, W.G., H.M.H. Braun and B.J.A. van der Pouw, 1982. Exploratory Soil map and agro-ecological zones map of Kenya, scale 1:1 000 000. Report E1, Kenya Soil Survey, Nairobi.

Stoorvogel, J.J., and E.M.A. Smaling, 1990. Assessment of soil nutrient depletion in Sub-Saharan Africa: 1983 – 2000. Winand Staring Centre report 28. Winand Staring Centre, Wageningen, The Netherlands.

Stoorvogel, J.J., E.M.A. Smaling and B.H. Janssen, 1993. Calculating soil nutrient balances in different scales. I. Supra-national scale. Fertilizer Research 35, 227-235.

Tittonell, P., B. Vanlauwe, P.A. Leffelaar, E.C. Rowe and K.E. Giller, 2005a. Exploring diversity in soil fertility management of smallholder farms in western Kenya. I. Heterogeneity at region and farm scale. Agriculture, Ecosystems and Environment 110, 149-165.

Tittonell, P., B. Vanlauwe, P.A. Leffelaar, K.D. Shepherd and K.E. Giller, 2005b. Exploring diversity in soil fertility management of smallholder farms in western Kenya. II. Within-farm variability in resource allocation, nutrient flows and soil fertility status. Agriculture, Ecosystems and Environment 110, 166-184.

Tittonell, P., P.A. Leffelaar, B. Vanlauwe, M.T. van Wijk and K.E. Giller, 2006. Exploring diversity of crop and soil management within smallholder African farms: A dynamic model for simulation of N balances and use efficiencies at field scale. Agricultural Systems: in press.

Van den Bosch, H., A. De Jager and J. Vlaming, 1998a. Monitoring nutrient flows and economic performance in African farming systems (NUTMON): II. Tool development. Agriculture, Ecosystems & Environment 71, 49-62.

Van den Bosch, H., J. Vlaming, M.S. van Wijk, A. de Jager, A., Bannink and H. Van Keulen, H., 2001. Manual to the NUTMON Methodology. Alterra/LEI, Wageningen University, The Netherlands.

Van den Bosch, H., J.N. Gitari, V.N. Ogaro, S. Maobe and J. Vlaming, 1998b. Monitoring nutrient flows and economic performance in African farming systems (NUTMON): III. Monitoring nutrient flows and balances in three districts in Kenya. Agriculture, Ecosystems & Environment 71, 63-80.

Vanlauwe, B., and K.E. Giller, 2006. Popular myths around soil fertility management in sub-Saharan Africa. Agricultural, Ecosystems and Environment 116: 34-46.

Wortman, C.S., 2001. Nutrient dynamics in a climbing bean and sorghum crop rotation in the central African highlands. Nutrient Cycling in Agroecosystems 61, 267-272.

Roles of ponds in integrated agriculture-aquaculture systems

Patricia N. Muendo[1], Jetse J. Stoorvogel[2], Marc C.J. Verdegem[1], El-Naggar Gamal[3] and Johan A.J. Verreth[1]
[1]Fish culture and Fisheries Group, Wageningen Institute of Animal Sciences, Wageningen University, The Netherlands
[2]Laboratory of Soil Science and Geology, Wageningen University, The Netherlands
[3]The World Fish Center (ICLARM), Africa and West Asia Regional Center, Egypt

Abstract

The effects of adding an aquaculture pond to existing farming operations on nutrient use efficiency and productivity were explored. Ponds can be used to cycle nutrients from agricultural by-products such as manures and composted grasses. The production from ponds using agricultural by-products was similar to that of ponds receiving formulated pellets. The largest fraction of the nutrients fed to ponds accumulated in the sediment. Pond sediments were rich in nitrogen and potassium, but only contained small amounts of soluble reactive phosphorous. Pond sediments supplemented with phosphorous produced a similar corn yield to standard inorganically fertilised plots. Starting from a database on nitrogen flows through Kenyan highland farms, a typical farm operation was defined for each agro-ecological zone. In the agro-ecological zones where annual temperature ranges allowed for the culture of tilapias, a tilapia pond receiving nutrients from on-farm by-products was integrated into the farming operations. Results showed that the nutrient depletion, typical for highland cultures in Kenya, is reduced, while more nutrients end up in harvestable products than in situations without aquaculture. The results are promising, but more work is needed on fine-tuning capital and labour availability for integrated aquaculture-agriculture systems to the needs or opportunities of farming households.

Keywords: fishpond, manure, sediment, maize, integration

Introduction

Aquaculture ponds can play an important role in nutrient cycling in mixed farming systems. Nutrients contained in by-products of terrestrial farming can contribute to the production of aquatic food. In turn, ponds can provide nutrient-rich irrigation water while nutrients stored in pond sediments can be utilised to fertilise terrestrial crops (Jamu and Piedrahita, 2002). As such, integration of aquaculture and agriculture can result in improved nutrient use efficiencies (Pillay, 1994). Unfortunately, fish, shellfish or shrimp are often considered the only beneficial products from aquaculture, thereby neglecting other benefits ponds provide in integrated aquaculture-agriculture systems (Pillay, 1994). In addition, formulated feeds are preferred above agricultural by-products as inputs to ponds because farmers believe formulated feeds boost pond production beyond that achievable with fertilisation only (Diana *et al.*, 1991). Hence, the potential of otherwise wasted agricultural by-products through pond aquaculture and its effects on nutrient use efficiencies in mixed aquaculture-agriculture farms, remains insufficiently explored.

The present study quantified the nutrients trapped in pond sediments and the effect the use of pond sediments as crop fertiliser has on the nutrient use efficiency of the overall aquaculture-agriculture (IAA) farming system. The fractions of nutrients from agricultural by-products applied to ponds ending up in fish biomass or accumulating in sediments were measured. The value of pond sediments as fertiliser for terrestrial crops was evaluated. Finally, the effects of incorporating pond aquaculture in Kenyan highland farming systems on nutrient use efficiency and overall farm productivity were explored.

Methods

In a first experiment, conducted between June and October 2002 in Abassa, Egypt, the trophic structure in organically fertilised and feed driven tilapia ponds was compared using multivariate analyses. The organic fertilisers evaluated were chicken manure and composted grass. Tilapias were stocked at one fish m^{-2} in ponds fertilised with chicken manure or composted grass and at two fishes m^{-2} in feed driven ponds. Fish were stocked at a higher density in the feed-driven ponds because it was assumed that the carrying capacity in feed driven ponds is higher than in organically fertilised ponds. Input rations of fertilisers or feed were such that in all treatments the nitrogen (N) input per fish was similar. In a completely randomised design, the three treatments were carried out in triplicate. Ponds were stocked with 18-20 g Nile tilapia (*Oreochromis niloticus*) grown over a 139 day culture period. In the feed driven treatment fishes were fed at a rate of 3% body weight per day. In the organically fertilised treatments, fertiliser doses corresponded to a similar N-input per fish as in the feed driven ponds. Fishes were sampled monthly by seining a minimum of 10% of the numbers stocked. Fertilisation or feeding was suspended when dawn oxygen levels dropped below 2 mg l^{-1} and resumed when dawn oxygen levels were restored above this level. Temperature, dissolved oxygen (DO) and pH were measured twice daily (6.00 and 15.00 hours), Secchi transparency once daily (12.00 hours). Primary productivity (PP) was measured bi-weekly using the free water method (Hall and Moll, 1975). Community respiration was assumed to be twice the decrease in mean DO between 18.00 hours and 6.00 hours the following day. Gross PP was calculated as the sum of gains in mean DO during the day, plus half of the community respiration. Other parameters measured in the water column on a bi-weekly basis were soluble reactive phosphorus (SRP; APHA, 1995), total ammonia nitrogen (TAN; APHA, 1995), nitrite-nitrogen (NO_2^--N; Boyd and Tucker, 1992), nitrate nitrogen (NO_3^--N; Boyd, 1979), total nitrogen (TN; Gross and Boyd, 1998), potassium (K; Page *et al.*, 1982) and chlorophyll *a* (APHA, 1995). Bi-weekly sediment samples were analysed for TN by the Kjeldahl method (Page *et al.*, 1982), phosphorous, organic carbon and potassium following procedures described by Buurman *et al.* (1996). Data were analysed through factor analysis to identify the principal ecological processes acting in the environments (Milstein, 1993).

In a second pond experiment, sediment and sediment nutrients accumulating during a 5 month tilapia growout cycle were quantified monthly by measuring changes in sediment depth, bulk densities and nutrient (N, P, K and organic carbon (OC)) concentration. The same analytical techniques as for the first experiment were used. Chicken manure or pelleted feed were the nutrient sources in ponds stocked with 1 or 2 tilapias m^{-2}. At harvest, the accumulated sediment was transferred to experimental corn plots. Corn production relying on fertiliser from either chicken manure or pellet fed plots was compared to production based on a standard inorganic fertiliser dose (positive control) or receiving no fertiliser at all (negative control).

To explore the feasibility of developing IAA systems in existing agricultural farming systems, different tilapia production scenarios were defined based on information found in literature. Calculated nutrient flows through the modelled tilapia production systems were integrated into a database of nitrogen flows though existing farming systems in Kenyan highlands. Four farming systems, each representing a different agro-ecological zone in the Kenyan highlands, were explored[2]. Nutrient balances of the resultant IAA systems were compared with those of the present agricultural farming systems (obtained from NUTrient MONitoring (NUTMON) survey data sets) and the changes in farm nutrient balances were quantified.

Results and discussion

In the first experiment, a specific growth rate of 1.7% body weight (bw) d^{-1} was realised in the chicken manure driven ponds compared to 1.1% bw d^{-1} in the other treatments ($P < 0.05$). Survival rates were similar between treatments ($P > 0.05$). Due to the double stocking density and N-input in the feed driven ponds production was nearly twice as high in the pellet fed ponds as in the organically fertilised ponds ($P < 0.05$) (Table 4.7). Factorial analysis demonstrated that primary production and decomposition processes are the driving forces behind tilapia production. The most important factor was related to algal biomass synthesis in the water column, which was significantly higher in the chicken manure driven ponds than in the pellet fed ponds ($P < 0.05$). Factor 2 suggested that DO, temperature and pH levels were higher in the chicken manure driven ponds than in the pellet fed ponds ($P < 0.05$). The higher pH levels in turn favoured the liberation of orthophosphate, which again was much higher in chicken manure

[2] These are the farms AEZ-2 to AEZ-5 from the previous chapter by Stoorvogel.

Table 4.7. Tilapia growth and yields in ponds fertilised with agricultural by-products (chicken manure and field grasses) and in feed driven ponds. Values are means of three replicates for each treatment (Muendo et al.*, 2006). Same letters (superscripts) within a row indicate no significant difference at the 0.05 level (a>b).*

	Treatment		
Parameter	Chicken manure	Composted grass	Pelleted feed
Mean weight at stocking (g)	18.5^{a}	18.3^{a}	19^{a}
Number stocked	200	200	400
Mean weight at harvest (g)	252^{a}	172^{b}	174^{b}
Survival (%)	41^{a}	48^{a}	54^{a}
Daily growth rate (g day^{-1})	1.7^{a}	1.1^{a}	1.1^{a}
Total biomass at harvest (kg $pond^{-1}$)	30.7^{b}	26.0^{b}	53.1^{a}
Net fish yields (NFY)			
kg $pond^{-1}$	27.0^{b}	22.3^{b}	45.5^{a}
kg ha^{-1}	1350^{b}	1117^{b}	2275^{a}
kg ha^{-1} $year^{-1}$ (assuming 2 seasons)	2700^{b}	2234^{b}	4550^{a}

driven ponds (Factor 3, $P < 0.05$). Compared to other treatments, the higher photosynthetic activity in the chicken manure driven ponds did not result in higher accumulation of nitrogen and organic carbon in pond sediments (Factor 4, $P > 0.05$). This was linked to differences in decomposition of organic matter between treatments (Factor 5, $P < 0.05$). Although the results of the factorial analysis are exploratory, the results concur with findings from Knud-Hansen *et al.* (1993) who suggested that with equal stocking densities and equal nutrient inputs, similar fish productions can be achieved in fertilised and feed-driven ponds.

To confirm these results, equal input levels and stocking densities were used in the second experiment. Fish survival and production was similar in ponds receiving comparable N-inputs either in the form of chicken manure or pelleted feed. About 125 – 300 kg N, 1.8 – 5 tons organic matter and 50 – 125 kg K accumulated per ha in pond sediments during the 5 month culture period (Table 4.8). The amount of available P that accumulated was negligible. Statistically, the amounts of N, OC, K and P that accumulated in pond sediment and that can be extracted for production did not differ between ponds fertilised with chicken manure and those fed with pellets ($P > 0.05$). In conclusion, up to 173 tons of pond sediment ha^{-1} (5 month period)$^{-1}$ accumulated in the ponds. Although questions remain as to the origin of the accumulated sediment, influent water and fertiliser or feed inputs are not the major sources. Nutrients released during mineralisation of organic inputs enriched the sediment, especially with nitrogen. The accumulated sediment is rich in nitrogen, exchangeable potassium and organic matter. As such, it has a high potential as a nitrogen and potassium fertiliser. An additional advantage of the use of pond sediment is its richness in organic matter, which also functions as a soil conditioner. In heavy clay and in light sandy soils, the organic matter can improve soil structure and increase the water holding capacity. The negligible accumulation of available phosphorus during the experiment suggests that the phosphorus accumulates in a non-available form. In future experiments, total phosphorus concentrations should also be measured to distinguish between available and non-available phosphorus.

Fertiliser requirements are highly situation (e.g. soil fertility and crop) dependent, so care must be taken not to generalise. Considering the Egyptian situation, the nutrients accumulated in an one ha pond over one growing season potentially meet the nitrogen and potassium requirement for over 1.5 ha of maize, provided it is supplemented with phosphorus fertiliser. When this is done, crop production was similar ($P > 0.05$) in plots fertilised with pond sediments and in plots fertilised with the standard inorganic fertilisers, but was significantly lower in unfertilised plots ($P < 0.05$) (Table 4.9). The results indicate that, pond sediment is an excellent crop fertiliser, regardless of the pond input types: chicken manure or formulated commercial feed (Muendo, 2006).

In the four agro-ecological zones of the Kenyan highlands considered in this study, nitrogen depletion is common. By adding the right size of semi-intensively managed pond to the farming system in each agro-economic zone nutrient depletion was reduced by 23 – 35%. Table 4.10 shows the N-balances of the present farming systems. The N-balances after inclusion of modelled aquaculture ponds are reviewed in Table 4.11. The results show that integration of an aquaculture component in existing farms provides an opportunity to utilise and store nutrients from agricultural by-products that would otherwise have been lost through leaching

Table 4.8. Sediment and sediment nutrient concentrations at the end of the culture period. CM = chicken manure; P = pellet; 1 = stocking density of one fish m^{-2}; 2 = stocking density of two fishes m^{-2} (Muendo, 2006).

	Treatments			
Parameters	CM-1	CM-2	P-1	P-2
Accumulated sediment quantity (tons $pond^{-1}$ cycle)	3.1	1.5	2.8	1.67
Nitrogen				
Concentration at harvest (g kg^{-1})	1.9	1.9	1.7	1.6
Quantity in accumulated sediment (kg $pond^{-1}$)	5.89	2.85	4.76	2.67
Quantity in accumulated sediment (kg ha^{-1})	295	143	238	124
Organic carbon				
Concentration at harvest (g kg^{-1})	14.5	12.5	11.7	12.7
Quantity in accumulated sediment (kg $pond^{-1}$)	45.0	18.8	32.8	21.2
Quantity in accumulated sediment (tons ha^{-1})	2.3	0.9	1.6	1.1
Available phosphorus				
Concentration at harvest (mg kg^{-1})	6.3	3.8	3.3	3.5
Quantity in accumulated sediment (g $pond^{-1}$)	19	6	9	6
Quantity in accumulated sediment (kg ha^{-1})	0.97	0.28	0.46	0.29
Potassium				
Concentration at harvest (mg kg^{-1})	72	77	74	73
Quantity in accumulated sediment (kg $pond^{-1}$)	2.23	1.16	2.07	1.22
Quantity in accumulated sediment (kg ha^{-1})	112	58	104	61

Table 4.9. Maize production in unfertilised plots, plots fertilised with inorganic fertiliser and in plots fertilised with pond sediment (Muendo, 2006).

Fertiliser type	Fertiliser (kg $plot^{-1}$)	Total plant biomass (kg $plot^{-1}$)	Grain weight (kg ha^{-1})
Pond sediment (from pellet fed ponds)	28.7	7.5 [a]	6.3
Pond sediment (from chicken manured ponds)	26.7	6.0 [a]	5.0
Inorganic fertiliser	20.6	5.3 [a]	4.4
None	11.2	2.5 [b]	2.1

[a,b]Values in the same column with no superscripts in common are significantly different. In case of no significant difference, no superscripts are given.

Table 4.10. Nutrient balances in kg ha^{-1} yr^{-1} of existing farming systems without fishponds in 4 agro-ecological zones (AEZ); PPU = primary production unit (crop fields); SPU = secondary production unit (livestock); RU = Redistribution unit (manure or garbage compost heaps); FP = fishpond. Adapted from Muendo (2006).

	AEZ 1				AEZ 2				AEZ 3				AEZ 4			
	PPU	SPU	RU	FP	PPU	SPU	RU	FP	PPU	SPU	RU	FP	PPU	SPU	RU	FP
Area (ha)	1.41	-	-	0	3.51	-	-	0	2.12	-	-	0	2.28	-	-	0
Inputs																
Mineral fertiliser	111	0	0	0	25	0	0	0	131	0	0	0	0	0	0	0
Organic fertiliser/feed	56	211	179	0	42	129	109	0	65	162	138	0	66	168	184	0
Atmospheric deposition	9	0	0	0	22	0	0	0	12	0	0	0	11	0	0	0
Nitrogen fixation	7	0	0	0	18	0	0	0	11	0	0	0	12	0	0	0
Sedimentation	0	0	0	0	0	0	0	0	0	0	0	0	0	0	0	0
Total IN	183	211	179	0	107	129	109	0	219	162	138	0	89	168	184	0
Outputs																
Crop/animal products	19	21	0	0	57	13	0	0	56	16	0	0	66	17	0	0
Crop residues	44	179	54	0	73	109	33	0	127	138	41	0	10	142	55	0
Leaching	67	0	126	0	112	0	77	0	74	0	96	0	41	0	129	0
Denitrification	35	0		0	62	0		0	31	0		0	4	0		0
Erosion	182	0	0	0	154	0	0	0	18	0	0	0	134	0	0	0
Total OUT	346	200	179	0	458	122	109	0	306	154	138	0	255	159	184	0
Net balance	163	11	0	0	-351	6	0	0	-87	8	0	0	-166	8	0	0
Depletion rate	116				100				41				73			

Table 4.11. Nutrient balance in modelled IAA systems with a fishponds integrated in four agro-ecological zone (AEZ) (Muendo, 2006); PPU = primary production unit (crop fields); SPU = secondary production unit (livestock); RU = Redistribution unit (manure or garbage compost heaps); FP = fishpond.

	AEZ 1				AEZ 2				AEZ 3				AEZ 4			
	PPU	SPU	RU	FP	PPU	SPU	RU	FP	PPU	SPU	RU	FP	PPU	SPU	RU	FP
Area (ha)	1.35	-	-	0.06	3.47	-	-	0.04	2.07	-	-	0.05	2.21	-	-	0.07
Inputs																
Mineral fertiliser	111	0	0	0	25	0	0	0	131	0	0	0	0	0	0	0
Organic.fertiliser/feed	18	211	179	91	20	129	109	56	37	162	138	70	27	168	185	94
Atmospheric deposition	8	0	0	0.4	22	0	0	0.2	12	0	0	0.3	11	0	0	0.3
Nitrogen fixation	7	0	0	0.2	18	0	0	0.1	11	0	0	0.2	12	0	0	0.2
Sedimentation & irrigation	121	0	0	87	89	0	0	76	54	0	0	9	108	0	0	65
Total IN	265	211	179	179	173	129	109	132	244	162	138	79	158	168	185	160
Outputs																
Crop/animal products	24	21	0	36	64	13	0	26	57	16	0	16	80	17	0	32
Crop residues	56	179	108	0	82	109	66	0	128	138	83	0	12	143	111	0
Leaching/pond drainage	73	0	72	9	118	0	44	7	75	0	55	4	51	0	74	8
Denitrification	38	0	0	18	65	0	0	13	31	0	0	8	5	0	0	16
Erosion/pond sediment	174	0	0	116	152	0	0	86	18	0	0	52	131	0	0	104
Total OUT	366	200	179	179	481	122	109	132	309	154	138	79	278	160	185	160
Net balance	-101	11	0	0	-308	6	0	0	-65	8	0	0	-120	8	0	0
Depletion rate(kg/ha/yr)	75				89				31				54			

during storage and application. In addition, by putting the pond at the lowest altitude in the farm, the pond also efficiently traps eroded nutrients. A weakness of the study is that it did not consider labour requirements. The assumption was made that labour will be available for pond management and pond sediment removal and application on land. Labour constraints are common in the Kenyan highlands, especially at ploughing/planting time. Future studies should take labour availability into consideration and cheap, quick and reliable methods to extract sediments from ponds need to be developed.

Conclusions

The study showed that equal production levels can be obtained from ponds receiving manures, crop residues or formulated pellets (Muendo *et al.*, 2006). Under the Egyptian conditions, pond sediments proved to be a good fertiliser for corn production, provided they are supplemented with phosphorous. Ponds are excellent nutrient traps, as on average 40-60% of the input nutrients accumulate in the sediment, where they can be stored for a long periods with minimal quality loss. When looking at the ideally configured integrated aquaculture-agriculture systems, the derived benefits are increased cycling of nutrients between farming components, and larger percentages of nutrients converted into harvested products. In each of the agro-ecological zones investigated, the net depletion of N was reduced after pond integration; hence the problem of nutrient depletion remains, although it is reduced. One aspect requiring further study relates to the costs of nutrients in terms of labour and cost. Quality fish pellets are expensive but cheap to transport, store and apply. Most farm households have no labour to spare, especially during the planting and cropping seasons. In the modelled IAA systems a lot of additional labour will be needed for fish production, collection and application of manure or crop residues, and extraction and use of pond sediments (Muendo, unpublished data; Pant *et al.*, 2004). If outside labour has to be hired the costs may be too high. In addition, input-output nutrient balances of individual activities in mixed farms should match in time. In some situations farmers also tend to specialise, reducing the number of farming activities rather than increasing them. In such a situation further integration could be achieved between farms or even between groups of farms (Edwards, 1998).

References

APHA (American Public Health Association, American Water Works Association and the Water Environment Association), 1995. Standard methods for the examination of water and waste water. American Public Health Association, Washington, USA. 1108 pp.

Boyd, C.E., 1979. Water quality in warmwater fish ponds. Auburn University Agricultural Experimental Station. Craftmaster Printers Inc., Alabama, USA. 359 pp.

Boyd, C.E. and C.S. Tucker, 1992. Water quality and pond soil analysis for aquaculture. Alabama Agricultural Experiment Station. Auburn University. 183 pp.

Buurman, P., B.V. Lagen and E.J. Velthorst, 1996. Manual for soil and water analysis. Backhuys Publishers Leiden, The Netherlands. 314 pp.

Diana, J.S., C.K. Lin, and P.J. Schneeberger, 1991. Relationships among nutrient inputs, water nutrient concentrations, primary production, and yield of *Oreochromis niloticus* in ponds. Aquaculture 92, 323-341.

Edwards, P., 1998. A system approach for the promotion of integrated aquaculture. Aquaculture economics and management 2, 1-12.

Gross, A. and C.E. Boyd, 1998. A digestion procedure for the simultaneous determination of total nitrogen and total phosphorus in pond water. Journal of the World Aquaculture Society 29, 300-303.

Hall, C.A.S. and R. Moll, 1975. Methods of assessing aquatic primary productivity. In: Leith, H. and Whittaker, R.H. Eds. Primary productivity of the biosphere. Springer-Verlag, New York, USA. pp. 19-53.

Jamu, D.M. and R.H. Piedrahita, 2002. An organic matter and nitrogen dynamics model for the ecological analysis of integrated aquaculture/agriculture systems II. Model evaluation and application. Environmental Modelling and Software 17, 583-592.

Knud-Hansen, C.F., T.R. Batterson and C.D. McNabb, 1993. The role of chicken manure in the production of Nile tilapia *Oreochromis niloticus* (L.). Aquaculture and Fisheries Management 24, 483 – 493.

Milstein A., 1993. Factor and canonical correlation analyses: basic concepts, data requirements and recommended procedures. In: M. Prin, G. Hulata and D. Pauly (eds.), Multivariate Methods in Aquaculture Research: case studies of tilapias in experimental and commercial systems. ICLARM Studies and Reviews Vol 20 Manila, Philippines, pp. 24-31.

Muendo, P.N., A. Milstein, A.A. van Dam, E. Gamal, J.J. Stoorvogel and M.C.J. Verdegem, 2006. Exploring the trophic structure in organically fertilised and feed driven tilapia culture environments using multivariate analyses. Aquaculture Research 37, 151–163.

Muendo, P.N., 2006. The role of fish ponds in the nutrient dynamics of mixed farming systems. PhD. thesis. Wageningen University, Wageningen, The Netherlands. 120 pp.

Page, A.L., R.H. Miller and D.R. Keeney, 1982. Methods of soil analysis 2nd edition. American Society of Agronomy. Soil Science Society of America, Madison, WI, USA. 1159 pp.

Pant, J., H. Demaine and P. Edwards, 2004. Assessment of the aquaculture subsystems in integrated agriculture-aquaculture systems in Northeast Thailand. Aquaculture Research 35, 289–298.

Pillay, T.V.R., 1994. Aquaculture Development, Progress and Prospects. Fishing news. Cambridge, USA. 182 pp.

Quantitative agro-ecological indicators and productive performance of Integrated Agriculture-Aquaculture systems in the Mekong Delta

Le Thanh Phong[1], Henk M.J. Udo[2], Martinus E.F. van Mensvoort[3], Anne A. van Dam[4], Le Quang Tri[1] and Akke J. van der Zijpp[2]
[1]Cantho University, Can Tho City, Vietnam
[2]Animal Production Systems Group, Wageningen University, The Netherlands
[3]Laboratory of Soil Science and Geology, Wageningen University, The Netherlands
[4]UNESCO-IHE Institute for Water Education, The Netherlands

Abstract

One aspect of farm sustainability is ecologically sound and productive farming. This study evaluates agro-ecological indicators and productivity of IAA farms with low, medium and high external input fish systems in the Mekong Delta of Vietnam. Eleven farms with Integrated Agriculture-Aquaculture systems in O Mon, Tam Binh and Cai Be districts with high, medium and low input fish systems respectively, were monitored from 2002 to 2004. The Ecopath model 3.0 was used to quantify a range of agro-ecological farms indicators. Based on species richness, functional agricultural diversity, actual efficiency, Finn's cycling index, sum of production, total system biomass, net system yield, animal ecotrophic efficiency and internal farm resource use, the medium input fish system appears superior to the high and low fish input systems in terms of sustainability. The surplus of N (nitrogen) recorded in all three systems represents an overload of nutrients and reflects a trend of farm intensification.

Keywords: Vietnam, IAA, Ecopath, sustainability

Introduction

In Asia, Vietnam is one of the countries where Integrated Agriculture-Aquaculture (IAA) farming systems have emerged over the last 30 years. In the Mekong Delta (MD) many farm households have integrated a fishpond into their production system. The potential benefits are not only risk aversion and a more even distribution of cash generating opportunities, but also more efficient use of resources as wastes from one component of an IAA system can be used as input into another component. Recently, economic globalisation has affected the economy of Vietnam and enhanced the intensification of agricultural activities through increased use of external inputs (Watanabe *et al.*, 2001). Constraints on the farm agro-ecosystem are likely to occur. A mass-balance model approach that allows quantification of ecological indicators can be used to evaluate such constraints. For aquatic systems, Polovina (1984) developed the Ecopath model which was modified by Dalsgaard and co-workers for farm systems (Dalsgaard and Oficial, 1998; Christensen *et al.*, 2005). This study uses Ecopath to evaluate agro-ecological indicators and productivity of IAA farms in three districts in the MD with low, medium and high external input fish systems and different agro-ecological conditions.

Materials and Methods

Early in 2002, an expert consultation and a participatory rural appraisal (Haverkort *et al.*, 1991) indicated the presence of three production systems, based on the dominant crop types and the intensity of fish culture, distributed over the MD. Subsequently, eleven selected smallholder IAA farms in three fresh water districts, were monitored on a weekly basis for a period of two years. There were (1) three farms with high input fish culture (HIF) and rice as main components in O Mon, (2) four farms with medium input fish culture (MIF) and rice as a main component in Tam Binh, and (3) four farms with low input fish culture (LIF) and fruit trees as main component in Cai Be. This fish culture input classification is based on the source of fish feed:

- In the LIF system fish are fed with on-farm crop residues, manure and human excreta.
- In the MIF system fish are fed with mainly pig and poultry manure, human excreta, crop residues, and some pellet feed (e.g. at fingerling stage).
- In the HIF system fish are fed with mainly pelleted feed, by-products from a fish processing factory and some manure or human excreta.

Fish in the HIF system is raised mainly for the export market, while in the MIF and LIF systems fish are produced for household consumption and the local market. Soil in the LIF area is fertile and mainly alluvial. Weather conditions are approximately similar in the three systems. Between 2002 and 2004, annual rainfall ranged from 1,154 mm in the MIF area to 2,271 mm in the LIF area; the high rainfall in HIF area was caused by an extremely high rainfall in this area in August 2004. Average temperatures were around 27 °C.

Modelling approach

The Ecopath model 3.0 (ICLARM, 1995) was used as a tool to quantify agro-ecological indicators of the IAA farms. At farm level, groups or stocks in the model are represented by annual crops, perennial trees, weeds/grasses, livestock, phytoplankton, and fish. The soil components were split into boxes according to the land use types of the farms. Flows represent bio-materials flowing into, out of and between boxes. The values of biomass (B), biomass accumulation (ΔB), production (P), consumption (Q), production/biomass ratio (P/B), consumption/biomass ratio (Q/B), harvest (H), and diet composition of groups/stocks for each component were calculated in Excel® according to Dalsgaard and Oficial (1998), and imported into Ecopath. The Ecopath programme quantified the agro-ecological indicators on the basis of annual farm data. For each monitored year in each farm an Ecopath model was estimated (i.e. 8, 8 and 6 models in the LIF, MIF and HIF systems, respectively). All annual input and output flows of the farm components were expressed in kg N per hectare of farm land (kg N ha^{-1} $year^{-1}$).

We used the following agro-ecological indicators:

- Finn's cycling index (ratio of total throughput that is recycled within a farm system, in percentage; Finn, 1980).
- Net system yield (net harvests in N which is exported from the whole farm).
- Sum of all production (sum of all materials produced by a farm whether harvested, added to the stock, or returned to the soil for decomposition).
- Total system biomass (total average standing biomass above ground of all farm groups/stocks).
- Animal ecotrophic efficiency (fraction of animal production that is harvested or used by other groups in the system).

In addition, in order to explore developments in the farms, we computed outside Ecopath:

- Species richness (counting the number of farmed and utilised species of crops, plants and animals within the farm).
- Functional agricultural diversity (using the Shannon-index: $H' = -\Sigma p_i \, Ln(p_i)$ (Magurran, 1988), with p_i group biomass proportion, dimensionless).
- Actual efficiency (ratio of total farm exports and total imports in feeds, fertilisers, biological N fixation, wet and dry atmospheric deposition and incoming irrigation water into a farm, in percentage).

Farm monitoring and data calculation

To gain insight in the bio-resource flows within a farm, all input and output flows of the farm components were recorded during weekly farm visits throughout the period 2002-2004. Soil was sampled at the beginning, in the middle and the end of the monitored period to a depth of 20 cm in rice fields, vegetable fields and fishponds, and 50 cm in homesteads and orchard beds. The total N (%) and bulk density (g/cm^3) were analysed for calculation of soil biomass. The values of dry matter (DM) and N contents of farm products and by-products (e.g. rice, vegetables, fruits, animal meat, and eggs) were mainly based on FCTUEA (1972) and FNRI (1990). The nutrient value of human faeces was derived from Suong and Nguyen (2000), and that of faeces of pigs, poultry and small ruminants was based on Yem *et al.* (2001). The values of DM and N of weeds and grasses were derived from Dung (1996). The biomass of phytoplankton and zooplankton was quantified from the average N content of fishpond water (Nhan, unpublished data), and the production value of phytoplankton in integrated rice-fish systems was based on Lightfoot *et al.* (1993) and Ruddle and Christensen (1993). We classified the various weed and grass species in a farm as one species; the same was done for phytoplankton and zooplankton.

To quantify the in-out nutrient flows of the farms we distinguished the soil detritus for the different land use types: rice fields, orchards, vegetable fields and fishponds. For exported N from the detritus (detritus losses) of rice field, orchards and vegetable fields we used the fractions 0.65 and 0.75 to estimate the N losses of inorganic and organic materials, respectively via volatilisation and denitrification (Dalsgaard, 1996; Dalsgaard and Oficial, 1998).

Based on averages of nutrient flows for semi-intensive tropical ponds, the denitrification, volatilisation and leaching accounted for 10% of total nitrogen inputs (Briggs and Funge-Smith, 1994; Lorenzen *et al.*, 1997), and loss of nitrogen from drainage water was 5% of total input nitrogen (Green and Boyd, 1995). Nitrogen fixation of 24 mg N m^{-2} day^{-1} (Acosta-Nasser *et al*, 1994) was applied to calculate the BNF of fishponds.

The N balance of each IAA farm was derived from the following equation: (Feed and Fertiliser inputs) + (BNF) + (Dry and wet deposition) + (Run-on with incoming irrigation water) – (Net harvest) – (Detritus losses) – (Other exports/losses). Loss of N via burning of rice straw on irrigated fields was based on Dobermann and Fairhurst (2000). N leaching from pond mud was calculated as a function of clay percentage of mud and mean annual precipitation (NUTMON, 2004). Losses from erosion, runoff and leaching of the farm soil were ignored.

Farm economy

Annual average economic performance of the farms for the two monitored years was calculated on the basis of a gross margin analysis. Farm gross returns were estimated as the sum of cash and non-cash returns of all farm products. Household consumption was not included in the calculations. Variable costs were the sum of cash costs and non-cash costs. Total costs equalled the sum of variable costs and general charges (e.g. taxes, local fees, transportation costs). Gross margins were calculated as gross returns minus variable costs. Net cash returns were calculated as cash returns minus cash costs. Net incomes were gross returns minus total costs. On-farm family labour and hired labour were in full-time equivalent. For comparison of economic values in 2004 the annual average inflation of 5.1% (Viet, 2004) was added to the farm economic indicators in 2002. One-way ANOVA was used to identify the differences between selected indicators in the three systems.

Results

Land use and farm components

The average farm size of monitored farms in the HIF, MIF and LIF systems was 2.90, 1.16 and 0.64 ha, respectively. The average orchard area was larger in the LIF system than in the MIF and HIF systems. Water surface of ponds and ditches was much larger in the HIF system. The ratio of water surface to orchard was 0.3, 0.5 and 1.25 for the LIF, MIF, and HIF systems, respectively. Among the LIF farms, three out of the four farms had an orchard-fish-livestock farm model; the fourth was an orchard-fish-livestock-rice-vegetable farm. In the second year this latter farm converted half of its rice field into a new orchard. All four MIF farms had five components, orchard-fish-livestock-rice-vegetables, but one farm had leased the rice field out during the second year because of financial constraints. One HIF farm practiced orchard-fish-livestock farming, the second orchard-fish-livestock-rice farming, and the third orchard-fish-livestock-rice-vegetables farming in the first year and orchard-fish-livestock-rice farming in the second year.

Mass-balance diagram

For each farm, Ecopath presented the quantified N flows in a diagram. Figure 4.4 shows the nutrient flows for a MIF farm in the second monitored year. The trophic levels are calculated by Ecopath based on the nutrient consumption/production of groups/stocks of crops/plants and animals, and soil detritus. The level 1 is assigned to nutrient producers (e.g. BNF) and soil detritus, the levels larger than 1 are assigned to nutrient consumers (e.g. crops/plants, animals, planktons) depending on their diet composition (Christensen *et al.*, 2005). In Figure 4.4 the level 1 is set to detritus of garden, rice, vegetables and the fishpond. Level 2 is set to crops/plants and phytoplankton and level 3 to animals. The Ecopath diagram shows the biomass (B), production (P), consumption (Q), import and harvest flows in the crops/plants (e.g. rice, vegetables, fruit trees, bamboo, weeds), phytoplankton, and animal (e.g. fish, pigs, chickens, ducks) boxes and the B, import, and internal flows (TI) for each soil box, and subsequently the agro-ecological indicators (Table 4.12).

Ecological indicators

Table 4.12 presents the mean annual values of the agro-ecological indicators for the three systems. Species richness in the MIF system was significantly higher than in the other two systems. The highest value for the functional agricultural diversity was also found in the MIF system. The annual inputs of inorganic fertilisers and feed were 229, 191 and 216 kg N ha^{-1}, corresponding to annual net system yields of 22, 51 and 50 kg N ha^{-1} in the LIF, MIF, and HIF systems, respectively. Based on the net system yield and total input flows (e.g. fertilisers, feeds, BNF, atmospheric deposition and incoming irrigation water) the actual efficiency was derived. The actual efficiency was higher for the MIF and HIF systems than for the LIF system. Sum of all production of the three systems was significantly lower for the LIF system than for the

Table 4.12. Annual mean values (± standard error) of agro-ecological indicators for the three external fish input systems.

	LIF	MIF	HIF	CV*
Species richness (n)	19 ^{b} ± 1	24 ^{a} ± 1	19 ^{b} ± 2	19
Functional agricultural div. (index)	0.94 ^{b} ± 0.17	1.37 ^{a} ± 0.04	0.91 ^{b} ± 0.08	29
Actual efficiency (%)	11.0 ± 0.02	20.0 ± 0.02	21.0 ± 0.05	47
Finn's cycling index (%)	43 ± 6	49 ± 4	43 ± 2	27
Net system yield (kg N ha^{-1})	22 ^{b} ± 4	51 ^{a} ± 8	50 ^{a} ± 11	53
Sum of all production (kg N ha^{-1})	156 ^{b} ± 13	282 ^{a} ± 40	226 ^{ab} ± 25	36
Total system biomass (kg N ha^{-1})	100 ± 10	111 ± 11	83 ± 4	27
Animal ecotrophic eff. (fraction)	0.75 ^{b} ± 0.04	0.90 ^{a} ± 0.04	0.92 ^{a} ± 0.04	13

*Coefficient of variation in percentage.
a, b Different superscripts denote significant differences between means within rows (P<0.05).

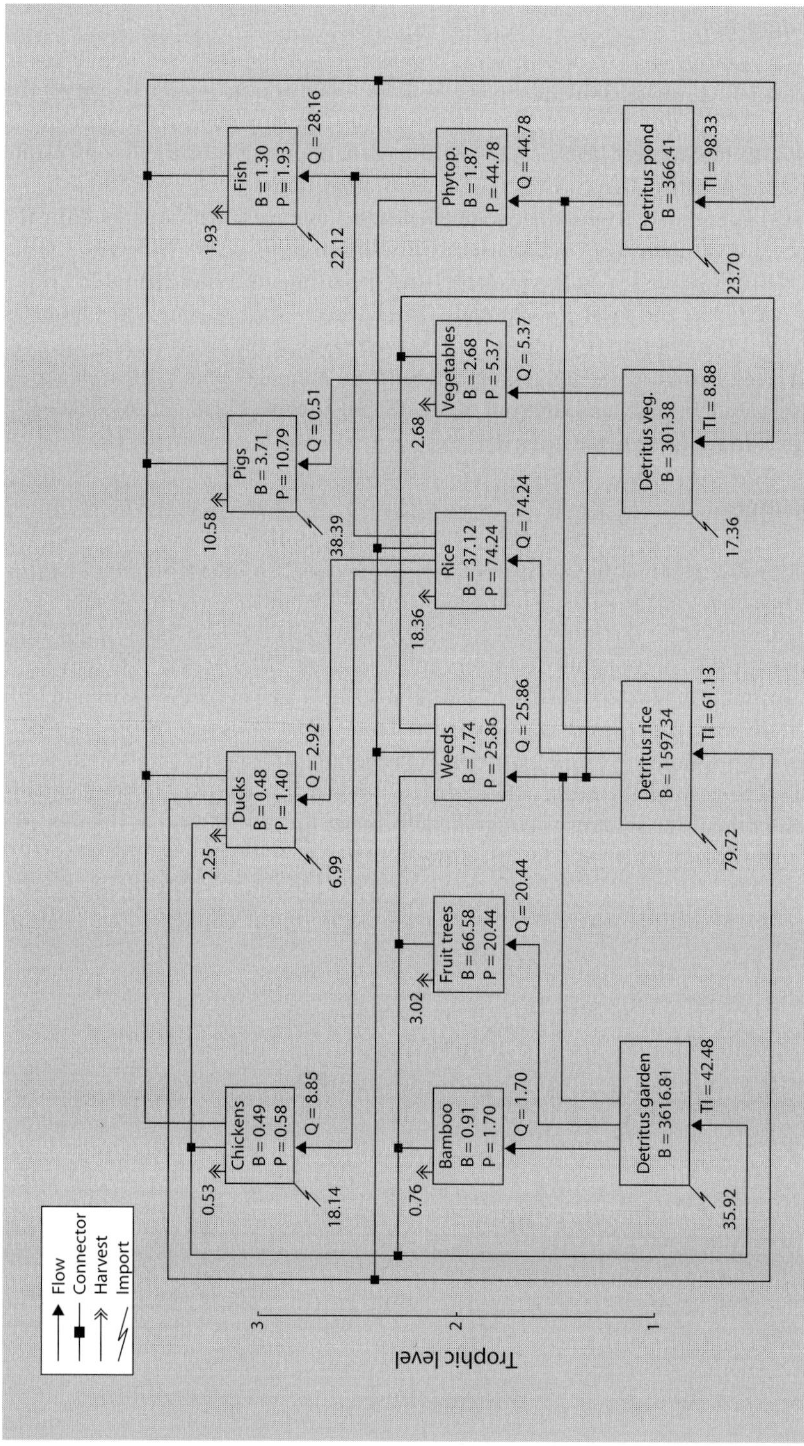

Figure 4.4. Ecopath flow diagram of one MIF farm in the second year (kg N ha^{-1}year^{-1}). (garden = orchard; veg. = vegetables; phytop.= phytoplankton; B: biomass, P: production, Q: consumption, TI: internal flow).

MIF and HIF systems. The important contribution of crop and weed biomasses led to a higher total system biomass for the LIF and MIF systems than for the HIF system. Nutrient cycling, expressed by Finn's cycling index, showed a higher value in the MIF system than in the LIF and HIF systems. Regarding animal production, the MIF and HIF systems showed higher animal ecotrophic efficiencies than the LIF system.

The high amount of inorganic fertilisers in the LIF system (159 kg N ha^{-1}) was mainly due to two farms using large amounts of fertilisers for fruit trees and vegetables. The quantity of imported feeds was highest in the HIF system (113 kg N ha^{-1}) because one farm had a large number (99) of pigs in the first year. In addition, the BNF was highest in the MIF system (46 kg N ha^{-1}) and the HIF (41 kg N ha^{-1}) because of a larger area for the rice field. The average positive N balance from the two monitored years was +105, +112 and +132 kg N ha^{-1} in the LIF, MIF, and HIF system, respectively (Figure 4.5).

N losses from rice straw burning and leached pond mud

Only one farm in the LIF system had a rice field, which was cropped once a year only. After harvesting, rice straw was used to mulch watermelon beds. In the MIF system all four farms had rice fields that were triple-cropped. Straw from the dry season rice crop was burned in the field, while in the wet season it was mainly used for mushroom cultivation. In the HIF system two farms had rice fields. Instead of burning the rice straw was left on the field for decomposition.

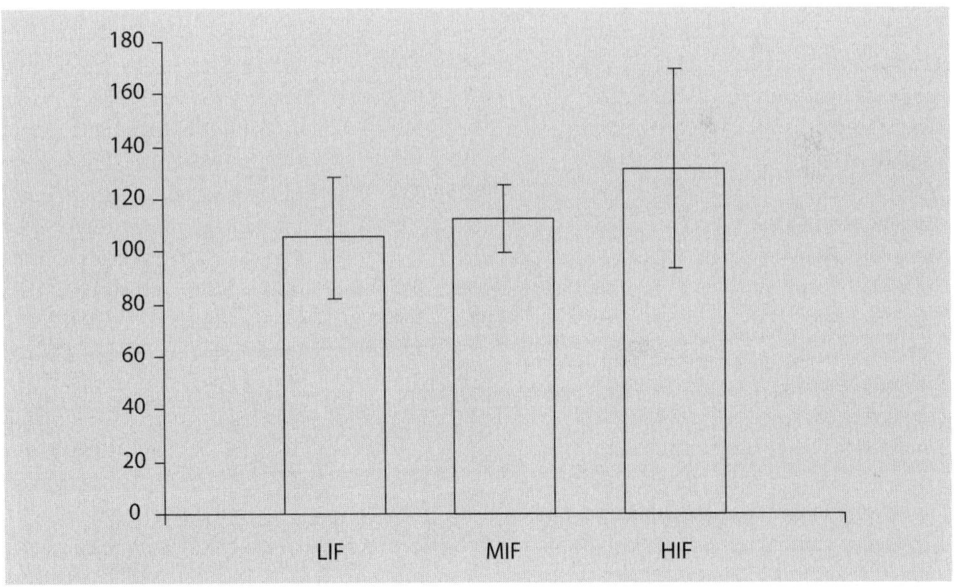

Figure 4.5. Annual N surplus (± se.) in the three input fish systems (not including N-losses from rice straw burning and leached pond mud).

Pond mud removal for hygiene of fishponds was practiced by one farm in the LIF system, three farms in the MIF system and three farms in the HIF system. Pond water was drained and a mud layer of 20-30 cm at the fishpond bottom was removed. This practice was recorded mainly around mid to end of the second monitored year. It caused an extra loss of N via leaching. Annual total N loss from burning and leaching in the MIF system (22 kg N ha^{-1}) was significantly higher than in the LIF (3 kg N ha^{-1}) and HIF (1 kg N ha^{-1}) systems. When those extra losses were corrected for the N balance in the three fish input systems, the annual N surplus was +102, +91 and +131 kg N ha^{-1} in the LIF, MIF, and HIF systems, respectively.

Farm economic efficiency

Farm gross returns, gross margins, net cash returns and net incomes tended to be higher in the HIF farms, but the values for these parameters were not significantly different for the three systems, due to the large standard deviations (Table 4.13). Annual non-cash returns and non-cash costs of LIF farms were lower than those in the MIF and HIF systems. Significantly lower on-farm family labour was recorded in the HIF farms; but its hired labour was highest.

Table 4.13. Farm economy (€ y^{-1}) and on-farm family labour in the three input systems (± standard error); applied exchange rate: €1 = VND 19,000 (2005).

	LIF	MIF	HIF	CV*
Cash returns	1218 ± 195	1888 ± 276	4870 ± 2405	125
Non-cash returns	181 ^{b} ± 28	481 ^{a} ± 60	336 ^{ab} ± 85	47
Gross returns	1399 ± 211	2369 ± 319	5206 ± 2416	111
Cash costs	666 ± 138	1151 ± 153	2276 ± 1118	113
Non-cash cost	80 ^{b} ± 14	212 ^{a} ± 27	135 ^{ab} ± 62	65
Variable costs	746 ± 145	1363 ± 165	2411 ± 1165	106
General charges	12 ± 2	18 ± 3	16 ± 4	47
Total cost	757 ± 146	1381 ± 167	2426 ± 1166	105
Gross margins	654 ± 84	1006 ± 190	2795 ± 1263	119
Net cash returns	553 ± 80	738 ± 170	2594 ± 1301	142
Net incomes	642 ± 83	988 ± 190	2780 ± 1262	120
Family labour	205 ^{ab} ± 23	258 ^{a} ± 17	156 ^{b} ± 28	29
Hired labour	11 ^{b} ± 3	18 ^{b} ± 3	80 ^{a} ± 26	105

*Coefficient of variation in percentage.
^{a,b}Different superscripts denote significant differences between means within rows (P<0.05).

Discussion

Ecological sustainability should be assessed by looking simultaneously at a range of performance indicators - no single indicator is adequate. In this study we used agro-ecological indicators that represent diversity, efficiency and productivity of agro-ecosystems. Diversity is generally related to stability: the higher the diversity the greater the stability of a system. Maintaining a high diversity is deemed important for risk spreading, genetic conservation of native species, and efficient resource use (Dalsgaard *et al.*, 1995). Functional agricultural diversity in the farms in the MIF system corresponds with the high species richness in this system (24 species). This can be seen as an ecological attribute different from specialised farming where only high economic value crops and animal species are exploited. A major difference between the three systems was that farms in the MIF farming systems commonly worked with more farm components (e.g. rice, fruit tree, vegetables, livestock, and fish) than those in the LIF system (e.g. fruit tree-livestock-fish) and HIF system (e.g. rice-fruit tree-livestock-fish). It seems that local economics and less fertile soil (e.g. low land, acid soil) in the MIF farms pushes the farmers to specialise and intensify their farms. Farmers in this region had rice as their staple crop and added other components over time. This gradually resulted in an increase of functional agricultural diversity (e.g. via high species richness and high biomass). Added to this, avoidance of risks from market fluctuations probably supported the process of diversification. Due to high soil level and fertility, fruit orchards were historically more important in the LIF system, which gives this system a better starting point to specialise in fruit production with the farmers' know-how of fruits. Rice was the base of farming in both the MIF and HIF systems. However, because of the proximity of a major market and An Giang province having a tradition of fish culture in cages for export, at the HIF farms high input fish farming increased and is still expected to increase further. At present, the importance of fish in the HIF system is growing and leading to IAA systems specialising in rice and fish.

Christensen *et al.* (2005) showed that Finn's cycling index strongly correlates with system maturity and stability. A system with higher capacity to recycle detritus has a higher ability to recover from perturbations (Vasconcellos *et al.*, 1997). The high N cycling in the MIF system may reflect a better use of farm internal resources due to integration of farm components. The species richness, functional agricultural diversity and cycling index all indicate that the MIF farms are more integrated than the LIF and HIF farms.

High net system yield from the MIF and HIF farms corresponds to their high actual efficiency. It implies that more N was extracted from the whole farm, resulting in better harvest in terms of N and also in higher resource use efficiency than on the LIF farm. Although annually a large amount of harvested fruits (e.g. 3,827 kg) was produced at the LIF farms, net system yield was low because of the low N and DM content of fruits.

Although the high value of sum of all production in the MIF system might underline the better farm management for crops and animals, the contribution of weed and plankton to production is considerable. High values of animal ecotrophic efficiency in both the MIF and HIF systems indicate a good utilisation of the animals produced during the monitoring period, but in all three systems the contribution of weeds/grasses to livestock production was low.

Weeds/grasses are not commonly used, and are left to decompose after weeding. Though this practice supports the maintenance of soil organic matter content, the further utilisation of weeds/grasses as animal feeds should be considered.

The external input of fertilisers for crops and feed for animals in the three systems was large when related to their net system yields. This represents a trend of intensification. The resulting high N losses could pollute the surrounding environment. The pond mud is commonly removed every 2 or 3 years. This low frequency causes loss of N through leaching from the pond before mud removal, and also after its removal when the mud is put on a heap in the garden and not incorporated evenly into vegetable fields or orchard beds.

The N from BNF of rice fields can be considered an important nitrogen source in the farm nutrient balance. Rice straw burning caused a significant N loss in the MIF system. This practice is a cost-effective method of straw disposal and helps to reduce pests and diseases that may occur because of re-infection from inoculums in the straw biomass (Dobermann and Fairhurst, 2000). It causes, however, an important loss of nutrients (e.g. almost complete loss of N) and contributes to atmospheric pollution. The incorporation of rice straw into the soil should be further considered to avoid extravagant nutrient losses.

The economic analysis confirms the trend of intensification. The high level of inputs in the HIF system is reflected in the variable costs, being 1.8 and 3.2 times higher than in the MIF and LIF system, respectively (Table 4.13). The higher gross margins, net cash returns and net income in the HIF may also be the result of more intensified farming. The significant higher annual harvested fish in the HIF system (1526 kg) compared with that of the MIF system (461 kg) and the LIF system (119 kg) reflects the intensification of fish culture. The high non-cash returns and non-cash costs in the MIF system imply that more internal materials with an opportunity cost value were used in these IAA farms. The high use of family labour in the MIF system also reflects the higher degree of integration of the many farm activities. The relatively high labour demand in the LIF system was due to the rotational cultivation of rice, watermelon and maize in one farm. The high quantity of hired labour in the HIF system was mainly due to rice and pig production, but also related to the fact that two of the household heads were public servants who consequently had limited time for farming.

Conclusions

Based on the agro-ecological indicators; species richness, functional agricultural diversity, actual efficiency, Finn's cycling index, sum of all production, total system biomass, net system yield, animal ecotrophic efficiency, and internal resource use the MIF farms appear superior to the HIF and LIF farms, in terms of the potential resource use for a sustainable farm development. The surplus N recorded in the three systems represents an overload of nutrients and reflects a trend of farm intensification in all three systems. The increase in farm integration can be considered as an economic and efficient strategy to reduce risks from market fluctuations. Although the market price of farm products and the costs of inputs played an important role, farmers in the three systems economically benefited from their farming activities. Regarding farm diversity and resource exploitation, however, farmers in the three systems should consider the use of weeds for ruminant animals, the trade-off of nutrient application, and the use of family labour (e.g. in the HIF system).

Acknowledgements

The study was conducted as a part of the cooperative research Project for Optimisation of Nutrient Dynamics and Animals for Integrated Farming (POND), funded by the Interdisciplinary Research and Education Funds (INREF) of Wageningen University and Research Centre, the Netherlands. We wish to thank the eleven farm households for their hospitality and keen participation in the study. We thank Mr. P.T. Loi, K.T. Hoang, N.H. Kiet and V.P.D. Tri (College of Agriculture and Applied Biology) for their enthusiastic support with field data collection.

References

Acosta-Nassar, M.V., J.M. Morell, and J.E. Corredor, 1994. The nitrogen budget of a tropical semi-intensive freshwater fish culture pond. Journal of the World Aquaculture Society 25, 261-269.

Briggs, M.R.P. and S.J. Funge-Smith, 1994. A nutrient budget of some intensive marine shrimp ponds in Thailand. Aquaculture and Fisheries Management 25, 789-8111.

Christensen, V., Carl J. Walters and Daniel Pauly, 2005. Ecopath with Ecosim: A user's guide. Fisheries Centre, University of British Columbia, Vancourer, Canada. 12 pp.

Dalsgaard, J.P.T., 1996. Tracking Nutrient Flows in a Multi-Enterprise Farming System with a Mass-Balance Model (ECOPATH). In: Morris, R.A. (ed.), Managing Soil Fertility for Intensive Vegetable Production Systems in Asia. Proceedings of an international conference, 4 to 10 November 1996, Tainan, Taiwan, pp. 333-343.

Dalsgaard, J.P.T. and R.T. Oficial, 1998. Modeling and Analyzing the Agro-ecological Performance of Farms with ECOPATH. ICLARM. Manila, Philippines.

Dalsgaard, J.P.T., C. Lightfoot, V. Christensen, 1995. Towards quantification of ecological sustainability in farming systems analysis. Ecological Engineering 4, 181-189.

Dobermann, A. and T. Fairhurst, 2000. Rice: Nutrient Disorders and Nutrient Management. IRRI, Philippines.

Dung, N.N.X., 1996. Identification and evaluation of non-cultivated plants used for livestock feed in the Mekong Delta of Vietnam. MSc. Thesis. Swedish University of Agricultural Science. Department of Animal Nutrition and Management. Uppsala.

FCTUEA (Food Composition Table for Use in East Asia), 1972. U. S. Department of Health, Education, and Welfare and FAO, Food Policy and Nutrition Division. Rome, FAO. 334 pp.

FNRI (Food and Nutrition Research Institute, Philippines), 1990. Food composition tables: recommended for use in the Philippines. Food and Nutrition Research Institute, Department of Science and Technology, Manila.

Finn, J.T. 1980. Flow Analysis of Models of the Hubbard Brook Ecosystem. Ecology 61, 562-571.

Green, B.W. and C.E. Boyd, 1995. Water budgets for fish ponds in the dry tropics. Aquacult. Eng. 14, 347–356.

Haverkort, B., J. van der Kamp and A. Waters-Bayer (Eds.), 1991. Joining farmers' experiments: experiences in Participatory Technology Development. Intermediate Technology Publications. 269 pp.

ICLARM (International Center for Living Aquatic Resources and Management), 1995. ECOPATH 3.0. A software for the construction and parameterization of mass-balance trophic models of aquatic and terrestrial ecosystems. ICLARM.

Lightfood, C., P. Roger, A.G. Cagauan and C.R. Dela Cruz, 1993. Preliminary steady state nitrogen model of a wetland rice field ecosystem with and without fish. In: V. Christensen and D. Pauly (eds.), Trophic models of aquatic ecosystems. ICLARM Conf. Proc. 26, pp. 56-64.

Lorenzen, K., J. Struve and V.J., Cowan, 1997. Impact of farming intensity and water management on nitrogen dynamics in intensive pond culture: a mathematical model applied to Thai commercial shrimp farms. Aquaculture Research 28, 493–507.

Magurran, A.E., 1988. Ecological diversity and its measurement. Croom Helm, London.

NUTMON v3.0, 2004. NUTMON toolbox. Alterra/LEI/Envista, the Netherlands.

Polovina, J.J., 1984. An overview of the ECOPATH model. Fishbyte, 2, 5-7.

Ruddle, K. and V., Christensen, 1993. Energy flow model of the mulberry dike-carp pond farming system of the Zhujiang Delta, Guangdong province, China. ICLARM.

Suong, N.K. and L.D. Nguyen, 2000. Produce biogas by anaerobic fermentation technique (in Vietnamese). Agricultural publisher, Vietnam.

Vasconcellos, M., S. Mackinson, K. Sloman and D. Pauly, 1997. The stability of trophic mass-balance models of marine ecosystems: a comparative analysis. Ecological Modeling 100, 125-134.

Viet, V.Q., 2004. Inflation in Vietnam. http://zdfree.free.fr/diendan/articles/u143vqviet.html

Yem, V.H., P.Q. Tuan, N.T. Dao, 2001. Cultivation, Soil, Fertilizer, Variety (in Vietnamese). Vol. 1. Education Publisher, Thanh Hoa. 129 pp.

Watanabe, T. and T. Nagumo, 2001. Nitrogen flow in Can Tho Province. In: Proceedings of the 2001 annual workshop of JIRCAS Mekong Delta Project. November 27-29, 2001. Cuu Long Delta Rice Research Institute, O Mon, Can Tho, Vietnam, pp. 221.

Nutrient accumulation and water use efficiency of ponds in integrated agriculture-aquaculture farming systems in the Mekong delta

Dang K. Nhan[1], Le T. Duong[1], Le T. Phong[2], Marc C.J. Verdegem[3], Jetse J. Stoorvogel[4] and Johan A.J. Verreth[3]
[1] Mekong Delta Research Development Institute, Can Tho University, Vietnam; dknhan@ctu.edu.vn
[2] Department of Crop Sciences, Can Tho University, Vietnam
[3] Aquaculture and Fisheries Group, Wageningen University, The Netherlands
[4] Laboratory of Soil Science and Geology, Soil Research Centre, Wageningen University, The Netherlands

Abstract

To further develop integrated agriculture-aquaculture (IAA) farming systems, better integration of ponds with terrestrial crops and human activities should be considered. In the Mekong delta, improving resource use efficiency is the main factor influencing the adoption of IAA farming. In the Mekong delta on-farm by-products or residues, particularly livestock manures, contributed about 80% of the total food nitrogen (N) inputs of the pond. Pond water use efficiency was low, due to high water exchange rates practiced in manure-fed ponds. About 10-20% of the total food N input was recovered in harvested fish, while most of the remaining fraction accumulated in pond sediments. Multivariate analysis distinguished three pond systems with different levels of water exchange, waste and home-made feed utilisation. The trials showed that adjusting water exchange rates to fertilisation, especially from pig manure, contributed to the reduction of environmental pollution, while fish production was almost maintained or improved strongly, depending on the system. A trial to prune the longan trees showed that the mixed fruit-fish system can only be judged in the long run: with fish providing the main income while trees are young, but fish becoming secondary once the fruit production becomes good.

Keywords: integrated agriculture-aquaculture; water use efficiency; nutrient accumulation; Vietnam

Introduction

Concern for sustainable aquaculture development is increasing (Naylor *et al.*, 2000). Aquaculture relies on a wide range of natural resources to produce marketable products while discharging wastes into environment. The higher the demand for the natural resources, the more negative the impacts on the environment (Beveridge *et al.*, 1997). In developing countries, however, poor farmers are hardly concerned about the environment if they do not achieve their basic needs, i.e. food and income. Therefore, the challenge is to increase their agricultural (including aquaculture) productivity while safeguarding the environment (Edwards, 1993). Integrated farming with efficient nutrient recycling was suggested as one of the priorities for sustainable aquaculture development (Edwards, 1998).

In the Mekong delta, aquaculture is a fairly new activity (Rothuis, 1998). The government has promoted the development of aquaculture through on-farm integration of terrestrial and aquatic farming activities with fish, crustaceans, fruits, terrestrial animals and rice as the principal components (Nhan *et al.*, 2004). Such integrated farms will be further referred to as integrated agriculture-aquaculture (IAA) systems. The important characteristic of IAA farming is nutrient recycling within the system (Little and Muir, 1987). To develop the aquaculture sub-system, therefore, interactions between the pond and the terrestrial crops, and the overall productivity of the whole system, rather than of single components, should be considered (Edwards, 1998).

Previous studies reported that pond fish farming is water-intensive and that harvested fish recover a small fraction of nutrient inputs. Pond fish production consumes more water than most of the terrestrial crops, even when water losses are minimised (Boyd and Gross, 2000). Harvested fish recover approximately 11-27% nitrogen (N), 26-65% organic carbon (OC) and 30-32% phosphorus (P) of the total food inputs in intensive ponds, and 5-25% N, 2% OC and 5-18% P in semi-intensive ponds (Avinmelech and Lacher, 1979; Sinha, *et al.*, 1980; Boyd, 1985; Edwards, 1993; Acosta-Nassar *et al.*, 1994; Green and Boyd, 1995). The remaining fraction of nutrient inputs mostly accumulates in pond sediments or discharges. If this occurs in the pond of the IAA systems in the Mekong delta, this better pond management could give dual benefits, maximising the overall economical benefit to farmers while minimising adverse impacts on the Mekong River ecosystem due to excessive nutrient discharges.

In this paper, we describe the major reasons for the adoption of IAA farming by farmers, different fish production systems, linkages of pond nutrient input management with farming households' livelihoods options. Furthermore it looks at agro-ecological conditions, nutrient accumulation and water use efficiency of the pond of IAA systems in the Mekong delta. Finally, impacts of proposed technology interventions on fish production and pond nutrient use efficiency are analysed and evaluated by stakeholders.

Methods

After a Participatory Community Appraisal (Nhan *et al.*, 2007), a baseline study, and a State-of-the–System workshop (Duong *et al.*, 2004) a broad range of parameters was monitored from 10 selected farms at three sites, to identify research issues and to serve as a baseline (Table 4.14). The monthly datasets or the monthly averages were analysed with a factor multivariate analysis on pond properties and nutrient management (Nhan *et al.*, 2006).

From results obtained in the 2nd year, participating farmers, other farmers, extension workers, and researchers agreed that the last trial should aim to evaluate affects of commercial feed addition, of pond fertilisation, and of pruning fruit tree branches to reduce the effect of shading on fish yields and economic profitability. A split-plot design was applied with input level as the

Table 4.14. Characteristics of the ponds monitored at farm level.

No.	Research site	Area (m^2)	Depth (m)	Width (m)
1	Cai Be	652	0.6	2.6
2	Cai Be	1,327	0.7	1.9
3	Cai Be	329	0.7	9.9
4	Cai Be	624	0.6	2.3
5	Tam Binh	222	0.7	5.0
6	Tam Binh	1,241	0.9	23.1
7	Tam Binh	1,584	0.7	35.2
8	Tam Binh	1,011	0.8	5.1
9	O Mon	960	1.8	13.0
10	O Mon	483	1.5	7.2

main plot and pruning fruit branches as the sub-plot. Each pond system was divided into two parts: with and without pruned branches. Two ponds were used with high input and another two were used with low inputs for fish. For each of the trees, about 17-20% of the tertiary branches that shade the pond were cut. Fish species and stocking rate were unique among the ponds. The fish species were mrigal (*Cirrhina mrigala*, 45%), Nile tilapia (*Oreochromis niloticus*, 25%), giant goramy (*Ospronemus exodon*, 20%), silver carp (*Hypophthalmichthys molitrix*, 5%), and river catfish (*Pangasianodon hypophthalmus*, 5%). The stocking rate was 1.5 fish m^{-2}. The orchards were in full production. To assess the effect on the system we monitored several parameters at two levels of N input: 80 and 600 kg N ha^{-1} $year^{-1}$. A sensitivity analysis was applied to the results to estimate the effect of price levels of longan between 1,800 VND kg^{-1} and 4,000 VND kg^{-1}. Fish price was fixed at 10,000 VND kg^{-1}. The treatments in two other trials, both carried out on three farms, were: additional application of manure (120 to 180 kgN ha^{-1} $year^{-1}$) next to the commercial feed (pelleted feed, fish meal and rice bran), or of 43 to 57 kg N ha^{-1} $year^{-1}$ using DAP and urea. Combining data obtained from the three years' monitoring, a yearly dataset was constructed and multiple linear regression analysis was used to determine interactions of fish yields and economic profitability with nutrient inputs.

Results

Reasons for the adoption of IAA farming by farmers

Analysing farmers' opinions using participatory appraisal and multivariate approaches, Nhan *et al.* (2007) found four major reasons influencing the adoption of IAA farming in the Mekong delta, in the order of importance as follows: (1) improving on-farm resource uses, (2) increasing farming income, (3) safeguarding the environment, and (4) improving nutrition for the family. Particularly, farmers tended to adopt IAA farming to maximise the benefits of the whole farming system. The goal to improve resource use efficiency and income, rather than to protect the environment and to meet nutritional needs, could attract farmers' attention to develop new IAA technologies and to adopt IAA farming.

Pond systems, nutrients input and accumulation

The pattern of IAA pond farming and its' nutrient management are determined by terrestrial crop farming practices, which are in turn closely associated with agro-ecological conditions and livelihood options of farm households. Using a multivariate approach, Nhan *et al.* (2006) identified three indicative pond farming systems: low water-exchange-rate ponds in fruit-dominated area (system 1), low water-exchange-rate ponds in a rice-dominated area receiving home-made feed (system 2), and high water-exchange-rate ponds in rice-dominated areas receiving wastes (system 3).

In the fruit-dominated area, where the topography is relatively high and fertile soils are predominant, farmers cultivate fruit crops intensively while they grow fish in sub-optimal conditions (system 1; Table 4.15). Ponds are narrow, shallow, and shaded by canopies of fruit trees; the food input levels and the fish yield were low, but the system retained N from rivers and canals (Table 4.16). The farmers focus on fruit production to generate income and gave little or no or priority to pond farming. In the rice-dominated areas, where less fertile soils are predominant, farmers grow fish in wide or deep ponds receiving home-made feed or livestock manure inputs (systems 2 and 3; Table 4.15). Farmers pay more attention to fish farming for income generation (system 2), or they focus on development of livestock farming and use the pond primarily to dispose of wastes (system 3). In system 2, the concentration of dissolved oxygen (DO) was very low in the morning because of higher inputs of manure and low water exchange which also enhanced algal development and water turbidity; the system lost nutrients to rivers and canals (Table 4.16). The total N budget of the intensive aquaculture systems (3) was the highest but most of the N and P inputs were lost to outside rivers and canals due to frequent water exchange.

By-products or residues of the terrestrial components in the IAA systems are principal food inputs for fish, contributing about 80% of the total food N inputs (including inorganic fertilisers). Livestock manures are an important nutrient source, contributing about 70% of the total food N inputs in the fruit-dominated system 1 and the rice-dominated high water-exchange-rate system 3. In the rice-dominated low water-exchange-rate (system 2), farmers combine on-farm manures, by-products from rice fields (snails, crabs and rice residues) and purchased feed to fertilise the fishpond.

The organic carbon (OC) accumulating in the sediments is the major sink within the pond (Table 4.15). Frequent water exchange increases both nutrient loss and downstream pollution of the environment. Harvested fish recovered about 11% of the total food N inputs in the fruit-dominated system 1 and about 19% in the rice-dominated systems 2 and 3. Large amounts of N and other nutrients accumulated in all three systems, showing that fish is not the main output if the pond mud is made available for the fruit trees.

Pond optimisation strategies: a perspective for sustainable IAA farming

Increasing the animal manure application to ponds in system 1 from 83 to 471 kg N ha^{-1} yr^{-1}, while reducing the water exchange rate from 7.9 to 2.6 % pond volume day^{-1}, improved the fish yield significantly (Table 4.16). Nutrient in pond water, concentration and amount of

Table 4.15. Mean quantities (kg N ha^{-1} yr^{-1}) and portions of the total food N inputs (%) of food inputs and harvestable outputs of the pond for three systems.

Sources	System 1		System 2		System 3	
	Quantity	%[1]	Quantity	%	Quantity	%
Food inputs						
Livestock manures[2]	166	70	557	49	901	70
Rice by-products[3]	5	2	348	31	100	8
Falling fruit residues[4]	27	11	0	0	0	0
Human waste[5]	3	1	0	0	211	16
Inorganic fertilisers	10	4	5	1	3	1
Purchased feed	26	11	226	20	65	5
Total food inputs	236	100	1135	100	1281	100
Outputs						
Fish	26	11	215	19	140	19
Sediment	383	162	1081	95	774	61
Crop irrigation	16	7	18	2	2	1

[1]As % of the total food inputs.
[2]Included urine, 3 rice residues, snail and crab collected from rice fields, 4 leaves and fruits, 5 mainly night soil and urine.

Table 4.16. Fish yields with standard error (±) and bottom mud nutrient accumulation (kg ha^{-1} $year^{-1}$) in the ponds in year 1, farmers' practice, and year 2 with technical interventions.

Items	System 1		System 2		System 3	
	1st year	2nd year	1st year	2nd year	1st year	2nd year
Fish yields						
Stocked (x1000)	0.5 ± 0.1	1.2 ± 0.4	6.2 ±3.8	5.1 ± 0.9	3.8± 1.2	4.1 ± 0.7
Wild	75 ± 40	73 ± 44	477 ± 247	247 ± 129	235 ± 80	230 ± 129
Nutrient accumulation						
Nitrogen (x1000)	0.3 ± 0.1	0.4 ± 0.1	0.9 ± 0.32	0.6 ± 0.1	0.6 ± 0.05	0.6 ± 0.3
OC (x1000)	4.3 ± 0.3	7.4 ± 1.9	1.0 ± 4.1	8.9 ± 1.6	6.9 ± 0.4	7.6 ± 2.2
Phosphor	73 ± 8	127 ± 56	458 ± 217	219 ± 49	213 ± 31	321 ± 153

nitrogen, phosphorus and organic matter deposited in pond mud also increased, especially P and OM content (Table 4.17).

Table 4.17. Means of nutrient content in the pond management parameters contributing most to the changes in system 1 in the 1st and the 2nd year (yr), in kg ha⁻¹ day⁻¹ (trial 1).

Variables	For N		For OC		For P	
	1st yr	2nd yr	1st yr	2nd yr	1st yr	2nd yr
Inflow of water	1.46	0.28			0.15	0.03
Animal waste	0.26	1.34	0.93	8.55	0.18	0.47
Crop residues input	0.03	0.01			0.01	0.00
Home-made feed	0.06	0.05	2.23	0.47	0.05	0.02
Outflow of water			1.33	0.75		
Human waste			0.00	0.63		

Reducing the water exchange rate from 1.5 to 0.6 volume% per day, decreasing input of manure while supplementing with inorganic NPK fertiliser and increasing stocking ratio of silver carp and Nile tilapia improved water quality and increased primary productivity (Table 4.18). The changes increased DO in the morning and reduced water turbidity. Fish yield was slightly reduced (Table 4.16), but nutrient input was reduced by half.

In system 3, the water exchange rate reduced from 21.4 to 5.3 volume% per day, while the input to the pond of animal manure and supplementary feed increased (Table 4.19). Results show marked increases in DO, algae, and primary productivity and a decrease of nutrient loss through water exchange. Fish yield improved significantly (Table 4.16).

In the last trial at intensive horticulture farms 17%, on average, of the tertiary branches of the longan trees were pruned. The pond surface shared one-fourth of the total orchard area. The net fish yield increased due to pruning by 107 and 280 kg ha^{-1} yr^{-1} in systems with, respectively, 80 and 600 kg N ha^{-1} yr^{-1}. On the other hand, pruning decreased the longan yield on average to

Table 4.18. Means of nutrient content in the pond management parameters contributing most to the changes in system 2 in the 1st and the 2nd year (trial 2), in kg ha⁻¹ day⁻¹. Waste & HM feed = Animal waste and home-made feed.

Variables	For N		For OC		For P	
	1st yr	2nd yr	1st yr	2nd yr	1st yr	2nd yr
Inflow of water	0.34	0.06			0.01	0.01
Waste & HM feed	0.75	0.60	10.71	8.51		
Irrigation water	0.03	0.00			0.004	0.001
Human waste			0.00	0.64	0.00	0.02
Outflow of water			0.50	0.29		

Table 4.19. Means of nutrient content in the pond management parameters contributing most to the changes in system 3 in the 1st and the 2nd year, in kg ha^{-1} day^{-1} (trial 3).

Variables	For N		For OC		For P	
	1st yr	2nd yr	1st yr	2nd yr	1st yr	2nd yr
Inflow of water	9.21	0.95	8.93	2.16		
Home-made feed	0.21	4.99	2.18	19.92		
Human waste	0.87	0.13			0.25	0.04
Crop residues input			2.14	3.54		
Animal waste			10.37	4.72	0.76	0.26
Outflow of water					1.19	0.27

860 kg ha^{-1} yr^{-1} if farmers can practice three fruit crop seasons in 2 years. Sensitivity analysis with various levels of current market prices show negative effects of pruning branches of longan trees at lower N input (80 kg N ha^{-1} yr^{-1}) for ponds in all cases. For ponds fertilised with 600 kg N ha^{-1} yr^{-1} the pruning is advantageous if the longan prices fluctuate from 1,800 VND kg^{-1} to 3,000 VND kg^{-1} and fish price is fixed at 10,000 VND kg^{-1}. The higher the prices for longan the lower the returns for farmers; farmers will lose money when longan prices are above 3,000 VND kg^{-1}.

Results from the multiple linear regression analysis show interactions of fish yield and economic profitability with pond nutrient inputs. They indicate that fish yields are positively correlated with commercial feed (pellets, fish meals, rice bran, broken rice), golden snail and/or crab, livestock and human waste inputs, and stocking rates, and negatively correlated with pond water exchange rates and fruit-plant residues. Fish production is economically profitable and positively correlated with inputs from livestock, human wastes and inorganic fertiliser, and negatively correlated with commercial feed inputs and stocking rates (Figure 4.6). The results imply that farmers should maximise integration between pond with crop, animal components and human activities for better use of on-farm available nutrient resources for higher profitability.

Opinions from farmers and extension staff

Farmers and extension workers stressed the importance of integrated systems for socio-economical and ecological sustainability. Some experiment results should be put in the context of the system, for example concerning the pruning of fruit trees: when longan trees are young production of fish and vegetables is a way to buffer farmers' incomes. The stakeholders raised several questions requiring further research, e.g. the effect of a bio-digester in IAA. According to the participants, additional information would be needed on the appropriate fertilisation and fish species in ponds, on environmental issues and on the economic optimisation planning and implementing of IAA.

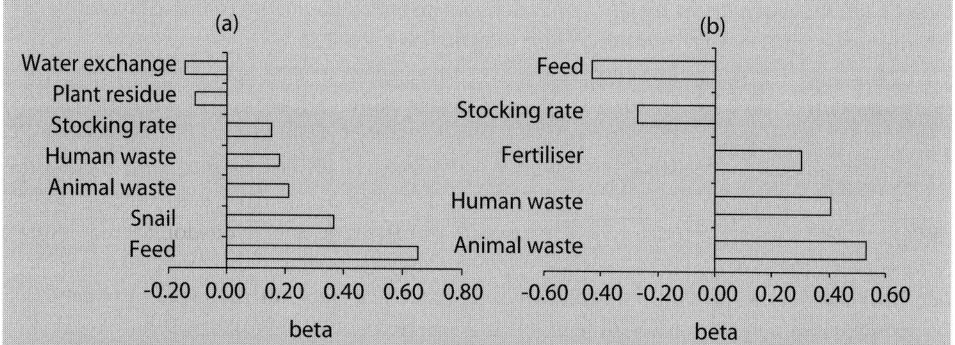

Figure 4.6. Standardised partial regression coefficients (beta-weights) for significant pond nutrient inputs in the models of fish yields (a) and economic profitability (b).

Farmers may not adopt IAA systems for three reasons: (1) the inappropriateness of technology, (2) either insufficient land holding or poor access to extension services, (3) limited farm management, and (4) a fear of conflicts associated with pesticide use on crops (Nhan *et al.*, 2007). The actions required to support the development of IAAS focus on the improvement of the information network (Table 4.20). The aquaculture know-how of most farmers in the Mekong Delta is still limited. The pond-dyke system is quite complicated to manage. For more efficient resource use farmers need not only techniques but also management knowledge and skills to enhance the interactions between the IAA components. The availability of micro-credit might make IAA systems accessible to resource poor farmers.

Table 4.20. Suggestions for improving the implementation of IAA according to the groups of stakeholders involved.

Issues	Government officials	Extension workers	Farmers
Good planning and implementation	+	+	+
More extension staff at village level		+	+
Local coordination of extension		+	
Training farmers on farm management		+	
Effective credit programs	+	+	
Establish community learning centres		+	
Effective agricultural services	+		

Discussion and conclusions

Conflicts for water resources may increase due to the growing shortage of water for industry, domestic and agricultural demands in some Asian countries, including Vietnam (FAO/NACA, 1995). Sustainable aquaculture should be water-efficient (Boyd and Gross, 2000), especially as aquaculture has to meet increasing demands for food supplies and income generation (Beveridge *et al.*, 1997). For aquaculture in the Mekong delta, high water exchange rates are practiced either to refresh the water in the orchard trenches (the fruit-dominated system) or to dilute the organic load in waste-fed ponds (the rice-dominated). To produce one kilo of fish about 170-300 m^3 are required, while in extensive stagnant ponds on average 45 m^3 water is used per kilo of fish produced (Verdegem *et al.*, 2006). Farmers use water lavishly because water scarcity is not a problem. The trials showed that adjusting water exchange rates to fertilisation, especially from pig manure, contributed to reduced environmental pollution.

A large fraction of added nutrients accumulated in the sediments. These are recycled as fertilisers for crops within the IAA farm. Due to flood protection dykes, the annual alluvial siltation decreased and the pond mud can contribute to maintaining soil fertility and fruit quality. This suggests that the benefit of the pond in IAA systems is broader than fish. In rain-fed agro-ecologies, the added economic benefit of the pond mainly comes from terrestrial crops relying on pond water, especially during dry periods (Prein *et al.*, 1996; Pant *et al.*, 2005). In contrast, in the Mekong delta where there is no water scarcity, the added economic benefit of the pond comes mainly from terrestrial crops through the pond sediments.

At present, IAA farming is mostly perceived from a narrow point of view, implying on-farm integration for increasing the variety of farming commodities produced at the farming household level (Nhan *et al.*, 2004). The main aim is to minimise risks from biotic and abiotic impacts and to sustain livelihoods for subsistent farmers. In the future IAA farming, however, would specialise and present fewer opportunities for on-farm integration in the future, due to forces of requirements of markets, technologies and income generations. IAA farming, therefore, should be placed in a wider context, implying off-farm integration for diversifying commercially-oriented production systems, income sources of the rural households and rural economic growth at the community level. The off-farm integrations include nutrient flows between farms, or between a farm and agro-industrial activities (Little and Muir, 1987). For example, an IAA farm with commercial pond farming might link to another with commercial livestock farming to receive the appropriate doses of nutrients at the right time to maximise the productivity and profitability of both farms, or it might link to another with commercial crop production to recycle pond mud or effluents. With this broader point of view, IAA farming might also create more jobs or services for poor rural households. Furthermore, poor rural people might also benefit from the provision of low-cost fish (Wohlfarth and Schroeder, 1979).

The pattern of IAA systems differs according to agro-ecological conditions and livelihood options of farming households, confirming observations from China and Thailand (Lo, 1996; Pant *et al.*, 2005), and this affects pond optimisation strategies. In the fruit-dominated area, reducing water exchange rates and improving fertilisation, by adding manures or inorganic fertilisers, improved not only fish yield but also more nutrients accumulated in the sediment.

In the rice-dominated area, where farmers tend to intensify pond farming or to use the pond primarily for disposing of wastes, manure input levels should be optimised, taking into consideration pond water quality, effluents discharged, fish production and sediment accumulation. Stocking appropriate fish species might be suggested as a way to better utilise the pond ecosystem. In addition, the seasonal farming calendar should be organised in a way that the availability of on-farm residues or by-products from the terrestrial crops can meet input needs for the pond (Prein, 2002), though this might be difficult to realise for small-scale farmers (Little and Muir, 1987).

In the Mekong Delta, to further the development of sustainable IAA farming, improved resource use efficiency at both farming household and community levels needs to be considered. Optimised pond nutrient accumulation and water use efficiency are important issues for management strategies of the pond, and between the pond and other IAA terrestrial subsystems. In the Mekong Delta, IAA systems will remain a target for food security, income generation and agricultural sustainability. The development of the IAA farming needs a system approach (Edwards, 1998). The pond needs to integrate with the terrestrial sub-systems as much as possible to maximise the multiple role the pond is fulfilling in IAA farming.

Acknowledgments

The results presented in this paper are mostly originated from a study conducted as part of the cooperative research project 'Improved resource use efficiency in Asian integrated pond-dyke systems' (INCO Pond-Live, ICA4-2000-20034), funded by the European Commission with Wageningen University and Research Centre (Aquaculture & Fisheries Group, Department of Animal Science, The Netherlands) and Can Tho University (Mekong Delta Development Research Institute; Vietnam) as project partners.

References

Acosta-Nassar, M.V., J.M. Morell, J.E. Corredor, 1994. The Nitrogen budget of a Tropical semi-intensive freshwater fish culture pond. Journal of the World Aquaculture Society 25, 261-270.

Avnimelech, Y., M. Lacher, 1979. A tentative nutrient balance for intensive fish ponds. Bamidgeh 31, 3-8.

Beveridge, M.C.M., M.J. Philips and D.J. Macintosh, 1997. Aquaculture and the environment: the supply of and demand for environmental goods and services by Asian aquaculture and the implications for sustainability. Aquaculture Research 28, 797-807.

Boyd, C.E., 1985. Chemical budgets for channel catfish ponds. Transactions of the American Fisheries Society 114, 291-298.

Boyd, C.E., and A. Gross, 2000. Water use and conservation for inland aquaculture ponds. Fisheries Management and Ecology 7, 55-63.

Edwards, P., 1993. Environmental issues in integrated agriculture-aquaculture and wastewater-fed fish culture systems. In: Pullin, R.S.V., Rosenthal, H., MacClean, J.L. (Eds.), Environment and aquaculture in developing countries. ICLARM conf Proc. 31, pp. 139-170.

Edwards, P., 1998. A system approach for the promotion of integrated aquaculture. Aquaculture Economics and Management 2, 1-12.

FAO/NACA, 1995. Regional Study and Workshop on the Environmental Assessment and Management of Aquaculture Development (TCP/RAS/2253). NACA Environment and Aquaculture Development Series No. 1. Network of Aquaculture Centres in Asia-Pacific, Bangok.

Green, B.W. and C.E. Boyd, 1995. Chemical budgets for organically fertilized fish ponds in the dry tropics. Journal of the World Aquaculture Society 26, 284-296.

Duong L.T., D.K. Nhan, J.C.M. Verdegem, D.C. Little. 2004. State-of-the-Systems (SoS) report: The VAC systems in the Mekong delta region in VietNam (bilingual English and Vietnamese).

Little, D.C. and J. Muir, 1987. A guide to integrated warm water aquaculture. Institute of Aquaculture, University of Stirtling, Stirtling, UK. 238 pp.

Lo, C.P., 1996. Environmental impacts on the development of agricultural technology in China: the case of pond-dike ('jitang') system of integated agriculture-aquaculture in the Zhujiang Delta of China. Agriculture, Ecosystems and Environment 60, 183-195.

Naylor, R.L., R.J. Goldburg, J.H. Primavera, N. Kautsky, M.C.M. Beveridge, J. Clay, C. Folke, J. Lubchenco, H. Mooney and M. Troell, 2000. Effect of aquaculture on world fish supplies. Nature 405, 1017-1024.

Nhan, D.K., L.T. Duong, N.V. Sanh and J.C.M. Verdegem, 2004. Development of "VAC" Integrated Farming Systems: a view of system and participatory approaches. In: Yamada, R. (Ed.), Development of integrated agiculture farming systems in the Mekong Delta, Vietnam. Tre Publishing House, pp. 101-125.

Nhan, D.K., A. Milstein, M.C.J. Verdegem, R.H. Bosma and J.A.V. Verreth. 2006. Food inputs, water quality and nutrient accumulation in integrated pond systems: a multivariate approach. Aquaculture 261, 160-173.

Nhan, D.K., L.T. Phong, J.C.M. Verdegem, L.T. Duong, R.H. Bosma, D.C. Little. 2007. Integrated freshwater aquaculture, crop and livestock production in the Mekong Delta, Vietnam: determinants and the role of the pond. Agricultural Systems (in press).

Pant, J., H. Demaine and P. Edwards, 2005. Bio-resource flow in integrated agriculture-aquaculture systems in a tropical monsoon climate: a case study in Northeast Thailand. Agricultural systems 83, 203-219.

Prein, M., 2002. Integration of aquaculture into crops-animal systems in Asia. Agricultural system 71, 127-146.

Prein, M., J.K. Ofori and C. Lightfoot, 1996. Research for the future development of aquaculture in Ghana. ICLARM Conference Proceedings 42, 94 pp.

Rothuis, R., 1998. Rice-fish culture in the Mekong Delta, Vietnam: constraint analysis and adaptive research. PhD thesis, Katholieke Universiteit Leuven, Belgium, 113 pp.

Sinha, V.R.P., H.A. Khan, D.P. Chakraborty and M.V. Gupta, 1980. Preliminary observations on nitrogen balance of some ponds under composite fish culture. Aquacultura Hungarica (Szarvas) II, 105-116.

Verdegem, M.C.J., R.H.Bosma and J.A.V. Verreth, 2006. Reducing water use for animal production through aquaculture. Water Resources Development 22, 101-113.

Wohlfarth, G.W. and G.L. Schroeder, 1979. Use of manure in fish farming - A review. Agricultural Wastes 1, 279-299.

Chapter 5

Sustainability, livelihoods and markets

Planet, profit and people: assessing sustainable development

Akke J. van der Zijpp
Animal Production Systems, Animal Science Group, Wageningen University, The Netherlands

Abstract

This paper puts the fishpond in the context of sustainable development, and presents a methodology for its assessment, based on an analytical framework for poultry housing systems and procedures provided by literature. The problems encountered in assessing sustainability of the integrated Agriculture-Aquaculture systems in the Mekong delta in Vietnam are discussed.

Keywords: sustainability, indicators, poultry, smallholder, Mekong Delta

Introduction

In the late eighties awareness of the limits of our natural resources was growing. Much earlier the Club of Rome had already warned the world with a doom scenario. The fertility of soils was decreasing caused by soil mining in most Sub-Saharan countries of Africa and Asia and by soil and water pollution in Western Europe, Eastern China and Thailand, and the Eastern states of the USA. The first signs of climate change had emerged and loss of biodiversity was universal. Forests in South America, Africa and Asia were disappearing at an alarming rate. Not only were natural resources at stake but also poverty and hunger were still widespread. Education was still not a common public good and poor health and reproductive diseases were common. Income for most people in the world continued to be too low for an acceptable livelihood. While migration became problematic for western and developing economies alike.

In 1987 the World Commission on Environment and Development reflected upon the above described state of the world and initiated the term sustainable development (SD). Despite its lack of a precise definition we share the basic meaning of sustainable development: That is that people secure sustainable development by covering the needs of the present generation without jeopardising the capability of future generations to cover their needs. SD concerns issues of ecology, economics and social-cultural well-being. These issues have been popularised as the planet, profit and people concepts. As many changes move in a negative direction concerns have been expressed about the driving forces of these processes. A change towards more balance in the ecological, economic and socio-cultural aspects is required for the world, national, regional and farm systems. Not only are these system levels connected, but also the justification for a fairer distribution between generations has to be made, given the definition of sustainable development.

In this paper a methodology for assessment of SD will be presented. Bell and Morse (1999, 2003) provided the analytical framework. De Boer and Cornelissen (2002), Cornelissen (2003) and Mollenhorst (2005) applied the procedures for different housing systems for egg layers.

I will pay attention to the problems encountered for assessing sustainability of the Integrated Agriculture-Aquaculture (IAA) systems in the Mekong delta in Vietnam.

Assessment of sustainable development

As changes occur over time from a variety of sources affecting farming systems, it is not easy to assess sustainability. Assessment is linked to a fixed point in time in a particular setting. When a study is finished the results may have to be interpreted in the light of even more recent changes. It is therefore important to be aware of the major and critical issues that affect systems and cause changes. These issues have to be determined and methods can be participatory in nature, introducing experimental and subjective judgment. Issue identification by scientists based on literature, primary and secondary data is another way of solving this problem, but will not completely erase the subjectivity of selection. A combination of both inputs may be the optimum for issue identification and selection, provided a proper stakeholder analysis has been carried out and all stakeholders expressed their views on the case. For example, farm income is generally accepted as an issue for economic sustainability and techniques for quantification are easily accessible, but views on relevance of animal health or welfare for sustainability may diverge and cannot easily be extracted from scientific literature.

In order to select issues of concern it is important to establish a basic and well defined concept of the system and the context of the society being studied. The boundaries of the farming system being studied have to be established and understood for later interpretation of the results. Some issues will be internal to and important for the farming system itself, other issues are external with large influence on the viability of the system. In the IAA systems in the Mekong delta the components of the system are varied: rice, fruit trees, horticulture, fishponds, pigs and poultry (chickens and ducks). Within the system knowledge about the land use differences in terms of inputs and intensity may be important. For example, growing rice three times per year can decrease soil fertility in the rice fields. And one could argue whether marketing activities are inclusive of the farming system. It will depend on definitions for labour as to what is included. This decision may affect other flows in the system like nutrient management, feed, energy, and so on. Before any research can start, or participatory workshops organised for issue identification and selection, the farming system and its boundaries have to defined and understood and agreed by the participants.

The next step in the process will be the identification of sustainability issues. For participatory approaches a stakeholder analysis is required and representatives of the stakeholder groups have to be invited. There is always a risk that not all stakeholders are represented, let alone that they can equitably express their views well. The process and outcomes of a participatory workshop will vary. This is the reason why a study of secondary data and literature can substantiate the results of the participatory workshop. If needed primary data can also be collected and analysed, this will markedly slow down the process and may cause loss of dynamics in the participatory course. Issues that can be expected are related to income, costs of inputs and prices of products, labour productivity and labour income, land prices, environmental impacts, labour availability and health as the farmers are aging, animal health control and productivity, governmental policies and regulations, subsidies and export opportunities.

When the identification and definition of relevant economic, ecological and societal issues has been completed selection and quantification of suitable sustainability indicators (SI) follows. These indicators have to be (1) relevant, i.e. they have to express something about the issue, (2) simple, i.e. they have to be understandable for users, (3) sensitive and reliable, i.e. they have to react to changes in the system, and different measurements must lead to the same outcome, (4) it must be possible to determine a target value or trend and (5) data have to be accessible. The search for indicators is sometimes difficult, but for many issues creative and satisfactory solutions can be found. Economic issues are often expressed as labour income, and for environmental impact nutrient balances or Life Cycle Assessment methods are used (for eutrophication and acidification and global warming). Animal welfare is usually a conglomerate of parameters including health.

When data have been collected on farms, showing also the variation between SI's, the scores for the indicators have to be utilised separately or jointly. From single indicator scores and their variability, already the key problems and potential roads to improvement become visible. This result can be linked to the original result on the issues important for the farming system.

Once SI's have been collected the next step becomes the judgment of the level of the indicators. Which reference values are acceptable? Reference values may come from different sources. For some reference values legal or regulation minimum/maximum are available. Others have to be found in scientific literature, for example for the Animal Needs Index presenting the issue animal welfare. For labour income minimum wages can be used. But for some indicators, an average value or a value linked to a reference year, may be the only solution. It should be understood, that the choice of reference value can have distinct effects on the final outcome of the indicator.

Overall assessment requires aggregation of the value of single SI's, weighting of these indicators and considering practical usefulness of an overall judgment. There are neither uniform approaches to aggregation nor to compensation. Threshold values can also be applied resulting in systems becoming unacceptable. This will depend on the user of the information and for what purpose. In our studies of farming systems we usually work with single indicators or groupings of indicators to better understand the trade offs and the potential for improvement of the farming system. The acceptation of compensation or threshold values will change the relationship between the SI's. In this evaluation the dynamics are important as the context is changing over time. Mollenhorst (2005) presented the consequences of decisions regarding SI's for welfare and labour income for different egg layer housing systems. The battery cage becomes easily the worst performing system in a situation where welfare has to meet a minimum level, while all other indicators may meet the agreed standard reference level.

IAA systems in the Vietnamese Mekong delta

In the Vietnamese Mekong delta IAA farmers are changing in response to national policy like Doi Moi reform (return to market economy and land ownership by households), international markets and access to knowledge and technology. In our study farm household information with 90 structured questionnaires was collected from three districts. All farms had a fishpond, but varied with regard to the intensity of the fish input of the farming system and marketing opportunities. The districts are not subjected to high frequency of flooding. Most farmers belong to the medium size farm group: 0.3-1 ha. Very poor farmers have no farming land. In all categories off-farm and non-farm activities were common. Data were collected in 2002 and 2004. The farms consisted of two to five of the following components: rice, orchard, cash crops, livestock (pigs and poultry) and fishpond.

Stable components of the system were rice, orchard and cash crops. Significant increases were found in livestock and even more so in fish production, measured as land use intensity (in %), the ratio between the land use of each component and the total agricultural area.

Rice was produced one to three times every year. Fishponds were common in two villages, in the third village fish was produced in ditches. Poultry was kept for family consumption and pigs for sale. Orchards were an important income source in one village and mixed orchards were found in the other two. The rice field is the traditional source of food and feed and mulch for vegetables and mushrooms. Pig manure is applied in orchards and fishponds. Poultry are free ranging. Orchards offer weeds and wastes as feed for pigs, poultry and fish. Fishponds supply feed like water spinach, snails and crabs for pigs and poultry.

Farm gross return increased considerably in 2004 and so did off-farm and non-farm income in two districts. Over the two years gross margins were highest for rice, followed by orchard, fish, pig and cash crops, but were negative for poultry. This was caused by the AI outbreaks in 2003 and 2004. The farm management differences can be explained by three factors: first the distance of the farm from the district market, second the land use intensity of products like rice, fruit, cash crops and fish, and third the opportunity for off-farm and non-farm labour income.

The changes in the IAA systems were strongly related to opportunities outside the farming systems in terms of employment and marketing. The farmers responded with intensification of the aquaculture component and pig production. There is also a trend towards fewer components: specialisation. Nevertheless, farmers continued with rice production and orchards, it being hard to change activities in the short term. The consequences of these changes are important for potentially sustainable natural resource management which is efficient relative to specialised farming systems with large external inputs.

The IAA system shows that it has great flexibility in adaptation to national and international market development. In the long term this may lead to further specialisation in fish, pigs, poultry production or cash crops and to greater risks for income security. Land use is very much determined by its agro-ecological status and by food security concerns. Rice production will therefore continue and changes in orchard land use will depend on competition with other components, especially fish production. Fish production integrates with pig and poultry production in the medium intensity fish production ponds.

Important policy issues for farm management are technology development in pig, poultry and fish production and cash crops, accompanied by efficient management of natural resources (such as nutrients from feeds and fertilisers cycling between the components), and provision of knowledge and transparency in the national and international marketing chains. The regional government has to consider infrastructural works (roads and communication) for better access to markets. As off-farm and non- farm employment increases education becomes more important and opportunities for children in the poorer segment of society have to be created. Employment on farms will benefit from these developments, as well as the communities. The future sustainability depends on a complex mix of improved technology for higher productivity, better utilisation of nutrients and a higher income for more people within and outside the farming system.

Conclusion for IAA systems

The IAA farms in the Vietnamese Mekong delta are participating in the national and international markets for rice, fruit and fish. The farmers are responding by specialisation and by employment off-farm. The IAA farmers require extension support for understanding the marketing opportunities, for technical expertise for the components of intensification and for risk management and maintaining efficiency of resource management. Infrastructural improvements will remove constraints for farmers to sell products for cash income.

Challenges for assessment of sustainability of IAA systems

A number of challenges will have to be faced when assessing sustainability of IAA systems. First the system and boundary definition will determine the calculation of flows such as income, labour and nutrients. The nature of a very mixed farm makes it more difficult to collect the data as often money, labour and nutrients will be divided over the different components, usually four or five in the farming system. Different components will have different effects on utilisation of capital, labour and nutrients and on air, water and soil emissions. Off-farm income and marketing activities also have to be decided as to if they are part of the farm or not.

The importance of the different components is difficult to express per se in the system although the system characterisation clarified that rice production and the fruit orchards are more permanent features of the system, whereas aquaculture and livestock vary depending on market prices. An interesting question would be which of the flexible components contribute most to sustainable development of the IAA systems. The response would have to be based on people, planet and profit issues such as, labour availability and quality, income and investment, maintenance of soil fertility and water quality.

In Le Thanh Phong's PhD research project farmers and researchers collaborated successfully (Le Thanh Phong, personal communication). For assessing issues and SI's all stakeholders should be involved and review and select. Regarding the choice of indicators for nutrient cycling, work is ongoing with different methods. This part of the assessment will be useful, as farmers have expressed worries about soil fertility when growing rice three times per year. The final choice will be dependent on its practical use with farmers. Similarly it will be important to select the best indicator for income: labour income, added value. The availability of data may be the next hurdle. In many situations surveys will be necessary.

A very challenging part of assessment will be the definition of reference values. The easiest way out would be to take a year of reference and compare indicator values with this reference year. However, it will be more useful to research national and international standards if available and furthermore to enquire what is regional norm. Particularly for off-farm labour and for opportunity labour income this would be a fair approach.

Aggregation of the SI's has to be reviewed carefully, in terms of compensation and thresholds and its relevance for the users. Moreover the single SI's, but studied as a whole set, may be more informative when considering the dynamic environment in the Mekong delta. Marketing situations vary, labour opportunities are changing, investment from elsewhere occurs and amongst these dynamic changes farmers are adapting with, e.g., more pig and fish production. Which trade-offs have the farmers made, aware of conditions like avian influenza outbreaks, catfish exports to the USA, rising commodity prices and weighing trade-offs, more or less consciously? Which risks have they avoided or accepted?

The assessment methods for sustainability show that this is a participative process following a series of defined steps. At each step decisions have to be made. The quality of these decisions is determined both by scientific inputs and by debate about practical relevance for the IAA system. Assessing SD is therefore both about the science and art of the people, planet and profit paradigm. Its value is based on the interactive process in gaining understanding and on the exact outcomes. Taking part in, and being aware of, the SD process creates opportunities for knowledge utilisation, innovation and change at IAA farm level, for policy makers, for the extension, industry and scientists (all stakeholders). The assessment of sustainability at farm level provides the basis for assessments at the higher scales like district, province, and nation. Finally these assessments provide the basis for policy making and farm management.

References

Bell, S. and S. Morse, 1999. Sustainability indicators. Earthscan Publications Ltd. London. 175 pp.

Bell, S. and S. Morse, 2003. Measuring sustainability. Earthscan Publications Ltd. London. 189 pp.

De Boer, I.J.M. and A.M.G. Cornelissen, 2003. A method using sustainability indicators to compare conventional and animal-friendly egg production systems. Poultry Science 81, 173-181.

Cornelissen, A.M.G., 2003. The two faces of sustainability: Fuzzy evaluation of sustainable development. Thesis Wageningen University. 183 pp.

Mollenhorst, H., 2005. How to house a hen: assessing sustainable development of egg production systems. Thesis Wageningen University. 135 pp.

Reducing waste production from aquaculture in China by feed formulation and system management: an overview

Shouqi Xie, Wu Lei, Xiaoming Zhu, Dong Han and Yunxia Yang
State Key Laboratory of Freshwater Ecology and Biotechnology, Institute of Hydrobiology, Chinese Academy of Sciences, Wuhan, 430072 Hubei, P.R. China; sqxie@ihb.ac.cn

Abstract

The fast increase in aquaculture is exacerbating problems of environmental pollution. To reduce the production of waste in aquaculture, recent research in China has focused on species selection, diet formulation and feeding management. Energy budgets are used to evaluate a potential aquaculture species. The improvement of diet formulation includes balancing nutrients, reducing N and P contents, and reducing the use of trash fish and the use of supplements. The digestibility of different feed ingredients has been determined in some species, for a balanced diet formulation. The potential contribution to the reduction of N and P loadings in waste water of increasing palatability and digestibility, selecting suitable protein sources, and using lipid, carbohydrate, and enzymes has been demonstrated. Bio-energetic modelling has been used to predict the growth and nutrient requirements, to decrease feed input and to lower the waste production. It has been shown that in some fish, compensatory growth increases feed utilisation and decreases N loading. Polyculture increases the utilisation of fish feed and decreases nutrient loading of waste.

Keywords: aquaculture, China, waste water, feed, management

Introduction

Aquaculture in China has a history of more than 3000 years and its production has increased dramatically in the past 10 years. In 2005, the total production of aquaculture was over 30 million tonnes, of which about 60% is from freshwater and 40% is from seawater. In China there are over 2000 marine species and 1000 species of freshwater fish; the latter include about 140 freshwater species with potential for aquaculture. Carp are currently the main culture species, accounting for more than 60% of the freshwater production. Pond culture provides slightly more than 70% of the freshwater production.

With the fast increase in aquaculture, some problems have emerged. The most important of these is pollution, especially from intensive aquaculture. This pollution has 5 major consequences:

1. Excessive nutrient loading in lakes. It has been found that in cage culture in Datong Lake (114 km^2) where the annual input of fertilisers was about 8,500 tonnes, more than 90% was not used by the fish and was deposited in the lake (Ou *et al.*, 2003). In the late 1990s, about 80,000 tonnes of fertilisers containing 8,600 tonnes N and 4,000 tonnes P were deposited in Dongting Lake each year, calculation based on the efficiency of 10% (Boyd and Tucker,

1998). In 2001, the water in five sections of that lake was found to be eutrophic. With the increasing ratio of fertilisers to feed in pond culture, the N and P loading increased linearly (Chen, personal communication).
2. Poor growth and feed utilisation. The high feed conversion ratio of commercial aquafeed for most species (more than 1.5) contributes to the waste of scarce feed ingredients in China. It would be possible to reduce the N and P contents in some commercial aquafeeds by 30%. Some farmers have even used raw feed ingredients, resulting in low feed efficiency and high feed cost (Xie, 1997). Due to the high feed conversion ratio and related problems, the feed cost accounts for more than 60% of total aquaculture cost and has become a barrier impeding aquaculture development.
3. High nitrogen (N) and phosphorus (P) loading. The waste production including N and P from ponds is very high. It is contributing to the eutrophication of lakes and is even becoming a threat to the marine ecosystem (Dong *et al.*, 2000; Ning and Gu, 2004).
4. Poor fish meat quality. In China, fish from natural water bodies fetches 1.5 to 10 times the price of fish from aquaculture. Poor meat quality not only reduces the price of fish products, but also damages the aquaculture sector. Unexpected substances such as microcystins, dioxins, hormones and heavy metals resulting from poor water quality or bad aquafeed have also resulted in food safety problems (Xu *et al.*, 1999; Xie *et al.*, 2004; Chen and Xie, 2005; Chen *et al.*, 2005; Hui *et al.*, 2005; Jiang, 2005). We give two examples. The content of gossypol originating from cottonseed meal in the muscle of gibel carp (*Carassius auratus gibelio*) was found to have increased exponentially with dietary gossypol intake (Figure 5.1). Eutrophication of water bodies can result in microcystins accumulating in fish muscle and tissues; its content in Nile tilapia could even exceed the WHO safety level (Chen *et al.*, 2005; Zhao *et al.*, 2006a,b).
5. Fish diseases due to poor feed formulation and poor water quality. In 2001-2004, mortality from fish diseases was 10-16 billion every year. The treatment of fish diseases, such as the use of antibiotics, could introduce other problems (Li *et al.*, 2001).

In order to solve these problems in aquaculture, research in China focuses on species selection, diet formulation and feed management.

Species selection

In past years, the most important factors for the selection of aquaculture species were fast growth and easy management. At present, energy budget studies are included in the selection of aquaculture species in order to evaluate the potential waste production. Comparative studies of the energy budgets of *C. carpio, O. mossambicus, P. fulvidraco, C. auratus, M. chinensis* and *P. para* have shown that notwithstanding the high growth efficiency (33.3%) of tilapia, its waste production is still high (16.9%). A comparison of nine freshwater species showed that the DL strain of gibel carp, mandarin fish, and snakehead had high growth and low waste production (Cui and Liu, 1990; Xie, 1997; Liu, 1998).

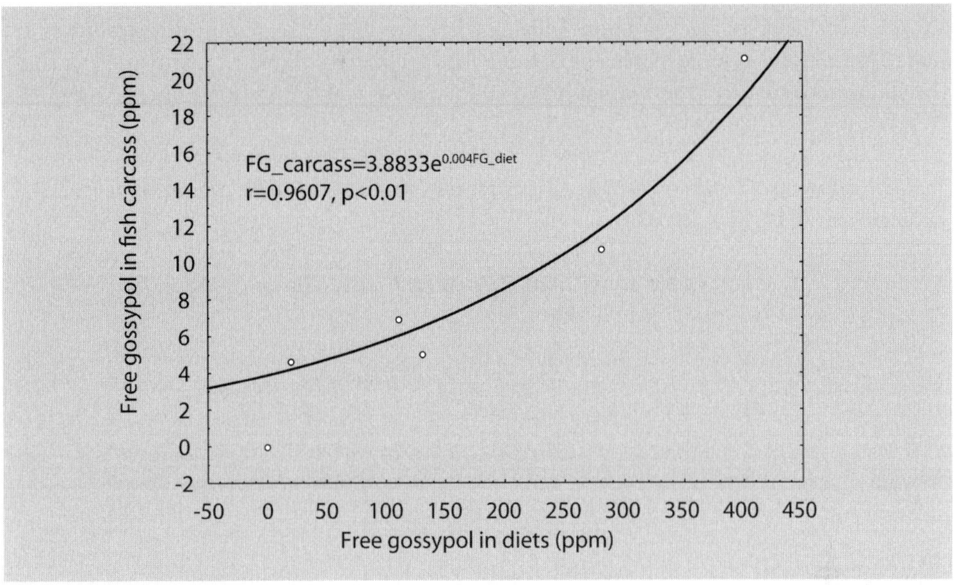

Figure 5.1. Relationship between free gossypol in carcass and diets (ppm) for gibel carp (Xie et al.*, unpublished).*

Diet formulation

Nutritional requirement is the base of diet formulation. Research has provided data on the requirements for protein, amino acids, lipid, fatty acids, minerals, vitamins, etc. (Li, 1990; Xie and Yang, 1995; Xie *et al.*, 1997; Yan, 2004; Zhou, *et al.*, 2006). The digestibility of different feed ingredients has been determined in some species (Lei, *et al.*, 1996). A balanced diet to produce 1 tonne of gibel carp reduced the excretion of ammonia and P by 26.6 and 70.2 kg respectively, while also decreasing the feed cost per kilogram fish produced (Lei *et al.*, personal communication). Feeding stimulants have been studied and used to increase the feed palatability (Xue *et al.*,2003; 2004) and this could reduce the feed waste.

In order to reduce the N and P loading, further research on aquafeed has focused on topics such as the lipid and carbohydrate content to decrease the dietary protein, on the content of fishmeal, on suitable protein sources, and on enzymes to increase the feed utilisation (Xue *et al.*, 2004; Yang *et al.*, 2004a, b). Increased dietary lipid and/or carbohydrate level increased protein retention and decreased N loading of waste; different dietary carbohydrates showed different effects (Tan, 2005). Increased dietary carbohydrate / lipid level also improved DM digestibility in gibel carp and Chinese long-snout catfish (Tan, 2005). The replacement of fishmeal by soybean meal decreased P loading by 15% (Zhang, personal communication). The replacement of trash fish by a formulated diet improved the growth rate and feed utilisation for rockfish (*Sebastes schlegeli*) and decreased nitrogen loading (Yan, 2004). Gibel carp showed different N and energy utilisation when fed with different sources of dietary protein (Xie *et al.*, 2001). Inclusion of meat and bone meal in gibel carp diet decreased the digestibility of dry matter, protein and P, while increasing N and P loading linearly (Table 5.1, Figures 5.2 and 5.3). The addition of citric acid and vitamin D could decrease N and P loading (Zhang, 2006).

Table 5.1. Apparent digestibility coefficient (ADC) of experimental diets in gibel carp (in %: means ± SE). Adapted from Zhang* et al. *(2006).*

Diet	ADC-DM	ADC-protein	ADC-energy	ADC-P
Control	69 ± 0.8 [a]	92 ± 0.7 [a]	80 ± 0.3	39 ±0.7 [a]
MBM_{20}	70 ± 0.7 [a]	92 ± 0.7 [a]	80 ± 1.4	36 ± 0.9 [a]
MBM_{40}	68 ± 0.4 [ab]	89 ± 0.3 [b]	80 ± 0.8	27 ± 1.2 [b]
MBM_{60}	67 ± 0.5 [b]	88 ± 0.3 [b]	80 ± 0.3	23 ± 1.4 [c]
MBM_{80}	62 ± 0.2 [c]	84 ± 0.7 [c]	78 ± 0.9	20 ± 1.6 [cd]
MBM_{100}	63 ± 1.0 [c]	83 ± 0.7 [c]	78 ± 0.1	18 ± 1.3 [d]

*Figures in the same column with different superscripts are significantly different (*P*< 0.05). MBM_{20}, MBM_{40}, MBM_{60}, MBM_{80}, MBM_{100} means 20%, 40%, 60%, 80% or 100% of dietary fish meal protein were replaced by meat and bonemeal (MBM) protein.

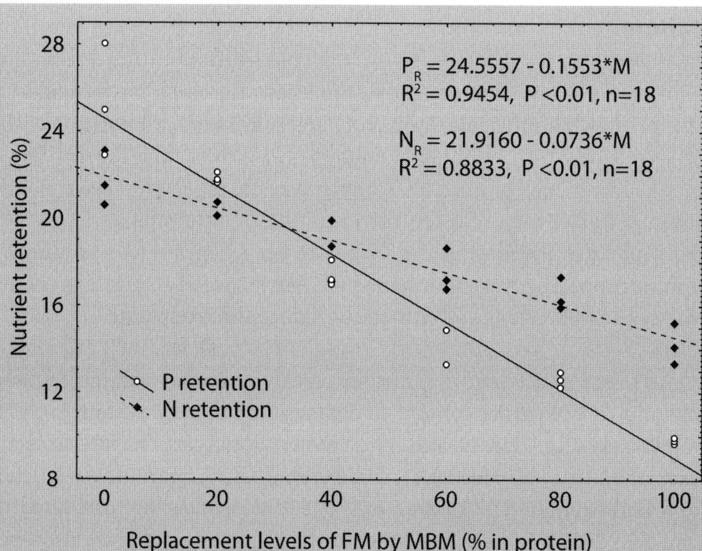

Figure 5.2. Relationship between P and N retention rate and dietary MBM protein levels in gibel carp fed test diets for 11 weeks (Zhang et al.*, 2006).*

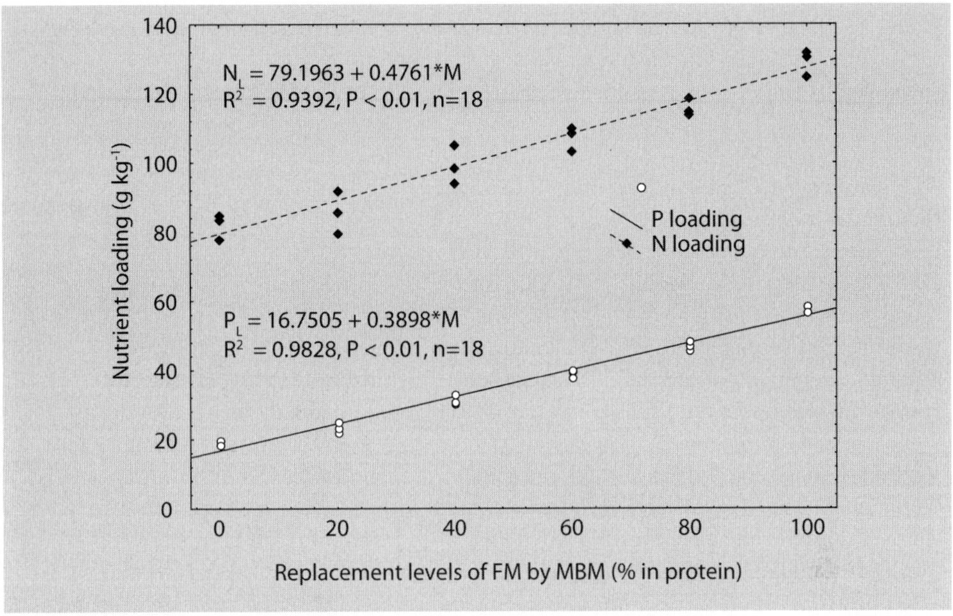

Figure 5.3. Relationship between P and N loading and dietary MBM protein levels in gibel carp fed test diets for 11 weeks (Zhang et al.*, 2006).*

The results of a modelling study showed that a balanced feeding regime could save about 0.86 tonnes of feed and reduce ammonia excretion by 31 kg when producing 1 tonne of gibel carp (Zhou, 2002; Han, 2005). The production of 1 tonne of Chinese long-snout catfish would save 0.27 tonnes of feed and reduce ammonia and P excretion by 21.7 and 7 kg respectively (Table 5.2).

Aquaculture management

Suitable management could also help to decrease the waste production in aquaculture. Such management includes the location of the culture system (cages), feeding regime, compensatory growth and polyculture.

Guo and Li (2003) have reported that, to respect the auto-cleaning capacity, the distance between net cages in freshwater lakes along the Yangtze River should be more than 50 metres. The feeding regime (including feeding frequency, feeding rate, and feeding rhythm) was studied, and bio-energetic modelling was included in order to predict the growth and feed required to build a dynamic feeding regime. In some fishes, a compensatory growth period could increase feed utilisation and decrease nitrogen loading (Zhu *et al.*, 2004, 2005).

In China and certain developing countries, semi-intensive culture is still very popular. In this culturing system, fish use some of the available natural food, thereby decreasing the need for

Table 5.2. Results of the inside validation of an N and P emission model for Chinese long-snout catfish.

Parameters	Observed	Predicted by bio-energetic models	Predicted by growth models
Body weight (g)	18.98	18.98	18.98
Feed intake (mg/fish/d)	283.1	283.1	-
Faecal production (kJ/fish/d)	0.285	0.325	-
Excretion of nitrogen (mg/fish/d)	9.40	9.87	-
Faecal nitrogen (mg/fish/d)	0.68	0.75	-
Waste nitrogen (mg/fish/d)	10.08	10.62	7.56
Nitrogen ingested (mg/fish/d)	18.48	18.48	13.25
Excretion of phosphorus (mg/fish/d)	1.44	0.98	-
Faecal phosphorus (mg/fish/d)	1.75	1.99	-
Waste phosphorus (mg/fish/d)	3.19	2.98	3.96
Phosphorus ingested (mg/fish/d)	5.71	5.71	5.65
Waste nitrogen/Nitrogen ingested (%)	54.53	57.49	57.07
Waste P / Phosphorus ingested (%)	55.77	52.10	70.15

Sources: Zhou, 2002; Han, 2005.

nutrients from a formulated diet (Rahman and Verdegem, 2007). Polyculture, i.e. producing fish with different feeding habits in the same pond, could increase the utilisation of fish feed and decrease the loading of nutrients into the natural water bodies (Dosdat, 2001).

Bio-energetic models have been used to estimate the contribution of natural food and the nutrient required through supplements. It has been found that daily natural food consumption estimated by direct diet sampling of stomach contents in September was 1.12% body weight/day in dry weight or 121.0 J/g/day in Nile tilapia weighing about 68 g in a no-feeding cage in the East Lake, when water temperature was around 30 °C. Estimated values (96.8–133.1 J/g/day) from the energetics model fitted the diet sampling value well (Xie, 1997; Xie *et al.*, 1998).

Conclusions

Given environmental pollution, food safety and the availability of aquatic feed ingredients, it is becoming increasingly urgent to maintain aquaculture sustainable. The inclusion of energy balance in the selection of fish species and in diet formulation can contribute to safeguarding the environment, while reducing feed cost. Adjusting the feed regime to obtain compensatory growth in some fishes, managing cage location, and using appropriate polyculture systems can also reduce waste production and improve the profit margins of aquaculture.

References

Boyd, C.E. and C.S. Tucker, 1998. Fertilization. In: Pond Aquaculture Water Quality Management. Kluwer Boston, Inc., USA, pp. 226-230.

Chen, J. and P. Xie, 2005. Tissue distributions and seasonal dynamics of the hepatotoxic microcystins LR and-RR in two freshwater shrimps, *Palaemon modestus* and *Macrobrachium nipponensis*, from a large shallow, eutrophic lake in subtropical China. Toxicon, 45, 615-625.

Chen, J., P. Xie, L. Guo, L. Zheng, and L.Y. Ni, 2005. Tissue distributions and seasonal dynamics of the hepatotoxic microcystins -LR and -RR in a freshwater snail (*Bellamya aeruginosa*) from a large shallow, eutrophic lake in subtropical China. Environ. Pollut. 134, 423-430.

Cui, Y. and J. Liu, 1990. Comparison of energy budget among six teleosts - III. Growth rate and energy budget. Comp. Biochem. Physiol. 97A, 381-384.

Dong, S., Pan, K., and U. Brockmann, 2000. Review on effects of mariculture on coastal environment. J. Ocen Univ. Qingdao. 30, 575-582.

Dosdat, A., 2001. Environmental impact of aquaculture in the Mediterranean: nutritional and feeding aspects. Environmental Impact Assessment of Mediterranean Aquaculture Farms. Cahiers Options Méditerranéennes vol. 55. CIHEAM-FAO, pp. 23–36.

Guo, L. and Z. Li, 2003. Effects of nitrogen and phosphorus from fish cage-culture on the communities of a shallow lake in middle Yangtze River basin of China. Aquaculture 226, 201-212.

Han, D., 2005. Modeling for feeding system and pollution evaluation in Chinese longsnout catfish. PhD thesis, Institute of Hydrobiology, Chinese Academy of Sciences, Wuhan, P.R. China.

Hui, Y., Y. Xu, Q. Tang, and W. Wang, 2005. Correlation between mercury concentrations in surface sediment and fish muscle in Ya-Er Lake, Hubei. J. Safety Envir. 5, 9-11.

Jiang, G., 2005. Developing healthy aquaculture to ensure the food safety. China Fish. 5, 66-68.

Lei, W., Y.X. Yang and X.Q. He, 1996. Apparent digestibility coefficient of crude protein and energy for six commercial feed ingredients for Chinese longsnout catfish. Acta Hydrobiol. Sin. 20, 113-118. (in Chinese with English abstract).

Li, J.S., 1990. Study on dietary zinc and copper nutrition on gibel carp (*Carassius auratus gibelio*) fingerlings. MSc thesis, Institute of Hydrobiology, Chinese Academy of Sciences, Wuhan.

Li, Z., K. Leng, J. Li, X. Li, and W. Wang, 2001. The quality of aquatic animal medicines and the supervision of medicine residues in aquatic products. Mar. Fish. Res. 22, 76-80. (In Chinese with English abstract).

Liu, J., 1998. Comparative studies on growth and energy budget of mandarin fish and snakehead fish. PhD thesis, Institute of Hydrobiology, Chinese Academy of Sciences, Wuhan, P.R. China.

Ning, F. and C. Gu, 2004. Actuality and countermeasure for pollution about net-cage aquaculture in Chongqing. J. Chongqing Techn. Business Univ. (Nat. Sci. Ed). 21, 543-548.

Ou, F., Y. Huang, Q. Tian and J. Zhang, 2003. Eco-environment situation and recovery strategies in inner lakes of Dongting Lake. Inland Fish. 12, 36-38.

Rahman, M.M and M.C.J. Verdegem, 2007. Multi-species fishpond and nutrient balance. In: A.J. van der Zijpp, J.A.J. Verreth, Le Quang Tri, M.E.F. van Mensvoort, R.H. Bosma and M.C.M. Beveridge (eds.), Fishponds in farming systems. Wageningen Academic Publishers, Wageningen, The Netherlands, pp. 79-88.

Tan, Q., 2005. A comparative study on dietary carbohydrate utilization by gibel carp and Chinese longsnout catfish. PhD thesis,, Institute of Hydrobiology, Chinese Academy of Sciences, Wuhan, P.R. China.

Xie, L.Q., P. Xie, K. Ozawa, T. Honma, A. Yokoyama and H.D. Park, 2004. Dynamics of microcystins-LR and –RR in the planktivorous silver carp in a sub-chronic toxic experiment. Environ. Pollut. 127, 431-439.

Xie, S., 1997. Supplementary feeding for cage-cultured tilapia: a bioenergetics modeling approach. PhD thesis, Institute of Hydrobiology, Chinese Academy of Sciences, Wuhan, P.R. China.

Xie, S.Q. and Y.X. Yang, 1995. On the formualtion of feed and feeding technique for Chinese longsnout catfish II. Methionine requirement of juveniles. In: Y. Liang and H. Liu (Eds.), Resources, Environment and Fishery Ecological Managment of Macrophytic lakes (Vol.I). Academic Press, Beijing, China, pp. 316-323. (in Chinese with English abstract).

Xie, S.Q., Y.B. Cui, Y.X. Yang and J.K. Liu, 1998. Effect of dietary protein level in supplemental diets on the growth of cage-cultured Nile tilapia in the East Lake, P.R. China. Asian Fish. Sci. 10, 231-238.

Xie, S.Q., X.Q. He and Y.X. Yang, 1997. Lysine requirement for longsnout catfish *(Leiocassis longirostris* Günther) juveniles. Acta Hydrobiol. Sin. 21, 1-7. (in Chinese with English abstract).

Xie, S.Q., X.M. Zhu, Y.B. Cui and Y.X. Yang, 2001. Utilisation of several plant proteins by gibel carp (*Carassius auratus gibelio*). J. Appl. Icthyol. 17, 70-76.

Xu, Y., W. Wu., Y. Zhang and Y. Zhang, 1999. Dynamics and long-term fate of residual hexachlorocyclohexane (HCH) in various types of environemtn in Ya-Er Lake area, China. Acta Hydrobiol. Sin. 23, 337-345.

Xue, M., S.Q. Xie and Y.B. Cui, 2004. Effect of a feeding stimulant on feeding adaptation of gibel carp *Carassius auratus gibelio* (Bloch), fed diets with replacement of fish meal by meat and bone meal. Aquacul. Res. 35, 473-482.

Xue, M., S.Q. Xie, Y.B. Cui and X.Q. Qian, 2003. Advances in studies on fish feeding stimulants. Acta Hydrobiologica Sinica 27, 639-643.

Yan, Q., 2004. Studies on lysine and methionine requirement and practical diets for *Sebastes schlegeli*. MSc thesis, Institute of Hydrobiology, Chinese Academy of Sciences, Wuhan, P.R. China.

Yang, Y., S.Q. Xie, Y.B. Cui, X.M. Zhu, Y.X. Yang and J.K. Liu, 2004b. Effect of replacement of fish meal by meat and bone meal and poultry by-product meal in diets on the growth and feed utilization of gibel carp, *Carassius auratus gibelio*. Aquacul. Nutr. 10, 289-294.

Yang, Y., S.Q. Xie, W. Lei, X.M. Zhu and Y.X. Yang, 2004a. Effect of replacement of fish meal by meat and bone meal and poultry by-product meal in diets on the growth and immune response of *Macrobrachium nipponense*. Fish Shellfish Imm. 17, 105-114.

Zhang, S., 2006. Utilization of phosporus of the diets containing meat and bone meal for gibel carp. PhD thesis, The Institute of Hydrobiology, Chinese Academy of Sciences, Wuhan, P.R. China.

Zhang, S., S. Xie, X. Zhu, W. Lei, Y. Yang and M. Zhao, 2006. Meat and bone meal replacement in diets for juvenile gibel carp (*Carassius auratus gibelio*): effects on growth performance, phosphorus and nitrogen loading. Aquacul. Nutr., 12, 353-362.

Zhao, M., S. Xie, X. Zhu, Y. Yang, L. Gan and L. Song, 2006a. Effect of inclusion of blue-green algae meal on the growth and accumulation of microcystin in gibel carp (*Carassius auratus gibelio*). J. Appl. Ichth. 22, 72-78.

Zhao, M., S. Xie, X. Zhu, Y. Yang, L. Gan and L. Song, 2006b. Effect of dietary cyanobacteria on growth and accumulation of microsystins in Nile tilapia (*Oreochromis niloticus).* Aquaculture, 261, 960-966.

Zhou, X., S. Xie, C. Xie, W. Lei, X. Zhu and Y. Yang, 2006. The utilization and requirement of dietary lysine for juvenile Gibel carp. Acta Hydrobiol. Sin., in press. (in Chinese with English abstract).

Zhou, Z., 2002. Establishment of feeding regimes for gibel carp (*Carassius auratus gibelio*): a bioenergetics modelling approach. PhD thesis, Institute of Hydrobiology, Chinese Academy of Sciences, Wuhan, P.R. China.

Zhu, X.M., S.Q. Xie, W. Lei, Y.B. Cui, X.Y. Yang and R.J. Wootton, 2004. Compensatory growth and food consumption in gibel carp, *Carassius auratus gibelio*, and longsnout catfish, *Leiocassis longirostris*, experiencing cycles of feed deprivation and re-feeding. Aquaculture 241, 235–247.

Zhu, X.M., S.Q. Xie, W. Lei, Y.B. Cui, Y.X. Yang and R.J. Wootton, 2005. Compensatory growth in the Chinese longsnout catfish, *Leiocassis longirostris* following feed deprivation: Temporal patterns in growth, nutrient deposition, feed intake and body composition. Aquaculture 248, 307-314.

Livelihood impacts of ponds in Asia-opportunities and constraints

David C. Little[1], Manjurul Karim[1,3], Danai Turongruang[1], Ernesto J. Morales[1], Francis J. Murray[1,2], Benoy K. Barman[1,4], Mohammad M. Haque[1,3], Nitai Kundu[5], Ben Belton[1], Golam Faruque[1,4], Ekram M. Azim[1], Faruk Ul Islam[1], Lindsay Pollock[1], Marc C.J. Verdegem[6], James A. Young[1], Will Leschen[1] and M. Abdul Wahab[7]
[1]Institute of Aquaculture, University of Stirling, Stirling, FK9 4LA, United Kingdom
[2]WorldFish Center, Jalan Batu Maung, Batu Maung, 10670, Penang, Malaysia
[3]Department of Aquaculture, Bangladesh Agricultural University, Mymensingh, Bangladesh
[4]WorldFish Center, Dhaka, Bangladesh and South Asia Office, House 22B, Road 7, Block F, Banani, Dhaka 1213, Bangladesh
[5]Institute of Wetland Management and Ecological Design, Salt Lake, West Bengal, India
[6]Aquaculture and Fisheries, Wageningen University, P.O. Box 338, 6700AH Wageningen, The Netherlands
[7]Department of Fisheries Management, Bangladesh Agricultural University, Mymensingh, Bangladesh

Abstract

Ponds are traditional multipurpose resources accessed by households and communities, and are increasingly being prioritised for aquaculture. High consumption of aquatic animals and declines in natural stocks has stimulated fish culture based on both stocked and natural seed across a broad spectrum of intensification. Management of a high proportion of ponds remains sub-optimal with respect to fish production because of conflicting uses, multiple ownership, and poor access to markets and information. Aversion to risk of drought and flood is also a factor that reduces interest in optimising fish yields through financial investment, but can enhance the importance of water storage and trapping of unstocked aquatic animals. The role of ponds as assets controlled by households and communities is reviewed. Ponds and their products embrace several types of assets and support a range of activities, both productive and social. Issues of access are critical to the potential for poorer people to sustain benefits. Ponds may reduce vulnerability and examples are presented of the mechanisms by which the resilience of households is enhanced with respect to withstanding shocks and seasonal fluctuations in food and income. Location, especially with respect to markets is identified as critical to the management strategies used by households. Key institutions that support the intensification of fish culture include dynamic networks of informal private sector service providers, many of whom are poor. Although the benefits of pond-based aquaculture have been associated with the resource-rich, poorer people are benefiting as producers, intermediaries and consumers.

Keywords: seasonality, livelihood, fishpond, multiple-use, vulnerability

Introduction

Small water bodies or 'ponds' have been a natural focus for human settlement, and are often common property resources with multiple users. 'Tanks', for instance, have been used for centuries in South Asia for domestic water supply and small-scale irrigation. More recently the specialisation of pond use for activities such as horticulture and aquaculture has increased, and in some cases associated changes in management practice have led to conflicting demands on the resource. These changes are often stimulated by opportunities for commercialisation but appear to be unsustainable in many instances, creating imbalances in access and the distribution of benefits among various groups within communities. On-farm ponds are also emerging as foci for agricultural diversification and the broader pluriactivity of households (Prein, 2002). Integrated pond-dyke cropping was a traditional response to high population densities in flood-prone agro-ecosystems in southern China (Ruddle and Zhong, 1988) but until recently was relatively uncommon elsewhere. The use of livestock waste as fishpond inputs has become common in tandem with the industrialisation of monogastric animal production (Little and Edwards, 2003). Although some report that integrated systems are less efficient in terms of nutrient, land and labour use than specialised fish culture (Micheilsons *et al.*, 2002), various degrees of integrated aquaculture-agriculture farming system have spread rapidly around the region over the last two decades, suggesting otherwise.

Pond based intensive, urban or export orientated commercial aquaculture is also well established and expanding rapidly in some parts of the region, with implications for the livelihoods of people in these areas. Impacts are likely to be variable however as the characteristics of, for example, intensive production of riverine catfish (*Pangasius* spp.) in Vietnam, are quite different from those of the valuable freshwater prawn (*Macrobrachium rosenbergii*) in Bangladesh.

It is common practice to identify aquaculture development in terms of its separateness from the surrounding fishery of wild stocks and other livelihood activities, and to associate farming fish with livestock and other types of agricultural production (e.g. Townsley, 1998). The stocking of ponds, especially with exotic species, has been used to distinguish 'farming' from fisheries of indigenous stocks. Although social and cultural differences between fish farmers and fishers have been identified, the reality for many rural people is a fishery-aquaculture continuum. Increasingly aquatic products derived both from hatcheries and natural stocks are managed together in aquatic systems under farmers' control. Development of ponds has played a critical role in this process. The location of ponds within or next to rice-fields and the importance of the rice-field based fisheries for food security and broader livelihoods in much of the South and South East Asia (Gregory and Guttman, 2002; Morales *et al.*, 2004) are major determinants of pond management practices in many rural households. Thus, accounts of the modern development of aquaculture in Asia which emphasise only the stocking of hatchery seed in purpose-built ponds are incomplete. In fact, ponds fulfil multiple roles and are part of a broader household managed aquatic resource base in which there is a continued reliance on naturally occurring aquatic animals in addition to stocked seed.

Many observers have reported that, as aquaculture can be resource intensive, it's practice tends to be dominated by the better-off (e.g. Lewis, 1997; Townsley, 1998). This view may have been informed by a narrow perspective on pond utilisation and its impacts at the community level. For example, recent studies in various regions of Bangladesh indicate that poorer households receive many benefits from fish production (Barman, 2000; AIT/DOF, 2001; Karim, 2006). The ponds of the poor are often ignored because of their very small size and tendency to be seasonal, but fulfil important livelihood functions none-the-less.

Ponds in Asia often have ancient roots and important traditional functions but have evolved in modern settings to deliver a variety of new livelihood outcomes. More recently, pressure on resources has resulted in dynamic changes in access to, and use of, the water ponds contain, and they have increasingly become part of more intensified food production systems through use for fish culture and the irrigation of surrounding crops. Increasing levels of integration within rain-fed farming systems are having fundamental impacts on livelihoods. The influences of broader development and demographic change, especially urbanisation, are particularly important. The physical and social characteristics of water bodies and their benefits and constraints are complex and highly location specific. This article focuses on inland areas of Asia[3]. The roles of ponds on inland livelihoods are reviewed based on evidence derived from a selection of recent research.

Methodology

This assessment was based on a series of field research projects conducted with a variety of partners over the last eight years in South and South East Asia. These include research in North West Province, Sri Lanka, on farmer-managed irrigation systems, in Bangladesh, Southern Vietnam and North East Thailand into the role of ponds as a focus for diversified farming systems, in West Bengal, India into the use of ponds, and at five sites throughout South and Southeast Asia into the role of non-stocked aquatic animals in smallholders systems. The paper first introduces the physical and social contexts of ponds before reviewing the role of ponds to household livelihoods.

The nature of 'ponds'

Physical

Water bodies defined by the term 'pond' range from small excavated holes in the ground little larger than open wells to much larger systems enclosed with dykes or levees. This definition also includes small impoundments created by damming natural stream courses. It excludes larger institutionally managed reservoirs but includes household, group and community managed systems that are typically up to tens of hectares in area. Ponds are typically shallow (<3m), and are temporary unless succession is managed to control encroachment by macrophytes and build-up of sediments that lead them to become wetlands and, finally, dry land. 'Ponded' water bodies are primarily lacustrine, although they may experience significant water exchange

[3] The implications of pond development on livelihoods in coastal areas are not reviewed, largely because potential synergisms within broader livelihood systems possible in inland contexts (storage of fresh water, for example) are less obvious. Furthermore, any analysis of coastal aquaculture is coloured by the phenomenon of widespread ecological and social change that accompanied the growth of shrimp farming.

depending on rainfall and management characteristics. The distinction typically made between on-farm and off-farm water storage resources becomes blurred when the hydrological linkages that typically occur are considered. Such linkages, moderated by rainfall and seasonal flood, support movements of nutrients and of aquatic life that are often difficult to control.

Seasonal ponds (small water bodies that dry out for a proportion of the year) can be more common than perennial water bodies (Figures 5.4 and 5.5b) and are often highly productive for a range of purposes. They are typically ignored by institutions promoting pond-based aquaculture however, and management complexity, particularly in relation to the stocking of hatchery seed, makes them highly underutilised for fish production. In one study of tanks in one district in Tamil Nadu, India the proportion of tanks stocked with seed was highly variable (Table 5.3). Definitions of perennial and seasonal water bodies are typically related to use. Tanks located in watersheds in the dry zone of Sri Lanka for example may be referred to as seasonal because of a lack of water for irrigation and yet retain enough water for continued production of fish (Murray, 2004). In contrast seasonal ponds in Bangladesh are considered as such based on the duration for which they can support fish culture.

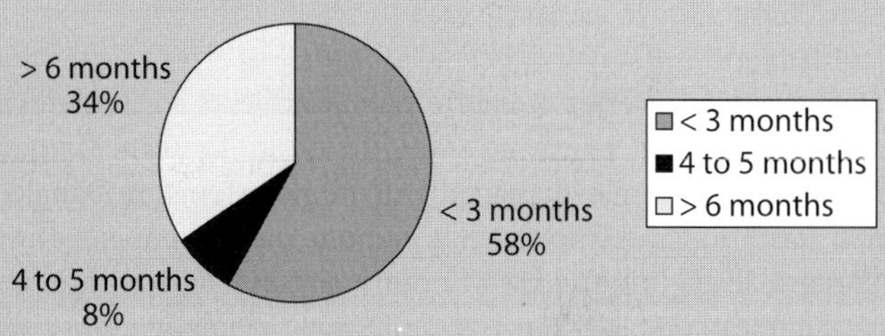

Figure 5.4. Water availability in seasonal tanks surveyed in Erode District, Tamil Nadu, India.

Table 5.3. Status of tanks stocked in Erode District, Tamil Nadu, India (Little and Pollock, 2003).

	Number of tanks		
Sub-District	Total identified	Stocked in the past	Currently stocked
Gobichettipalayam	17	12	10
Bhavani	31	31	19
Sathyamangalam	69	40	10
Perundura	16	4	1
Dharapuram	1	0	0
Erode	7	6	6
Kangeyam	14	-	-

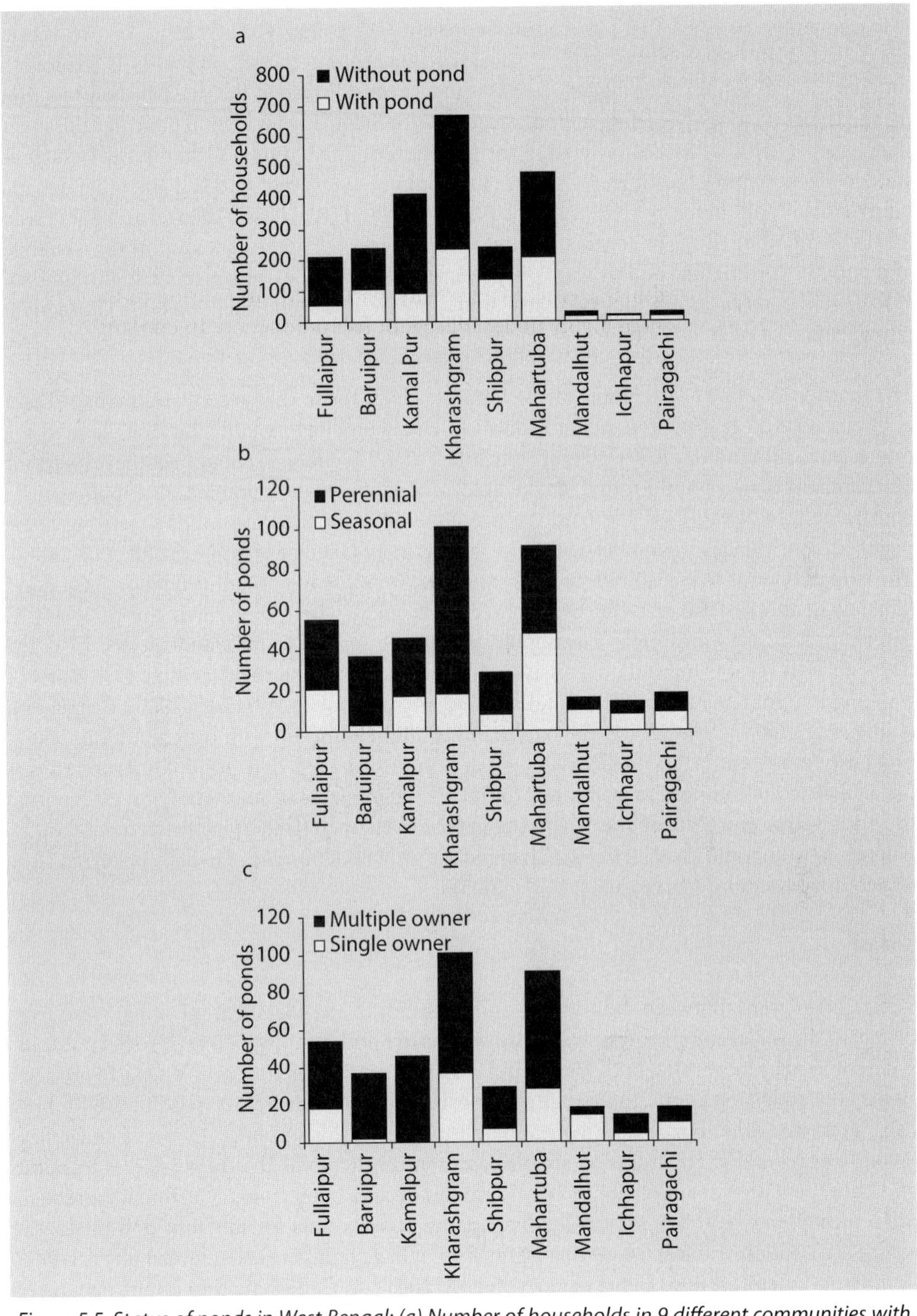

Figure 5.5. Status of ponds in West Bengal: (a) Number of households in 9 different communities with access to ponds, (b) number of seasonal and perennial ponds, and (c) number of ponds under single and multiple ownership (modified from Barman et al., *2005).*

It is frequently assumed that fish culture represents the primary motivation for excavating ponds, especially where the activity is subsidised or promoted, but the reality is often very different. Initial stimulus for digging ponds is in fact often linked to the need for construction fill and the avoidance of flooding in homestead development. The incidental pond development that results is an important motivation for adoption of aquaculture. Other benefits such as water collection for domestic purposes and the trapping or 'storing' wild fish and other aquatic animals also result in many areas, e.g. North East Thailand (AIT/SAC, 2003), Lao PDR (Bush, 2004), Cambodia. This is in contrast to sites where clusters of ponds, are specifically designed and built for commercial aquaculture purposes with the intent of optimising fish production. Construction costs and anticipated use undoubtedly affect the location and subsequent management of such 'fish' ponds.

The location of ponds in relation to the homesteads of those that access and manage them is often a critical factor influencing utilisation characteristics. A pond located close to the homestead will tend to attract greater user interest and is more likely to become a focus for development than a pond located 'in the field', where observation and labour inputs may be constrained. In North East Thailand the productivity of stocked fishponds located at some distance from villages was correlated to the frequency of farmers' visits (Arjinkit *et al.*, 2006). The siting of ponds may also reflect local rules and the value of agricultural land. Thus, pond location in the Red River Delta usually occurs in areas where rice yields are poor and is determined by the commune. 'Household ponds' are typically small and located near the homestead in rural Cambodia, whereas ponds located within ricefields, either as 'culture' or trap ponds dominate in North East Thailand and are common in West Bengal and Bangladesh (Little *et al.*, 2004; Table 5.4). Pond excavation is often bound up with broader development especially roads and access to markets. Bush (2004) observed that pond construction was particularly common close to roads and around urbanising areas in Southern Laos. He reported the greatest concentration of ponds as being in an area characterised by lack of perennial water, whereas in Cambodia and North East Thailand the growth of ponds has usually occurred close to perennial water bodies (Amilhat *et al.*, 2005).

Social

Strict distinctions between household and communal management of water resources disregard the social complexity of water use. Access to ponds for productive or social purposes is determined by local rules, both formal and informal, which together codify *de facto* property rights. These rights give rise to a range of opportunities and constraints to pond use for both fish culture and other purposes. Single and multiple ownership of ponds, owner-management or out-leasing and share cropping are all common mechanisms that make access to ponds and their benefits an option for poorer individuals and households. Customary practices often incorporate complex seasonally overlapping access rights for different pond uses. For example irrigation and fish harvesting rights from village tanks are often informally accorded to different interest groups in Sri Lanka (Murray, 2004). In Cambodia trap ponds may revert to open access or common property conditions when they are flooded. Ponds can be the foci for broader development, both on a household or community basis, resulting in a range of benefits to producers, intermediaries and consumers. A first transitional step in the continuum of common to private ownership often involves a move from community managed fisheries to simple community managed stocking enhancements. Further intensification is usually accompanied by increasing levels of enclosure, and a move from group to individual household or private commercial management - ensuring returns to individual effort commensurate with increased investment risk.

How livelihoods are impacted

Ponds and their products as assets

An analysis of impacts of ponds on livelihoods needs to consider "the capabilities, assets (stores, resources, claims and access) and activities required for a means of living" (Chambers and Conway, 1992) – it is necessary to consider both their physical and social dimensions. Ellis (2000) suggested that the combination of assets, activities and access determine the nature of the livelihood of either the individual or the household. Analysis that extends beyond producer households is required however if the role of ponds in relation to livelihood vulnerability or resilience is to be fully understood, since pond development can potentially undermine or enhance non-producer livelihoods and create impacts at a community or even wider geographical scale.

Carney (1998) defines five types of capitals that underpin household livelihoods as physical, natural, social, human and financial. Whereas unimproved water bodies and the water contained in ponds can be considered natural assets, constructed or improved ponds are clearly physical assets. Wild and self-recruiting stocks of fish and other aquatic animals that are often important components of harvested pond yield are a natural asset. Their continued importance not only appears to meet rural communities diverse needs but their integration within managed systems enhances local biodiversity with positive implications for non-pond owners in the vicinity. Ponds within rice-based farming systems enhance heterogeneity of landscape and create micro-habitats, challenging the perception that pond-based aquaculture inevitably leads to reductions in biodiversity (Little *et al.*, 1996). The harvest of ponds may meet both subsistence food needs (human) and be income generating or reduce expenses in purchasing food (financial). Producers' households often sell fish strategically in response their own needs and those of others, and to meet seasonal income demands (financial capital). Fish and other aquatic animals are core high quality dietary items, the micronutrient and essential fatty acid profiles of which have been associated with maintaining human health, maternal wellbeing and infant physical and mental development (human capital; Ruxton *et al.*, 2004). Ponds can be particularly important at times such as rice harvest when labour is constrained but social obligations are high. The relative efficiency of harvesting fish from ponds makes them an important asset even in areas where wild fish are abundant (Little, 1994). Ponds have been viewed as social assets and status symbols in parts of Africa (Harrison, 1994; Ruddle, 1996) and their role in Asia for supporting and improving social relations can be critical (Haque *et al.*, 2006). Improvement of social capital through networking with other pond operators or through group management of village ponds has been demonstrated by several projects aiming to promote pond-based aquaculture (Islam, 2002; Little *et al.*, 2004; Murray, 2004)

The physical and social assets embodied in ponds are interchangeable and influenced by management and seasonality. Audits of pond utilisation frequently indicate multipurpose use and sub-optimal fish production. Culture and capture of fish are usually subsidiary activities of rural people rather than dominating rural livelihoods. An assessment in three parts of South East Asia where ponds are common components of rice-dominated agro-ecosystems found that fish production was always a subsidiary activity among rice farmers, who prioritised crop, livestock and off-farm labour (Morales *et al.*, 2003). Other studies indicate that pond-based aquaculture can become a major, or dominant, activity where aquaculture has a stronger commercial orientation (Faruque *et al.*, 2006; Karim, 2006, Nhan *et al.*, 2007). This does not always correlate to the size of ponds, but may be more related to management intensity. Luu *et al.* (2002) reported that households in lowland areas of the Red River Delta, Vietnam, where ponds are smaller than in other agro-ecological zones, were more likely to consider fish culture as their primary activity.

The water retained in ponds and the productive and social activities based around them can be important, especially in water-stressed areas of Asia. Communities in dry zone Sri Lanka prioritise small ponds within watersheds more for irrigation, bathing, domestic use and livestock than their value as a source of fish (Murray, 2004). Households may trade off the benefits from retaining water for multiple uses to consumptive use for irrigation. In areas where aquaculture is well developed these priorities often change but can lead to other impacts. Intensification of aquaculture on marginal agricultural land can have adverse unforeseen consequences. Rice fields being converted to prawn ponds in South Bangladesh have led to increased landlessness for example and there may be externalities on other water users: In Rajshahi, Bangladesh farmers are depleting ground water to extend fishpond production through the dry season.

The use of ponds for fish production in the region is highly variable and mediated by a wide range of factors. For example, access to key inputs such as seed and nutrients can be affected by availability of credit, location and social status. Frequently access is mediated by institutions and social relations (Ellis, 2000).

Vulnerability and resilience

The impacts of ponds on vulnerability at the household and community level have been little studied to date. Few studies have focused on their role as multipurpose resources, as opposed as mono-commodity units. Pond development can also be counterproductive. Adger (2000), for instance, found that the development of coastal ponds for shrimp in Northern Vietnam had weakened social cohesion. Development of shrimp ponds by the better off placed increasing pressure on remaining mangrove resources available to poorer households. The mechanisms by which ponds can strengthen household resilience are related to longer term trends, shocks and seasonality. These are considered below.

Declines in fisheries, widely perceived to be an inevitable consequence of population growth and environmental deterioration, often stimulate political action towards promotion of aquaculture (Thompson *et al.*, 2002; Bush, 2003). However, in reality, fisheries remain critical resources for the poor in many parts of Asia. Although qualitative changes have certainly occurred through exploitation almost everywhere, Nguyen Khoa *et al.*, (2005) found floodplain fisheries remained robust and diverse in rain-fed areas outside of large-scale irrigation systems in Lao PDR. Other trends such as increased access to pumps and low cost synthetic nets have led to over-fishing of wild stocks, but coincidentally provide important inputs that improve the management of culture ponds. The intermittent or continuous harvest of culture ponds that such tools allow, have important implications for poor people as employees and consumers. Declines in the real cost of mechanised pond construction that have accompanied the expansion of road networks in the region, have led to increases in trap pond construction in some areas of North East Thailand. Such small, deep ponds undoubtedly increase the efficiency of trapping wild fish such as the Chevron snakehead (*Channa striata*) (Amilhat and Lorenzen, 2005) but, if appropriate management steps are in place, can also raise the possibility of increased yields of aquatic animals from surrounding ephemeral ricefields through expansions in the area of dry season refuges (Little *et al.*, 2004).

Broader changes in demography and trends towards urbanisation are also having important impacts on the role of ponds (Little and Bunting, 2005). Access to inputs and markets affect opportunities to commercialise pond use, some examples of which are given below. Ponds may cushion households from various types of shocks, improving their resilience to unforeseen events. Farmers raising pigs and fish in Central Thailand were more protected from price declines for pork than those without fish (Edwards, 1985). The physical structure of ponds may provide protection against otherwise catastrophic flood and drought, and pond raised fish are typically liquid assets that can be sold even at a small size if required; in this respect fish take on a similar role as small livestock within smallholder households (Little and Edwards, 2003). Ponds have particular importance in rain-fed marginal agro-ecosystems since they can smooth water availability during periods of seasonal shortage. Perennial water availability can broaden and increase the value of ponds greatly. For example ponds holding water year-round can be highly profitable for juvenile nursing prior to the onset of the rains and the corresponding jump in demand for fry (AIT/DOF, 2001). Maintaining surface water on-farm through storage in ponds is widely believed to reduce household vulnerability in climates with significant periods of irregular or no rainfall and where off-farm irrigation is unavailable. Smoothing consumption of fish and associated horticulture can in principle relieve hungry gaps common in such contexts but positive impacts on expenditure and income can also be important.

Producers, intermediaries, consumers

Producer households have been the conventional focus for understanding benefits from pond development, and a tendency for these to be better resourced than others in the community has been widely reported (e.g. Lewis, 1997). There is evidence that poorer households are becoming producers in many circumstances and that livelihood outcomes may be relatively more important for this group. For example, in a study of communities in Jessore, Bangladesh, where prawn production is important, Faruque *et al.* (2006) have observed that while the basic needs of poorer producers were improved after adoption (e.g. through better nutrition and clothing) better-off prawn farmers tended to upgrade housing, education for children and investment in other productive assets. Other studies have found that although absolute production and benefits were lower for poorer producers they were relatively more important when assessed as part of overall income. Although the better-off had larger ponds, poorer households managed their smaller units more productively in a study in North West Bangladesh (Haque *et al.*, 2006). Commercial pond-based aquaculture also generates employment among the poor. Significant benefits in terms of increasing demand for paid labour and higher labour rates for agricultural labourers are associated with clusters of aquaculture development around Mymensingh, Bangladesh (Faruque *et al.*, 2006).

The role of the poor as service providers, in terms of input provision, marketing networks or production labour has important impacts on pond productivity and, subsequently, through the market chain to consumers. Local reuse of by-products and wastes is normally facilitated by poorer people (Prax *et al.*, 2000; Little and Edwards, 2003). The development of aquaculture in rural and peri-urban areas typically generates a large number of opportunities in terms of both forward and backward linkages within marketing chains; seed production and trading supports the productivity of fish farmers and a choice of marketing channels tends to increase returns. This has led to poor fishers being able to improve their livelihoods through harvest of pond cultured fish belonging to others. In Bangladesh fishers often benefit from pond culture through both a partial share of the catch and cash payment. Growth in cultured fish availability and demand has also stimulated fishers to diversify into retail marketing in Bangladesh (Faruque *et al.*, 2006). In Northern Vietnam, women in producer households have been able to diversify and then specialise in trading fish (Phuong *et al.*, 2006). The development of local markets for food-fish and fingerlings has increased such opportunities in several parts of the region.

Factors affecting the importance of ponds to livelihoods

Several key factors affect the importance of ponds within livelihoods. These are assessed based on data from around the region. The role of ponds within the broader aquatic resource is examined before issues of access, investment and broader macro-level change are considered.

Ponds as part of broader aquatic systems

The high proportion of ponds that are constructed as integral parts of, or with strong linkages to, broader wetlands under household or community management indicates their increasingly important contribution to the sustained management of such resources. In Cambodia for example the incorporation of trap ponds as refuge areas for fish stocked in enclosed/fenced areas of pooled paddy land is being piloted as a collective management approach. This vision of ponds being *extensive* components of broader systems is contrary to a conventional, more linear view of aquaculture development i.e. ponds are stocked and managed as polycultures and develop inevitably towards monoculture as systems are intensified. Many commentators have speculated that this trend leads to a deterioration of surrounding aquatic biodiversity, which in turn negatively impacts on household nutrition (e.g. Roos *et al.*, 2002).

Such intensive pond management is limited in most of South and South East Asia and, more often, ponds contribute to biodiversity in several ways. One study has shown that even when farmer attitudes to unstocked species in fishponds were negative and efforts were made to prevent their entry, yields of such self-recruiting species remained significant (Little *et al.*, 2004). More often, households recognise the value of unstocked animals and are positive or neutral in their attitudes and behaviour toward them. Thus ponds (as part of linked rice agro-ecosystems in which fish are stocked and other aquatic animals enter or reproduce *in situ*) contribute to both 'planned' and 'associated' biodiversity' (Alteri, 1999). The compatibility of both stocked and unstocked species with meeting household needs has been confirmed in numerous recent studies on both a household (Yoonpundh, 1997; Islam *et al.*, 2003; Morales *et al.*, 2003) and group basis, (Garaway, 1999; Murray, 2004) and explains farmer behaviour that runs contrary to the advice of extensionists and scientists. Pond development has broader implications for enhancing biodiversity as the surrounding dykes provide small areas of upland on which perennial vegetation can thrive in seasonally flooded wetlands. Ponds thus contribute to a structural mosaic that can enhance overall farm productivity, diversity and stability (Surinteraseree, 1988, Little *et al.*, 1996). These roles may be most important in areas short of perennial water where natural habits for wild stocks of aquatic animals are few and particularly sensitive to over-exploitation.

Access

The basis on which access to ponds occurs (whether through single or multiple household ownership) is often indicative of the levels of intensification and integration likely to be employed in their management and of the production benefits likely to be enjoyed by users. Access to ponds tends to be highly heterogeneous in communities around Asia. Whereas research in Bangladesh has suggested that the major benefits from pond-based aquaculture have been enjoyed by better-off households (Lewis, 1997) in other places benefits appear to be more equitable. The proportion of households that own or access ponds is a useful indicator and this suggests that access to ponds in many South East Asian countries has expanded greatly. In the Red River Delta, Vietnam, up to 80% of households had ponds in the three irrigated agro-ecological zones surveyed by Luu *et al.* (2002); this contrasts with Little *et al.* (2004) who found lower proportions of households with ponds in other communes with a greater range of access to perennial water. Pond ownership frequency varies between less than 10% to more than 70% of households in North East Thailand (Table 5.4; Demaine *et al.*, 1999; Little *et al.*, 2004,) depending on specific agro-ecological conditions. At research sites in two provinces in Cambodia, ownership of ponds varied from less than 5% for trap ponds to nearly 90% for small household ponds in low-lying areas. In an area of West Bengal an audit of pond access revealed highly variable levels of access in nine communities (Figure 5.5a; Barman *et al.*, 2006). In Bangladesh, broader development and efforts by promoters appear to be increasing aquaculture as an opportunity among poorer landowners previously excluded from the activity. It appears that use of small ponds is now relatively common among poorer landowners as well as by the better-off in Mymensingh where pond-based aquaculture has become well-established (Karim, 2006). Better availability of juveniles of a range of fish species suited to small ponds, improved design of small ponds that integrate within rice-based systems and the high value of fish appear to support this trend in several parts of Bangladesh.

The basis of ownership and access can also be variable with both multiple and single ownership being common in West Bengal (Figure 5.5c) and Bangladesh. The outcomes suggest that

Table 5.4. Percentage of households that managed different farmer managed aquaculture systems (FMAS) in rural areas of South East Cambodia (SEC), North East Thailand (NET) and the Red River Delta, Vietnam (RRD) by category of well-being (Little et al.*, 2004).*

	SEC		NET		RRD	
System types	Poor (n=29)	Better-off (n=14)	Poor (n=12)	Better-off (n=17)	Poor (n=25)	Better-off (n=17)
Rice field	100.0	100.0	100.0	100.0	100.0	100.0
Culture pond	0.0	0.0	8.3	41.9	52.0	88.2
Household pond	72.4	78.6	0.0	0.0	0.0	0.0
Trap pond	17.2	7.1	91.7	88.2	0.0	0.0
Ditch	55.2	64.3	0.0	0.0	0.0	0.0

Note: Each column does not add up to 100 percent as households have multiple FMAS.

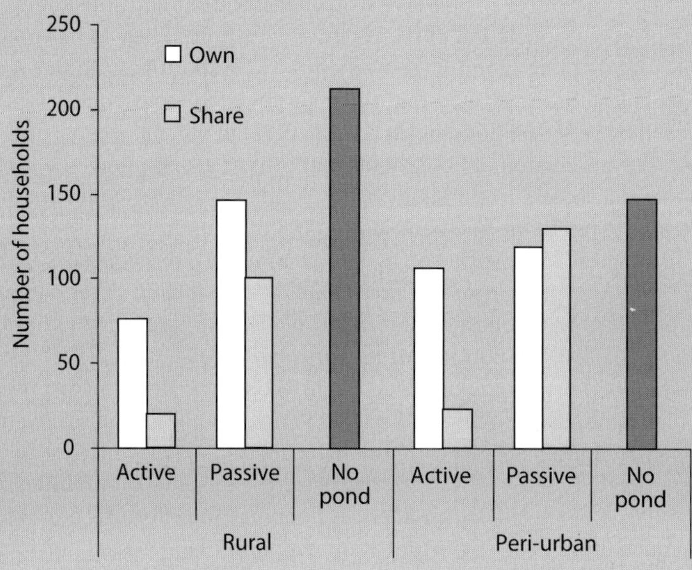

Figure 5.6. Status of pond ownership (individual (own) or shared (share)) in relation to use (active or passive) and location (rural or peri-urban) in Mymensingh district, Bangladesh.

whereas single ownership may prove to be a constraint to intensification and integration in West Bengal, the opposite was found in Mymensingh, Bangladesh where active integration of pond water use was more common among single household owned ponds (Figure 5.6).

In multiple ownership or common property contexts social complexity can undermine any intensification of pond management (Murray, 2004). The transaction costs required to organise collective effort for stocking, guarding and equitable benefit sharing are major, often overlooked challenges in such systems and need to be supported by well developed local institutions or poor sustainability and wider conflicts are a major threat. Lewis (1997) also found that joint ownership had been a disincentive to investment in aquaculture in Bangladesh leading to passive types of management, although the experience of NGOs indicates that these problems are not insurmountable.

Access to direct benefits is gained through share cropping and leasing in some areas and appear to reflect a growing importance of fish culture as a commercial opportunity. The practice is particularly common where clusters of juvenile fish production exist and where short-crop production can yield appreciable income. In West Bengal, poorer people tend to manage leased ponds more intensively than the better-off, who retain a higher proportion of their crop for subsistence (Barman *et al.*, 2006). The role of ponds and pond water in the development of social capital may be important; one example of this is poorer people accessing better off household ponds for the purpose of short-term cage culture for fingerling production in North West Bangladesh (Barman *et al.*, 2002). The relationships between access to, investment in, and benefit from ponds are important and are discussed below.

Investment

Understanding the behaviour of households investing in ponds is critical to identification of constraints and opportunities for pond utilisation. Community or group based management faces constraints that appear to become magnified as numbers and heterogeneity of stakeholders and size of resource increase. Thus the variability in sustained stocking and use of larger, leased ponds ('tanks') in Erode District Tamil Nadu, India, reflects the challenges faced in managing traditional common pool resources more intensively (Table 5.4). In an action research study in North West province Sri Lanka, only simple enhancements of small seasonal tanks by poor kinship groups living close to the resource proved successful (Murray, 2004).

For single household managed ponds the quality and availability of key inputs such as seed and nutrients has been identified as a major constraint to improving productivity (Little *et al.*, 2002; Litdamlong *et al.*, 2002; Edwards and Allen, 2001). These factors do not solely explain the wide range of input levels common among households with ponds in the same vicinity however. Similarly the level of integration that occurs between ponds and associated farming systems is often highly variable suggesting that ponds are viewed, and used, quite differently even within similar contexts. In a study in Mymensingh District, Bangladesh and Sisaket, North East Thailand the relative level of integration between ponds and associated agriculture (i.e. 'passive' or 'active' integration) tended to reflect the household resource base, with 'active' integrators tending to be relatively 'asset' rich with better access to information, credit and supportive institutions. At a lower level of intensification, pond based aquaculture has often not been dependent on credit for inputs (Lewis, 1997) although this may reflect the sub-optimal levels of feed and fertiliser used in such systems. Access to information is critical for take-up of intensified management of pond-based aquaculture but there is often little direct contact between formal promoters and farmers. The quality and consistency of local institutional support can affect decisions to invest in aquaculture. Communes with well developed commercial pond aquaculture near Hanoi often have supportive administrators with a background in aquaculture. Security of land tenure can also be a major factor; risk of non-renewal of land leases has affected interest in investment in Vietnam for example. Learning from other farmers and through informal knowledge brokers such as seed traders and other input providers are the routes through which most knowledge reaches farmers (Karim, 2006). This probably explains the geographical clusters of pond development that occur where pond management has been intensified, and in results of studies in North East Thailand and Bangladesh showing that proximity to urban areas has a major effect of pond-dyke design and management (Karim, 2006; Turongruang *et al.*, 2004).

Broader development and demography

Broad economic trends and the profile of producer households are both important to understanding the relative utility of ponds across Asia. A study of pond-owning households in Sisaket, North East Thailand found incomes to be dominated by off-farm earnings with less than 30% of total household income derived from agriculture (Turongruang *et al.*, 2004). Unsurprisingly, if household income was disaggregated by age, older people tended to dominate agricultural income. In contrast in Mymensingh, Bangladesh, more than 70% of household incomes are derived from farming. Such different profiles are likely to affect householder expectations of, and capacities for, pond development. In general, ease of access to alternative employment is a critical factor in changing agrarian livelihoods. Rapidly growing urban centres and associated transport links are providing a new level of dynamism in many places where aquaculture is expanding. Indeed, urbanisation has been identified as a key stimulus to aquaculture in Asia (Little and Bunting, 2005). Pond development, like many enterprises, often occurs in geographical 'clusters' to service the high demand in urban areas. Households located closer to urban areas appear to be more likely and able to intensify fish production. Karim (2006) found that communities with better linkages to inputs and markets tended to have higher yields. In Vietnam Luu *et al.* (2002) found that ponds in suburban communes were higher yielding than those located in rural areas and related this to use of larger, higher quality fingerlings, higher densities of fish and better management. In a study of three Provinces in southern Laos Bush (2004) found that although most producers were subsistence orientated those producing with an exclusive commercial orientation were closest to market towns. Pant *et al.* (2005) reported that a combination of the resource base and market accessibility affected the development of integrated pond culture in North East Thailand.

Motivations

Interest in developing ponds is rarely linked exclusively to their use for fish culture, although once established there is a tendency for the relative importance of fish culture to increase,

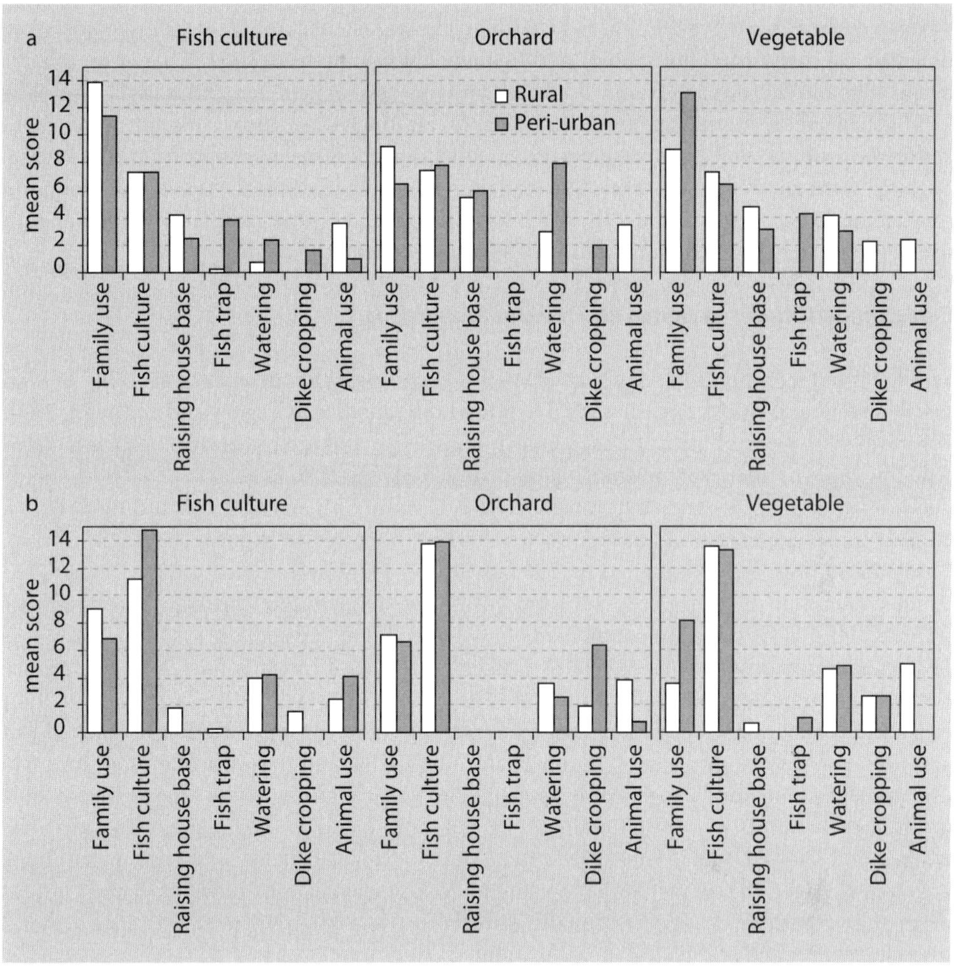

Figure 5.7. Importance of ponds to households focusing on fish, orchard or vegetable production in Mymensingh district, Bangladesh at time of development (top) and currently (bottom) (Karim, 2006).

whatever the original motivations. In Mymensingh, Bangladesh, an area where aquaculture has become increasingly commercially orientated in recent years (ADB, 2004), fish culture is now the most important reason for pond ownership (Figure 5.7). In Sisaket Province, North East Thailand, the primary motivation for pond construction was, and remains, on-farm water storage to support associated horticulture and despite availability of seed and other inputs, culture and trapping of fish remains secondary. Other studies in North West Bangladesh (AIT/DOF, 2000) and North East Thailand (AIT/DOF, 2000) indicated the subsistence orientation of most pond-based fish culture, although recent field work suggests this is also changing rapidly as marketing opportunities expand.

Taken together, this evidence suggests that meeting subsistence needs for fish remains a key objective for many rural households and that any level of intensification is based on relative availability and costs of fish from wild stocks and markets. Where demand is high, especially near urban areas, intensification is most likely, although households may become more sensitive to opportunity and labour costs as off-farm employment opportunities increase. Such off–farm factors are also critical in explaining household decisions regarding their pond management. Given this context there are several key factors that can explain differences in use of ponds and their role in household livelihoods around the region.

Explaining attitudes to ponds and their management

Limited interest in pond-based aquaculture has been linked to adequate availability of wild fish where natural stocks remain critical to household nutrition (Gregory and Guttman, 2002; Murray, 2004). Lack of institutional support, knowledge and credit has also been associated with low levels of interest in optimising fish productivity but there is evidence that households decisions to manage systems extensively are based more on perceived risk and uncertainty. Such attitudes are common among practitioners of low external input technologies (Tripp, 2006). Action research conducted by Turongruang *et al.* (2004) in North East Thailand led to the conclusion that risks of losses from both flood and drought constrain investment by farmers in the fish production component of integrated pond-horticulture systems. Risk of loss of stock and investment in feed/fertilisers and a requirement for early harvest at a small size through lack of water both undermine confidence. Bush (2004) described a similar situation in southern Laos. Ponds are still considered useful for extensive production of stocked fish and to trap and store wild stocks, however, and in NE Thailand these are often sold for cash. The perception of ponds as valuable assets is particularly strong, even among households locating their ponds in sub-optimal sites or pursuing off-farm employment. In situations where pond construction is subsidised or can be achieved at very low cost these perceptions are often stronger. In contrast, in Mymensingh, declining availability of wild fish and cultural aversion to consumption of other aquatic animals creates a much higher subsistence demand for cultured fish and stimulates intensification.

An important aspect of the innovation expressed by households in these different areas of Asia is that they focus not on fish culture alone, but rather on a 'package' which includes use of the dykes for associated horticulture and management of wild stocks of aquatic animals in and around the pond. In this way ponds have developed as part of an integrated livelihood system. The ease with which these activities complement off-farm labour is also often critical. The relationships between components of the package are all important and can explain why adoption is often partial (Smale and Heisey, 1993).The importance of off-farm activities to household livelihoods in many contexts has been found to affect the assessment of risk, investment and adoption in general (Kebede, 1992).

Income, food security, employment

Ponds can be important for reducing the seasonal income and consumption vulnerability common in rain-fed agricultural environments. Furthermore, employment opportunities are not only created in production but also in service roles prior to and after harvest regarding inputs and supply and marketing of products respectively particularly as culture intensifies. Unfortunately there is relatively little information on the intra-household impacts of ponds and intensification of pond-based activities.

Food produced in association with ponds needs to be placed in the context of overall diets if its implications for the quality of the diet are to be understood. Reliance on freshwater fish and other aquatic animals to complement diets based on rice and vegetables among Asian rice-growing households may appear self-evident but researchers have often failed to acknowledge contributions from the immediate rice agro-ecosystem, and these may also be under-appreciated by local people (Burlingame *et al.*, 2006). Recent field studies although typically limited by geographical scope have provided more precise evidence (e.g. Gregory and Guttman, 2002; Meusch *et al.*, 2003; Morales *et al.*, 2004). Formal data collection at national and international level, often based on simplistic aggregation of market data, has typically failed to capture the nutritional significance of managed aquatic resources. Data from monitoring household consumption has given new insights into the nutritional role of pond produced fish and other food products and, especially if households are disaggregated by wealth, into the relative importance to the poor. Households that 'actively' integrated fish and vegetable production together through use of pond water for irrigation produced more of both: 33% and 84% higher respectively than passive households in Bangladesh, for example (Karim, 2006). Production and sale of vegetables from households with ponds exceeded the level of that from households without ponds in Bangladesh. There was little difference in overall household consumption however (Figure 5.8; Karim 2006), since, as with fish, active integrating households tended to benefit mainly by selling vegetables rather than from eating

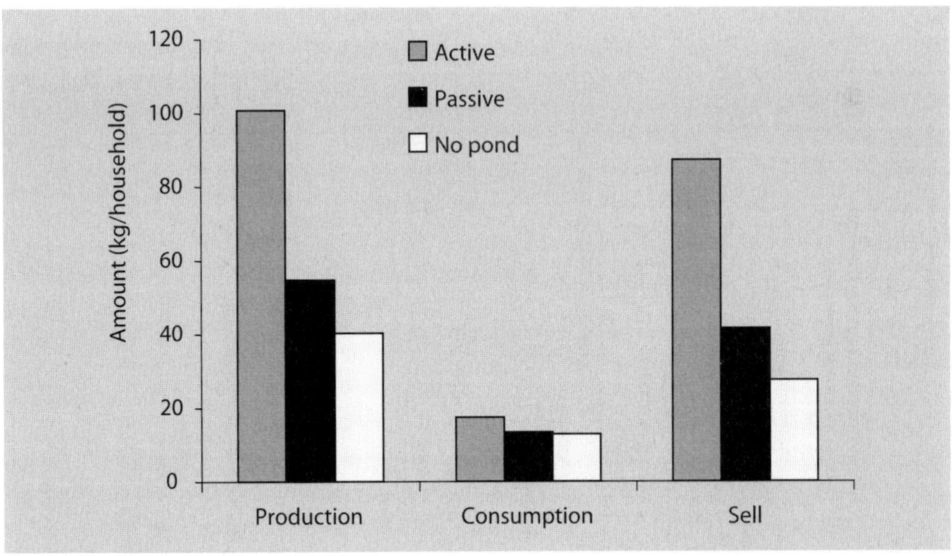

Figure 5.8. Vegetable production, consumption and sales for households with ponds practicing active and passive integration between pond and vegetable production and without ponds (No pond) in Mymensingh, Bangladesh (Karim, 2006).

more themselves. Bouis (2000) found that neither fish nor vegetable consumption increased in project households where aquaculture and commercial vegetable production had been extended in Bangladesh. He interpreted these results in terms of a lack of desire to increase consumption of vegetables.

Consumption of fish and meat/poultry is considerably higher in North East Thailand than Bangladesh but there are significant levels of seasonality in both countries, reflecting availability and price. Poorer people are relatively more dependent on aquatic food than the better-off for most of the year in Thailand. Both wealth groups are simultaneously reliant on fish and substitutes in Bangladesh, where ponds are a particularly important source of fish during the 'hungry' gap. Poorer producers consume less fish overall, and rely relatively more on wild fish from off-farm sources. In overall dietary terms, however, ponds are more important as sources of fish for home consumption for poorer than for better-off people.

The relative importance and complementarity of pond cultured fish for the producer household compared to fish purchased and caught from unmanaged stocks indicate that ponds contribute to improving food security (Figure 5.9). Catches from household ponds are low during periods when fish availability from unmanaged stocks is high.

Active pond owners in Mymemsingh, not only sell more fish and vegetables, but also consumed more wild fish than the other groups suggesting that income from home produced food encourages households to use this asset strategically (Karim, 2006). Clearly then, ponds are rarely the only source of fish for those who access them and their ownership meets a broad set of livelihood objectives. Prein and Ahmed (2000) highlight the importance of selling home produced fish to support the purchase of other food items. In contrast Bouis (2000) notes that

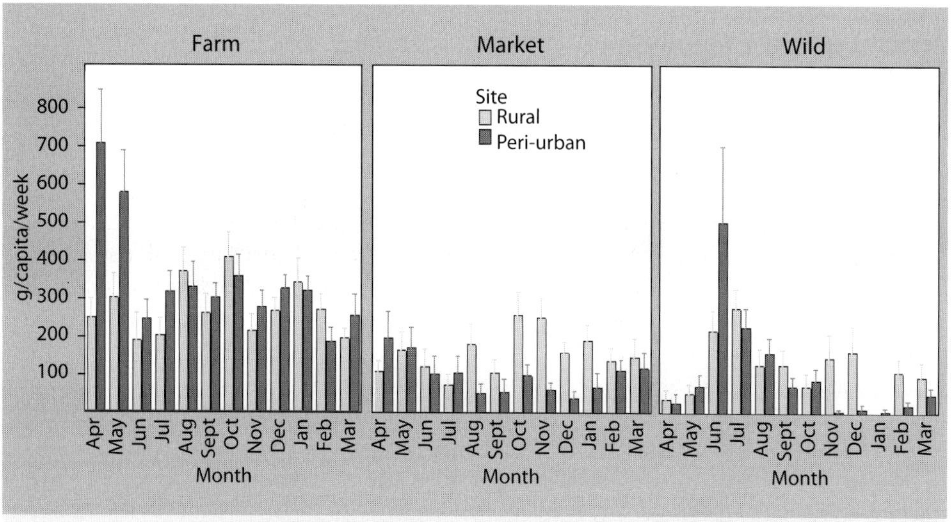

Figure 5.9. Consumption of freshwater fish by source (farm, market or wild) for households located in rural and peri-urban communities in Mymensingh, Bangladesh by month.

producers surveyed in Bangladesh tend to consume more of their own large fish and reduce purchase and consumption of smaller wild fish. He raised the possibility that as a consequence of small fish being more valuable nutritionally, diet quality might actually deteriorate among pond culturists, and that the culture of small fish should be promoted.

As pond culture expands and wild catch contracts, the changing size and origin of fish will have potentially important implications, especially for poor consumers. Evidence shows that markets for small cultured fish have expanded rapidly in many areas encouraging producers to manage their systems accordingly. The intermittent harvest of 'overstocked' ponds, while contrary to conventional extension advice, reduces risk, increases yield and results in larger quantities of small fish, which are affordable to the poor compared to a single harvest of larger sized fish (Barman *et al.*, 2004). Data from both Bangladesh and West Bengal suggests that substitution of low cost wild fish by small pond reared silver carp and tilapia is increasingly common (Thompson *et al.*, 2005; Barman *et al.*, 2006).

Pregnant women and young children are most at risk from deficiencies in critical micronutrients that fish, especially some species of small indigenous fish, and vegetables contain (Roos *et al.*, 2003). The protection these types of food offer against vitamin A and calcium deficiency is particularly important for young children (Ahmed, 1999, 2000). Fish are typically the major source of essential fatty acids which are important in the development of brain and neural tissue (Lauritzen *et al.*, 2001). Intra-household studies have shown that males, even young boys have disproportionately higher access to fish and other quality foods than women (Bouis, 2000; Faruque *et al.*, 2006). The relative impact of high quality foods to impoverished diets may therefore have important implications for livelihoods of both current and future generations. Inter and intra-household variation in access to such food on the basis of wealth and gender means that increased production on-farm is most likely to ensure basic consumption needs are met.

Employment

In areas where pond culture is well established and commercially orientated, its benefits are often less seasonal and result in more consistent income through the year. Poor people employed as labourers around the ponds or within marketing and processing networks particularly benefit from such patterns of production. There is evidence from areas of intensified pond production in Bangladesh and Vietnam which indicates that poor groups benefit through employment gains. In Bangladesh both agricultural workers and fishers have benefited from higher wages and greater and more diverse work opportunities that have arisen as a result of market-driven pond aquaculture (Faruque *et al.*, 2006). In Vietnam urban-driven aquaculture development has stimulated employment and enterprise in both upstream and downstream service sectors, with rapid growth occurring in seed production and distribution, marketing and, increasingly, processing of fish (Young *et al.* 2006). Observation also suggests that many poorer people become involved in aquaculture through opportunity rather than necessity. Some may adopt these livelihood strategies to alleviate seasonal unemployment, but involvement in aquaculture has allowed others to accumulate capital; either by becoming producers themselves and/or by entering into less risky value added forms of marketing. In Bangladesh examples of such strategies include fry traders becoming nursery operators (Haque *et al.*, 2006) and fishers becoming *gher* (integrated farm) and *depo* (wholesale market) owners (Faruque *et al.*, 2006).

Sustainability of pond-based livelihoods

Widespread pond construction at the household level may have overall negative impacts if ponds facilitate the resource capture of land, water, nutrients and aquatic animals by the better-off. Lewis (1997) identified this as a major issue in Bangladesh where access to water bodies by groups of poor, landless people, often facilitated by NGOs, is not always sustained. A key issue is if poor people can 'hold onto their niche', whether it be the individual use of ditches and small ponds or group access to larger water bodies that in practice are controlled by the local power structure. Furthermore, in places where the growth of a competitive commercially orientated aquaculture industry is clearly underway, the sustainability of fish sales from small ponds becomes questionable.

Comparison of the situation around Asia suggests that while pond-based diversification is currently making valuable contributions to farm-based livelihoods, there will be trade offs in the way ponds are managed if off-farm labour demand grows. As the relative importance of off farm employment increases and less time is spent on farm by the most active household members, connections with the natural resource base inevitably weaken (Ellis, 1998). In North West Bangladesh it was found that middle-class households whose participation in off-farm activities was limited tended to optimise their pond management. Whereas poorer people with less land were forced to seek off-farm employment and the better-off diversified into value added activities, both resulted in poorer pond production outcomes (Barman, 2004).

In evaluating the retention of pond-based aquaculture among its adopters in Guatemala, Lovshin *et al.* (2000) found reduced interest after ten years, in part because children had left home and remitted urban wages to support older family members left on the farm. Under these conditions there was less need for maintaining fish yields and less labour to do so. In contrast, the integration of horticulture and livestock around small, deep ponds is particularly popular with older people in North East Thailand, as a method to save time and labour. Daily subsistence needs for vegetables, herbs and spices can be met reducing time and effort to gather such products from a dwindling natural resource base (Little and Edwards, 2003).

The pace of overall development will clearly have an important effect on the retention of ponds within household livelihoods. A comparison of household profiles of those employed in urban aquaculture in four cities in South East Asia found that where the wider economy has been most dynamic (i.e. Bangkok compared to Ha Noi, Ho Chi Minh City or Phnom Phen) over the longest period, the relative education level of those engaged in aquaculture was relatively lower suggesting its declining attraction for the most economically and socially mobile. Pond-based activities have probably rarely eliminated underemployment but rather been part of a broader livelihood portfolio. Ruddle and Zhong's (1988) assessment of labour use found that traditional pond-dyke systems in the Zhujiang Delta, China, only ever employed around 50% of available labour. This availability of labour possibly contributed to its development as a major manufacturing hub, although perhaps inevitably leading to a decline in the relative importance of ponds to overall livelihoods; as little as half the original area of integrated pond-dyke now remains (Ho, 2006).

Pond-based aquaculture, often integrated with other components on the farm, has been frequently promoted as part of a low risk developmental strategy for improving and diversifying household food security and rural livelihoods, particularly in marginal rain-fed agro-ecosystems (e.g. AIT, 1994; Prein, 2002). In development terms it has therefore been located within the norm of using small farms as an entry point to poverty reduction (Ellis, 2000). Commercialisation and specialisation threaten such multipurpose roles of ponds however. The intensive production of fruit trees around fishponds reducing light and aquatic productivity in southern Vietnam was found to limit opportunities for close integration (Nhan *et al.*, 2007). The challenge of optimising both livestock and fish production cycles within integrated systems becomes problematic as, for example, in Central Thailand where recent Avian Influenza outbreaks led to drastic changes in pond management as use of poultry waste became restricted (Belton, 2006). Competition with cage culture has also emerged although this approach generally has a higher, less flexible production cost and carries significantly greater risk. In southern Vietnam, pond production of *Pangasius* spp. appears to be growing at the expense of traditional cage culture.

Export-driven opportunities are offering new opportunities for the use of ponds. The intensive river catfish industry in the Mekong delta, Vietnam, and extensive production of giant freshwater prawn in South East Bangladesh are examples of the rapid change that a global relative advantage can bring. In both cases significant local wealth creation has occurred throughout the complex production and marketing networks that have grown up around the industries. Longer term trends in environmental quality are a potential threat, although they are yet to have a critical impact on production success. Dependence on highly specialised distant markets also imposes a range of specific risks, such as the trade dispute between US and Vietnamese catfish growers. This led to significant hardship within the Vietnamese industry, but also proved its resilience as it went on to seek out and penetrate new markets (Young *et al.*, 2006).

Take home points

- The rapid spread of ponds throughout Asia has only partly reflected interest in farming fish. Improved water availability is a critical part of development for marginal agro-ecosystems and a good deal of 'pond' construction has occurred as an outcome of the growth of human settlements, flood protection and road networks.
- The contribution of ponds to aquatic food availability in Asia has increased substantially during the last decades with implications for household food security and, increasingly, for income generation.
- The management of ponds, and livelihood outcomes that result, reflect a heterogeneity in agro-ecosystems and broader resource systems in which they are located.
- Ponds are part of households' capital assets which are managed strategically to meet broader livelihood needs. In general the intensity and level of integration of fish culture with other on-farm components reflects local opportunities and constraints.

- Ponds may reduce vulnerability and enhance resilience in several ways. They may have particularly important roles in marginal, seasonal contexts.
- Poorer people benefit from pond development in a wide range of ways: as producers, intermediaries and consumers.
- There is level of convergence in ponds, both those well established as multipurpose resources, and purpose-built fishponds, towards more productive fish culture, reflecting decline in availability of wild fish and returns from fish sales.
- Ponds may enhance local biodiversity through their management at the interface of household and common property and a common desire to maintain and produce both stocked species and natural stocks.
- Urbanisation and urban markets are major drivers of aquaculture development and ponds are the major source of fish supply in many urban situations.
- Pond development occurs mainly through private sector mechanisms. Maintaining access to ponds that are common property resources may be critical if the poor are to benefit directly, and often requires external institutional support.

References

ADB, 2004. Small-scale freshwater rural aquaculture development for poverty reduction. Special evaluation study. Operations Evaluation Department, Asian Development bank, Philippines, 67 pp.

Adger, W.N., 2000. Social and ecological resilience: are they related. Progress in Human Geography 24, 347-364.

Ahmed, F., 1999. Vitamin A deficiency in Bangladesh: a review and recommendations for improvement. Publ. Health Nutr. 2, 1-14.

Ahmed, F., 2000. Anaemia in Bangladesh: a review of prevalence and aetiology. Publ. Health Nutr. 3, 385-393.

AIT, 1994. Partners in development, the promotion of sustainable aquaculture. Asian Institute of Technology, Bangkok, Thailand, 98 pp.

AIT/DOF, 2000. Fish seed quality in Northeast Thailand. State of the System report. Aquaculture Outreach Programme, AIT, Thailand, 31 pp.

AIT/DOF, 2001. Fish seed quality in Northwest Bangladesh. State of the System report. Aquaculture Outreach Programme, AIT, Thailand, 37 pp.

AIT/SAC, 2003. Improved resource use efficiency in Asian integrated pond-dike systems. State of the system report, Thailand, Asian Institute of Technology, Bangkok, Thailand, 26 pp.

Alteri, M.A.1999. The ecological role of biodiversity in agroecosystems. Agriculture, Ecosystems and Environment 74, 19-31.

Amilhat, E., E.J. Morales, A.J. Immink, D.C. Little, K. Lorenzen, F.U. Islam, I. Karapanagiotidis, 2005. Self-recruiting species (SRS) from farmer managed aquatic systems: their role in rural livelihoods. AFGRP-DFID summary report, 12 pp.

Amilhat, E. and K. Lorenzen, 2005. Habitat use, migration pattern and population dynamics of chevron snakehead *Channa striata* in a rainfed ricefarming landscape. J. Fish. Biol. 67, 23-34.

Arjinkit, C., R.H. Bosma, D.Turongrouang, D.C. Little, and M. Verdegem, 2006. On-farm trials of improved nursing and fattening sex-reversed gift tilapia in Sisaket province, Northeast Thailand. In: Proc. Aqua 2006, Florence, abstract no 992. accessed on 26-10-2006 http://www.was.org/meetings/abstractData.asp?Abstractld=11270.

Barman, B.K., 2000. Assessment of Nile tilapia (*Oreochromis niloticus*) Seed Production and Grow-out Systems for Small-scale Farmers in Northwest Bangladesh. PhD Dissertation. Asian Institute of Technology, Thailand, 246 pp.

Barman, B.K. and D.C. Little, 2004. Household nutrition of street workers in northwest Bangladesh: focusing on the role of fish in their diets. Working paper. Institute of Aquaculture, University of Stirling UK, WorldFish Center Bangladesh, and South Asia Office Bangladesh.

Barman, B.K., D.C. Little and J. Janssen, 2004. Tilapia and Common carp Fingerling Production in Rice Fields of Farmers in Northwest Bangladesh: The Adoption and Dissemination Strategies. Working Paper 10. Institute of Aquaculture, University of Stirling, UK and WorldFish Centre Bangladesh, p. 43.

Barman, B.K., D.C. Little and N. Kundu, 2005. Studies of fish markets, hatcheries, nurseries and grow-out farmers in West Bengal, India, Working Paper, Institute of Aquaculture, University of Stirling UK, Institute of Wetland Management and Ecological Design India and WorldFish Center Bangladesh.

Barman, B.K., D.C. Little and N. Kundu, 2006. Marketing of aquatic products and its impacts on livelihoods of people in West Bengal, India. Paper presented in a seminar on Aquaculture Production and Marketing 2020, Dhaka 20 March 2006. Bangladesh.

Barman, B.K., A. MacNiven and D.C. Little, 2002. Comparison between the performance of overwintered and new season silver carp, *Hypophthalmichthys molitrix* in an on-farm grow-out trial with small-scale farmers in Northwest Bangladesh. Working Paper. Institute of Aquaculture, University of Stirling, UK and Northwest Fisheries Extension Project (NFEP-2) and Asian Institute of Technology, Thailand. 12 pp.

Belton, B., 2006. An assessment of the sustainability of tilapia culture in central Thailand. MSc Dissertation, University of Stirling, UK, 118 pp.

Bouis, H.E., 2000. Commercial vegetable and polyculture fish production in Bangladesh: their impacts on household income and dietary quality. Food and nutrition Bulletin, 21, 4, 482-487.

Burlingame, B., R. Charrondiere and M. Halwart 2006. Basic human nutrition requirements and dietary diversity in rice-based aquatic ecosystems. Journal of Food Composition and Analysis 19, 770.

Bush, S.R., 2003. Give a man a fish. Contextualising living aquatic resource development in the lower Mekong Basin. Australian Mekong Resource Center. Working Paper Series No. 8.

Bush, S.R., 2004. A political ecology of living aquatic resources in Lao PDR. PhD dissertation. University of Sydney, Australia, 359 pp.

Carney, D., 1998. Implementing the sustainable rural livelihoods approach. In: Carney D. (ed.), Sustainable Rural Livelihoods, What Contributions Can We Make? DFID, London, UK. pp. 3-23.

Chambers, R. and G. Conway, 1992. Sustainable Rural Livelihoods: Practical Concepts for the 21st Century. IDS discussion paper No 296 IDS, UK.

Demaine, H., N.L. Innes-Taylor, D. Turongruang, P. Edwards, D.C. Little and J. Pant. 1999. Small-scale aquaculture in Northeast Thailand. A case study from Udorn Thani.Studies. In Agricultural and Aquatic Systems 2. Aquaculture and Aquatic Resources Management Program, AIT, Thailand. 86 pp.

Edwards, P., 1985. Pigs over fish-ponds. Pig International 15, 8-11.

Edwards, P. and G. Allen, 2001. Review of feed and feeding for inland aquaculture in the Mekong region Countries, ACIAR.

Ellis, F., 1998. Livelihood diversification and sustainable rural livelihoods. In: Carney D. (ed). Sustainable rural livelihoods. What contribution can we make? DFID, UK. pp. 53-66.

Ellis, F., 2000. Rural livelihoods and diversity in developing countries. Oxford University Press, UK, 273 pp.

Faruque, G., D.C. Little, J.A. Young and M.M. Dey, 2006. Impacts of the production and marketing of freshwater aquaculture products on stakeholders in Bangladesh. Paper presented in a seminar on Aquaculture Production and Marketing 2020, Dhaka 20 March 2006, Bangladesh.

Gregory, R. and H. Guttman, 2002 The ricefield catch and rural; food security. In: P. Edwards, D.C. Little and H. Demaine (eds.), Rural Aquaculture. CABI Publishing, UK. pp. 1-14.

Garaway, C.J., 1999. Small waterbody fisheries and the potential for community-led enhancement: case studies in Lao PDR. PhD thesis, T.H. Huxley School for the Environment, Earth Sciences and Engineering, Imperial college of Science Technology and Medicine, University of London, UK.

Haque, M.M., B.K. Barman, D.C. Little and M.A. Wahab, 2006. Fish seed production in irrigated rice-fields: Livelihood impacts and factors affecting adoption and rejection process at household level. Paper presented in the International Seminar on Aquaculture Production and Marketing: 2020 held at BRAC Inn, Dhaka, Bangladesh, 20 March 2006.

Harrison, E., 1994. Aquaculture in Africa: socio-economic dimensions. In: Muir, J.P. and R.J. Roberts (eds.). Recent Advances in Aquaculture V. Blackwell Science, UK, pp. 240-300.

Ho, M.W. 2006 Circular economy of the dyke pond system. Accessed on 20/11/06 at http://www.i-sis.org.uk/DykePondSystem.php.

Islam, M.M., 2002. An Analysis of the Role of Extension Methodology Poverty Reduction: A Comparative Study of Aquaculture Extension Programmes in the Northwest Fisheries Extension Project (NFEP) Command Area, Bangladesh. PhD Thesis, University of Wolver Hampton, UK, 300 pp.

Islam, F.U., A. Immink, K.C. Shaha, S. Islam, A. Masud and D.C. Little, 2003. Self-recruiting species in aquaculture-their role in rural livelihoods: a case study from South-central and North-west Bangladesh. In: Wahab M.A., S.H. Thilsted and M.E. Hoq (eds.), Small Indigenous Species of Fish in Bangladesh: Culture Potentials for Improved Nutrition and Livelihoods, pp.103-110.

Karim, M., 2006 The livelihood impacts of fishponds integrated within farming systems in Mymensingh District, Bangladesh PhD Dissertation, University of Stirling, 351 pp.

Kebede, Y., 1992. Risk behaviour and new agricultural technologies: the case of producers in the Central highlands of Ethiopia. Quarterly Journal of International Agriculture 31, 269-284.

Lauritzen, L., H.S. Hansen, M.H. Jorgensen, K.F. Michaelsen, 2001.The essentiality of long chain n-3 fatty acids in relation to development and function of the brain and retina. Prog. Lipid Res. 40, 1-94.

Lewis, D., 1997. Rethinking aquaculture for resource-poor farmers: perspectives from Bangladesh. Food Policy 22, 533-546.

Lithdamlong, D., E. Meusch and N. Innes-Taylor. 2002. Promoting aquaculture by building the capacity of local institutions: developing seed supply networks in the Lao PDR. In: P. Edwards, D.C. Little and H. Demaine (eds.), Rural Aquaculture. CABI Publishing, UK. pp. 155-166.

Little, D.C., 1994. The development of small-scale poultry-fish integration in Northeast Thailand: potential and constraints. In: J.J. Symoens and J.C.Micha (eds.). Proceedings of a seminar `The management of integrated freshwater agro-piscicultural ecosystems in tropical areas', 16-19th May 1994, Brussels, Belgium.Technical Centre for Agricultural and Rural Co-Operation (CTA), Royal Academy of Overseas Sciences. Belgium. pp. 265-276.

Little, D.C. and S.W. Bunting, 2005. Opportunities and constraints to urban aquaculture, with a focus on south and southeast Asia. In: Costa Pierce, C.A., P. Edwards, D. Baker and A. Desbonnet (eds.), Chapter 3 Urban Aquaculture, CABI, Publications, UK.

Little, D.C. and P. Edwards, 2003. Integrated Livestock-Fish Farming Systems: The Asian experience and its relevance for other Regions. Inland Water Resources and Aquaculture Service and Animal Production Service. Food and Agriculture Organization, Italy. 177 pp.

Little, D.C. and L.J. Pollock, 2003. Integration of Aquaculture within Irrigation Systems, Final Technical Annex (Aquaculture), DFID KAR Project R7123, Institute of Aquaculture, University of Stirling, Scotland, U.K. 92 pp.

Little, D.C., K. Lorenzen, E. Amilhat, E. Morales, A. Immink, F. Ul Islam, I. Karapanagiotidis, 2004. Self recruiting species in aquaculture: their role in rural livelihoods Final technical Report, AFGRP/DFID, UK. 73 pp.

Little, D.C., A. Satapornvanit and P. Edwards, 2002. Freshwater fish seed in Asia. In: P. Edwards, D.C. Little and H. Demaine (eds.), Rural Aquaculture. CABI Publishing, UK. pp. 185-196.

Little, D.C., P. Surintaraseree and N.L. Innes-Taylor, 1996. Fish culture in rainfed ricefields of Northeast Thailand. Aquaculture 140, 295-321.

Luu, L.T., P.V. Trang, N.X. Cuong, H. Demaine, P. Edwards and J. Pant, 2002. Promotion of small-scale pond aquaculture in the Red River Delta. In: P. Edwards, D.C. Little and H. Demaine (eds.), Rural Aquaculture. CABI Publishing, UK. pp. 55-76.

Lovshin, L., N. Schwartz and U. Hatch, 2000. Impacts of Integrated Fish Culture on Resource Limited Farms in Guatemala and Panama: An Ex-post Evaluation. Research and Development Series No. 46. Agricultural Experiment Station, USA, 29 pp.

Meusch, E., J. Yhoung-Aree, R. Friend and S. Funge-Smith, 2003. The role and nutritional value of aquatic resources in the livelihoods of rural people. A participatory assessment in Attapeu Province, Lao PDR. FAO & IUCN. Thailand. accessed on 15/11/06 http://www.iucn.org/themes/wani/pub/FAOIUCNLaoResourcesLivelihoods.pdf

Micheilsons, C.G.J., K. Lorenzen, M.J. Philips and R. Gauthier, 2002. Asian carp farming systems: towards a typology and increased resource use efficiency. Aquaculture Research 33, 403-413.

Morales, E.J., D.C. Little and H. Demaine, 2003. Participatory approaches to define the role of self-recruiting species in aquaculture on rural livelihoods. In: Wahab, M.A., S.H. Thilsted and M.E. Hoq (eds.), Small Indigenous Species of Fish in Bangladesh: Culture Potentials for Improved Nutrition and Livelihoods.

Morales, E.J., D.C. Little, A.J. Immink, E. Amilhat, H. Demaine, A. Yakupitiyage and K. Lorenzen, 2004. Contribution of Self-recruiting species (SRS) produced in farmer-managed aquatic systems (FMAS) in rural areas of southeast Asia to food consumption. FAO Technical Expert Meeting on Aquatic Biodiversity, it's nutritional composition, and human consumption in rice-based systems, December 8–10, 2004. FAO Regional Office for Asia-Pacific (RAF) in Bangkok, Thailand.

Murray, F.J. 2004. Potential for Aquaculture in Community-Managed Irrigation Systems of the Dry-Zone, Sri Lanka: Impacts on Livelihoods of the Poor. PhD dissertation. University of Stirling. 502 pp.

Nhan, D.K, L.T. Phong, M.J.C. Verdegem, L.T. Duong' R.H. Bosma and D.C. Little, 2007. Integrated freshwater aquaculture, crop and livestock production in the Mekong Delta, Vietnam: Determinants and the role of the pond. Agriculture Systems.

Nguyen Khoa, S., K. Lorenzen, C. Garaway, B, Chamsingh, D.J. Siebert and M. Randone, 2005. Impacts of irrigation on fisheries in rain-fed rice-farming landscapes. Journal of Applied Ecology 42, 892–900.

Pant, J., H. Demaine and P. Edwards, 2005. Bio-resource flow in integrated agriculture-aquaculture systems in a tropical monsoonal climate: a case study in Northeast Thailand. Agricultural Systems 83, 203-219.

Phuong, N.T.D., P.A. Tuan, N.T.H. Tien, P. Bau, H.K. Diep and N.T. Tan, 2006. Report on Baseline and Monitoring Survey in aquatic producer households in peri-urban Hanoi. Accessed on 15/11/2006 at http://www.papussa.org/publications/bandm_surveyreport_hanoi.pdf. 59 pp.

Prax, J., D.C. Little, P.V. Sao and P.A. Tuan, 2000. Seed production and distribution network in Mao Dien community, Red River Delta, Northern Vietnam. Working Paper. Institute of Aquaculture, University of Stirling, UK.

Prein, M., 2002. Integration of aquaculture into crop-animal systems in Asia. Agricultural Systems 71, 127-146.

Prein, M. and M. Ahmed, 2000. Integration of aquaculture into smallholder farming systems for improved food security and household nutrition. Food and Nutrition Bulletin 21, 466-471.

Roos, N., S.H. Thilsted and M.A Wahab, 2002. Culture of small indigenous fish species in seasonal ponds in Bangladesh: the potential for production and impact on food and nutrition security. In: P. Edwards, D.C. Little, and H. Demaine (eds.), Rural Aquaculture, CAB International, U.K., pp. 245-252.

Roos, N., M. Islam and S.H. Thilsted, 2003. Small fish is an important dietary source of vitamin A and calcium in rural Bangladesh. Int. J. Food Sci. Nutr. 54, 329-339.

Ruddle, K., 1996. Household and community factors affecting development of small-scale fish farming in Africa. In C. Bailey, S. Jentoff and O. Sinclair (eds.), Aquacultural development, social dimensions of an emerging industry, Westview Press, USA, pp. 211-232.

Ruddle, K. and G.F. Zhong, 1988. Integrated agriculture - aquaculture in south China: The dike-pond system of the Zhujiang Delta. Cambridge University Press. USA, 173 pp.

Ruxton, C.H.S., S.C. Reed, M.J.A. Simpson, and K.J. Millington, 2004. The health benefits of omega-3 polyunsaturated fatty acids: a review of the evidence. J. Hum. Nutr. Diet 17, 449-459.

Smale, H. and P. Heisey, 1993. Land allocation inn HYV adoption models: an investigation of alternative explanations. American Journal of Agricultural Economics 76, 535-546.

Surinteraseree, P. 1988. Rice-fish culture systems: a survey in Northeast Thailand. MSc. dissertation, Asian Institute of technology, Thailand, 117 pp.

Thompson, P., N. Roos, P. Sultana and S.H. Thilsted, 2002. Changing significance of inland fisheries for livelihoods and nutrition in Bangladesh. In: Kataki, P.K. and S.C. Babu (eds.), Food systems for improved human nutrition: linking agriculture, nutrition and productivity. Binghampton, New York: The Haworth Press, Inc., pp. 249–317.

Thompson, P.M, A.K.M. Sultana and F. Khan, 2005. Aquaculture extension impacts in Bangladesh: a case study from Kapasia, Gazipur. Penang, Malaysia: The WorldFish Center.

Townsley, P., 1998. Aquatic resources and sustainable rural livelihoods. In: Carney, D. (ed.), Sustainable rural livelihoods. What contribution can we make?.DFID, UK, pp.139-154.

Tripp, R., 2006. Self sufficient agriculture. Labour and knowledge in small-scale farming.Earthscan, UK, 242 pp.

Turongruang, D., F.J. Murray and D.C. Little, 2004. Interim assessment of PONDLIVE baseline study. Paper presented at the 7th Asian Fisheries Forum, December 2004.

Yoonpundh, R., 1997. The reproductive biology of snakeskin gourami (*Trichogaster pectoralis* Regan) and intensification of seed production. PhD dissertation. Asian Institute of Technology, Thailand.

Young, J.A., N.P. Son, and T.V. Binh, 2006. Positioning Pangasius on the global market - a case of diversification through adversity? Paper presented at the seminar on Aquaculture Production and Marketing 2020, 20 March 2006, Dhaka, Bangladesh.

Environment, governance and global markets

Simon R. Bush and Peter J.M. Oosterveer
Environmental Policy Group, Wageningen University, The Netherlands

Abstract

As the intensity of fish farming in South East Asia has expanded in response to rising global demand for fish, the resilience of the coastal environment and effectiveness of management practices to ensure economic efficiency, environmental protection and social equity have come into question. Trade that once involved a stepwise progression through multiple global, national, local scales are now organised flexibly and largely beyond the reach of the state. Innovative 'governance' arrangements are needed to respond to the increasingly global nature of fish trade while remaining relevant to the local conditions under which actual production takes place. Using two case studies from South East Asia this paper addresses how local and global governance arrangements influence the livelihood decisions of fish farmers and how better decisions can be made that facilitate both environmental and socially sustainable production. The paper concludes that more attention must be given to both the formal and informal arrangements over local production practice and trade must be seen in its global context, considering different state and non state actors at multiple scales.

Keywords: market-chains, environment, governance, aquaculture, South East Asia

Introduction

Worldwide, coastal fisheries are placing substantial pressure on the ecological functioning of coastal zones, with production between 1989 and 1998 doubling to a current estimated value of around $38 billion. As fish emerge as one of the most widely traded commodities and the intensity of resource use has expanded, the resilience of these coastal areas and their related management practices have come into question. One of the strongest drivers of change, particularly in the coastal zones of South East Asia has been shrimp aquaculture, which has grown significantly with global demand. To illustrate, the production of the Penaeus shrimps (*P. vannamei* and *P. monodon*), which comprise around 80% of total shrimp production, increased from 1,618 t in 1950 to 3,679,753 t at a value of over US$14 billion in 2004 (FAO, 2006). In some countries of South East Asia this has led to a loss of 50-80% of mangrove area in 50 years (Valiela *et al.*, 2001). Shrimp aquaculture is now increasingly recognised as a key driver in the loss of biodiversity, pond productivity and the viability of the livelihoods that depend on these coastal systems.

As the practices of shrimp farmers in South East Asia gradually become embedded within global systems of production, the control and management over their resources, collectively termed governance, is also expanding to include networks of state and non-state actors at multiple spatial and political scales. The most 'fluid' connection between local and global scales is through trade, so understanding how market relations facilitate commodity flows provides an entry point to understanding the pressures faced by farmers and fishers in both their 'local' and 'global' contexts. This paper addresses how decisions made both above and below the level of the state influence decisions over production practices, and how these existing systems can be used to improve environmental and social sustainability in aquaculture production.

Aquaculture production remains one of the most challenging primary production systems to understand because of the complex interrelations between inputs and outputs and the diversity of actors involved. Applying the commodity chain literature to the specific context of fisheries, Thorpe *et al.* (2005) use the heuristic of the 'fishery chain', to describe a set of scaled relations that can be conceptualised as a series of vertical 'linkages' between ecosystem functions, modes of production, processing, marketing and distribution. As fish are grown, processed and transferred to local, national and international markets they pass through a series of scaled networks along the chain, each with formal and informal norms, rules and regulations that control and manage activities and social relations. Producers and traders alike are located first within communities with interrelated sets of social, cultural and familial norms, practices and expectations; second, within formal state legislation; and third, within emerging combinations of formal and informal rules and regulations set by non-government institutions. Together these scaled networks influence the flow of the commodity up the chain to consumers and information down the chain to producers. The following presents some of the challenges of promoting environmentally and socially sustainable aquaculture production. Of particular interest is how governance arrangements above the state set production standards and how diverse and complex governance arrangements below the state control the ability of producers to respond. Two case studies highlight the interactions of fish farmers with the complex mix of global and local trade dynamics and governance arrangements.

Aquaculture production and trade

Fish and fish products offer particular challenges for trade. The highly perishable nature of the product means that producers and traders must either facilitate its quick transfer or invest in freezing infrastructure or other preservation techniques. When chains are short the need for preservation is less, but as chains extend larger investments in preservation technologies are required. Fish processing was originally an artisanal practice to preserve fish for postponed consumption or for trade to distant markets. As markets have grown traditional local products, such as fish oil and fish sauce, are produced on an industrial scale and exported around the globe. This process has resulted in a gradual 'upscaling' of local production to service global markets, so that its intensity and volume has outstripped the carrying capacity of many areas for meeting production in a sustainable manner. Moreover, in the context of the global market, producers have to respond to specific consumer demands, in terms of quantity and taste, but also in terms of health and safety. The ability of farmers to respond to these rapid changes is limited by a combination of environmental, social and economic factors, including the supply of seed, feed, the availability of and access to land and water, and the control of disease.

The global flows of commodities cannot be seen out of their local contexts, where diverse relationships and practices surround interactions between different actors involved, and politics and local histories contribute to a wider context for decisions over resource exploitation. The interactions at the local level within the commodity chain can be termed artisanal, meaning that trade practices are based on the negotiation skills of individuals. Local systems also determine the specific context in which fish farmers make decisions over production that affect their livelihoods and the environment, including decisions over market access. Under these conditions producers rely on a series of social and familial relationships based on trust to reduce risks and uncertainty. Trade relationships in artisanal trade networks are varied, including patriarchal, debt-tied relations while the prices in non-tied markets are generally higher. Fish farmers seek a range of relationships that helps them to sell their produce and obtain market information, with the aim of meeting their families' financial and nutritional needs, sometimes balancing lower market prices with access to credit, emergency cash flows, feed, equipment and a guaranteed market for their fish (Platteau and Abraham, 1987). The livelihood decisions taken by farmers are directly influenced by both the structure and character of the integrated global and local trade networks. What is best for a farmer's livelihood may be provided for through trade of their product, but this does not mean that the market necessarily provides pressure to improve environmental performance, especially when such improvements may in fact increase rather than reduce an individual's risk and uncertainty.

Governance

The position of fish farming within ever expanding international markets is best conceptualised through the emergence of the global network society (Castells, 1996). In the global network society the focus on local production practices is being replaced by a more dynamic understanding of the interrelations of production within global networks. Economic activities and social practices in the 21st century are less organised on the basis of material transactions and face-to-face interactions and more via exchanges and interactions without geographical contiguity. Therefore to understand individual practices at the local level attention must also be given to the global level, as individuals build their identity and alter practices by orienting their lives to this level. Contemporary governance needs to reconcile global with local dynamics, while being conscious of the continuous transformations in the relationships between both levels. In the network society, a nation-state can no longer be understood as a sovereign, independent entity because all states are increasingly less able to control social and economic processes within its borders. These decreasing possibilities for governmental regulation of fish production and trade combined with growing concerns among consumers result in a demand for more adequate arrangements. In recent years a growing number of innovative governance arrangements has emerged. There is a shift from centralised systems of decision making through the state, to networked dynamics of formal and informal negotiations among numerous individuals and organisations at multiple scales, including private enterprises, non-government organisations (NGOs) and consumers. The role of the nation state is not completely diminished as government departments may provide technical support, create legal and regulatory arrangements, fund and carry out research for primary production, processing or trade. Thus governance responses relate to the new dynamics of interconnecting production and consumption through a mosaic of both formal and informal networks, and through the horizontal and vertical dimensions of commodity chains.

Governing the 'fish chain' means balancing a series of often competing goals: enhancing development of coastal areas, while maintaining a predictive fish supply to global markets, ensuring the sustainability of resources in already stressed coastal environments, and securing a minimal level of food and income security for local populations. Arrangements can be endogenous: carried out by producers and traders themselves; or exogenous, facilitated by national and international agencies (Thorpe *et al.*, 2005). Endogenous arrangements could mean better coordination to facilitate quick transfers, better use of ice and freezing technologies, or collective actions to maintain private standards of operation or quality guarantees. Exogenous activities at the national level can be direct government interventions through taxation, subsidies and infrastructure development, and beyond the national level multilateral organisations can set particular norms, rules and standards to guide production and post-production practices. Harnessing the global dynamics of trade and information exchange can facilitate a process of 'chain reversal': the transfer of information down the chain back from consumers to producers through a series of tools such as labelling, standards and consumer guides. Oosterveer (2005) describes the combination of such processes as innovative 'global food governance' which represent the wider shift in relationships between fish consumers and producers beyond simple market-based relationships by introducing different, specific, and changing concerns in new arrangements. Consumer oriented governance mechanisms are becoming a tool for 'political consumerism' in an era where governments seem less able and willing to intervene actively in economic processes.

Local governance arrangements over aquaculture focus on individual decision making with the aim of maximising outputs. These decisions may be embedded within a combination of local customary arrangements on access to land and water, as well as emerging issues of access to inputs, the environmental consequences of feed, genetic leakage, impacts on other sectors and local aspects of ponds (Pullin and Sumaila, 2005). Unlike fishers, fish farmers are able to at least partly control production and thereby able to respond to supportive as well as coercive forms of governance (rules, laws and regulations) in order to balance livelihood needs and issues such as cleaner production. Regardless of their specific strategies, fish farmers have links with global markets while embedded within farming communities and local trade networks. The challenge is to understand how individuals make decisions faced with both the supportive and coercive influences of these scaled networks. The two case studies that follow highlight different governance aspects of trade that influence fish production, farmers' livelihoods, and the environment

From consumers down: EU-Thai nitrofuran case

The European shrimp market has always been considered more particular than the USA or Japanese markets and, due to consumer pressure, has recently become even more demanding due to rising environmental and food safety concerns. These concerns include sustainability and control in fish farming, regulation of antibiotic use, application of ethical employment standards, traceability throughout the supply chain, no use of genetically modified feed ingredients or gene technologies in shrimp breeding, sustainability in the use of fishmeal, attention to issues of animal welfare, and no use of dioxins, PCBs, heavy metals, agrochemicals and irradiation. Despite these concerns, the EU remains the world's largest market importing about 50% of all shrimp traded internationally.

The 'nitrofuran' incident can be singled out as an example of the impacts of global market relations on local production practices. Nitrofuran, a group of antibiotics used to inhibit bacterial growth, is a cancer-causing chemical and has been banned by most countries including the European Union (EU), where since 1994 the presence of nitrofuran is forbidden in shrimp and other food products. In practice the technology used to detect antibiotics had a limited sensitivity of 5 ppb and so set a 'de facto' standard in view of the exporting countries. However, technological improvement and the introduction of new testing practices lowered the detection threshold to 0.05 ppb and in February 2002 EU customs officials discovered the presence of nitrofuran in shipments of Thai shrimp. In reaction the EU decided to test all shrimp imports from Thailand and several other Asian countries. As a direct result, shipments of frozen shrimp from Thailand to the EU immediately fell from 7,000 t in 2001 to 1,850 t in 2002 and less than 700 t in 2003.

Thai shrimp produced from aquaculture links producers and consumers at very large distances and creates impacts at different scales. Environmental impacts may occur locally where shrimp ponds are constructed and managed, while the demand for fishmeal to feed the shrimp and the collection of broodstock has much wider impacts. Understanding local practices can therefore only be achieved by linking them to the transnational shrimp chain, including the financial and informational aspects. The most important pressure to reduce the environmental impact of shrimp production in Thailand has, for example, not been of domestic but more of foreign origin. Under pressure from European concerns about the effects of antibiotics the Thai government introduced a national Code of Conduct for Sustainable shrimp farming. This code was developed to obtain a framework for the shrimp farming industry's goal to take up its environmental, social and economic responsibility (Nissapa *et al.*, 2002). If all actors involved in the shrimp production chain abide by this code, the final product can be labelled as 'Thai Quality' shrimp. The French supermarket chain, Carrefour, intends to use this quality assurance scheme and sell these shrimps under its own private brand. The Thai shrimp producers also introduced another scheme, Good Aquacultural Practices (GAP), as the Code of Conduct may be too challenging for some fish farmers. This GAP program is simpler and has two main objectives: that shrimp farmers effectively meet the official hygienic standards, and that no prohibited antibiotics are used. These innovative arrangements show that contemporary governance cannot be organised by conventional governmental regulation alone. Production areas, structures of trade and places of consumption may move swiftly without national governments being able to control them. Effective governance of global shrimp provision has to combine various governmental structures and non-governmental actors at different levels to link in with global shrimp chains as well as establish connections between the local dynamics of production and consumption.

From producers up: power-based artisanal production in the Mekong Delta

The case of Ap Cho village in Vietnam exemplifies the complexity of customary production and trade arrangements faced by small scale fish farmers in marginal coastal areas, highlighting local governance.[4] Ap Cho is located within Hiep Thanh commune at the confluence of the Mekong and South China Sea in Tra Vinh province Vietnam. The village is comprised of 135 families of which only 30 have agricultural land gradually converted from mangrove over the last fifteen years. Those households that do not have land either fish in inshore areas or provide labour for surrounding land owners who commercially farm shrimp and tilapia. The vulnerability of the community was compounded by tropical cyclone Linda in 1997 which destroyed all but a few of the village's coastal fishing boats, leading to the death of many of the community's male fishing population. In response the government provided low interest loans for families to recover their fishing boats. Due to the depleted status of the fishery some of the families invested in shrimp production, following other communities in the surroundings. However, families within Ap Cho have been slow to adopt shrimp aquaculture, due to the high investment costs, lack of land ownership and past investment in coastal fishing.

Fish farmers in Ab Cho sell to domestic and international markets through different channels, starting with either local traders, agents and processing companies. These different market channels not only facilitate the trade of produce, but also supply farmers with technical and market information, on which they base their production decisions. Who a farmer sells to is determined by a combination of personal relations, wealth and social status. Some farmers enter into debt tied relations which forces them to sell only to their creditor, thereby fixing prices and their ability to react to alternative sources of market and technical information. In return, farmers benefit from access to often short term emergency credit for health, education or productive inputs such as feed. As such, processing companies selling directly to international markets may find it difficult to access supply from these debt tied farmers. Far from comprising rational, transparent networks and flows of information, as exhibited in the nitrofuran case of Thailand, fish farmers in Ap Cho are located within socially embedded local trade and credit networks through which information and market access is controlled by local elites, traders and processing companies.

The rapid increase of shrimp culture systems has led to the emergence of new governance arrangements dealing with access to land and water, as well as information on production techniques. The state's promotion of mangrove clearance, land tenure changes and the shift in practices from fishing to aquaculture has caused disruption in existing rules and regulations over the use of the coastal environment. Indeed, the limited ability of individual farmers to react to changing rules and regulations has been an important factor in hampering both increased and more efficient production. For example, after the storm the government provided farmers with access to finance and some households borrowed up to 20 million VND (equivalent to approximately US$1400) over three years. Much of the technical assistance from government has been unsuitable, focusing on intensive aquaculture techniques in areas where, according to village representatives, farmers only have the financial capacity to farm extensively. The result has been the continued outbreak of diseases and poor harvest results, which has meant that some farmers have failed in all eight attempts to harvest shrimp and have little chance of repaying their debt. As a further measure district and provincial authorities have started

[4] This case study was part of a study funded by Oxfam America in the Lower Mekong Basin. For more details see Bush (2006).

a series of micro-finance initiatives, lending up to 1 million VND (or US$70) at a time for alternative activities such as livestock. The support provided by the government has been a proactive intervention to address poverty alleviation in the Mekong Delta using technologies that have been developed and transferred internationally. However, production practices have not been tailored to the specific local conditions due to poor information flows and lack of responsiveness between departments, commune officials and farming communities. The continued failure has meant that farmers have lost confidence in external support as they continue to default on debt repayments and lose their land, limiting their willingness to plan or seek out alternative information on either technical expertise or market trends.

The case of Ap Cho shows how before accessing either global or domestic markets, farmers must negotiate a series of rules and norms that regulate their access to finance, as well as market information and technology. The ability of a farmer to make decisions over production practices that secure their own livelihoods in a way that minimises environmental damage to coastal environments is primarily dependent on customary and state governance arrangements. The potential of global governance arrangements to influence production practices must therefore engage the complex, socially embedded conditions in which farmers operate.

Conclusions

The inclusion of a social science perspective provides valuable insights to the mechanisms and contexts in which decisions over production practices are made. If we refer in aquaculture governance to maximise social justice, environmental sustainability and economic development for coastal areas, greater emphasis needs to be given to multiple nested networks in which fish farmers, fish traders, and suppliers exist. For governance to be effective solutions must be sought that are inclusive of the demands of consumers, retailers, international traders, domestic traders, and producers. It is therefore imperative to reflect on the character and influence of local governance arrangements in order to contextualise what influence global governance arrangements can have over local production practices of fish farmers. Conversely, while remaining firmly within the daily struggles and material realities of South East Asian fish farmers, recognition must also be given to the influence that global scale regulation can have over these local production practices. As such, the attention being given to global governance arrangements, such as the EU's response to nitrofuran, must be supplemented with intensive engagement with the rules, norms and regulations under which farmers operate, including access to land, technology, market information, finance and trade. Reflecting the complex mix of state and non-state actors in the governance of fish farming is an important precursor for improved and targeted technical assistance for farmers, as well as for fostering more socially and environmentally equitable flows of commodities, information and capital.

Including the multitude of state and non-state actors involved in the extended vertical (see the nitrofuran-case) and horizontal (see the Vietnam-case) networks in the development and implementation of governance arrangements constitutes a complicated challenge. In this regard, the rapid introduction of non-state governance arrangements, such as third party labelling as a market-based tool to promote sustainable fish production, is drawing particular attention. Because they rely on market information these arrangements are restricted to the extent they can include informal social networks, or government institutions. Therefore, the main challenges for developing adequate contemporary fish governance arrangements are as follows:

- Give due respect to the continued role of the state, particularly in regulating overall production practices and providing information.
- Recognise the complexity of the stakeholders above and below the state and increase understanding between what consumers and retailers want and what is feasible given the realities of local conditions of production.
- Recognise that markets alone cannot balance the various environmental, economic, social, and political interests.

References

Bush, S.R., 2006. Livelihoods, markets and governance: market access and value adding for small-scale fishers in the Lower Mekong Basin. Oxfam America, Phnom Penh, Cambodia.

Castells, M., 1996. The information age: economy, society and culture: Volume I. Blackwell, Oxford.

FAO, 2006. Fisheries global information system (FIGIS), http://www.fao.org/figis/servlet/SQServlet?file=/usr/local/tomcat/FI/5.5.9/fi5/webapps/figis/temp/hqp_68650.xml&outtype=html. (Accessed June 2006)

Nissapa, A., S. Boromthanarat, B. Chaijaroenwatana and W. Chareonkunanond, 2002. Shrimp farming in Thailand: A Review of Issues. EU INCO-Dev project PORESSFA No. IC4-CT-2001-10042.

Oosterveer, P., 2005. Global food governance. Wageningen University, The Netherlands.

Platteau, J.P. and A. Abraham, 1987. An inquiry into quasi-credit contracts: the role of reciprocal credit and interlinked deals in small-scale fishing communities. J. Dev. Stud. 23, 461-490.

Pullin, R.S.V. and U.R. Sumaila, 2005. Aquaculture. In: J. Kooiman, M. Bavinck, S. Jentoft and R.S.V. Pullin (eds.), Fish for Life: Interactive governance for fisheries. Amsterdam University Press, Amsterdam, pp. 93-107.

Thorpe, A., D. Johnson and M. Bavinck, 2005. The system to be governed: Introduction. In J. Kooiman, M. Bavinck, S. Jentoff and R.S. Pullin (eds.), Fish for Life: Interactive Governance for Fisheries. Amsterdam University Press, Amsterdam, pp. 41-44.

Valiela, I., J.L. Bowen and J.K. York, 2001: Mangrove forests: one of the world's threatened major tropical environments. BioScience 51, 807-815.

Chapter 6

IAA systems and sustainable development

Research approaches to support development

Peter Edwards
Asian Institute of Technology, 593 Soi Lad Prao 64, Bangkok 10310, Thailand, pedwards@inet.co.th

Abstract

There is a need to use a systems approach to complement the traditional reductionist approaches of conventional science to ensure that development is both socially and environmentally sustainable. The systems approach is essentially three things: the philosophy or concept of holism, a study of the whole system; a physical system such as an aquaculture system located in a hierarchy of systems from organism, through enterprise and farm, to community, region, nation and the world; and methods for identifying and solving problems or identifying and exploiting opportunities. As most real world problems are complex, they require interdisciplinarity, most effectively provided by teams of natural and social scientists. A general, seven step systems methodology is outlined and a discussion of farming systems research and extension, with examples from aquaculture.

Keywords: farming systems research, interdisciplinarity, participatory research

Sustainable development

There is increasing recognition that development must be sustainable. There are many definitions of sustainability, but a widely accepted one is that of the Brundtland Report as published in a book 'Our Common Future' by the World Commission on Environment and Development (WCED) in 1987:

> "sustainable development is development that meets the needs of the present without compromising the ability of future generations to meet their own needs".

It contains two key concepts: the limitations of the environment to meet present and future needs; and the unsatisfied needs of the world's poor. The definition implies that there is a single correct definition but a "precise and absolute definition of sustainability, and therefore sustainable agriculture is impossible" … and it is "important to clarify what is being sustained, for how long, for whose benefit and at what cost, over what area and measured by what criteria" (Pretty, 1995). According to Pretty (1995), sustainable agriculture is not an imposed package but is a "process for learning and perpetual novelty" as conditions and knowledge change.

Unfortunately, short-term needs and objectives such as increased productivity, employment and income and quick returns to investment usually dictate policy as well as practice with insufficient attention to sustainability. However, non-negotiable natural laws will eventually force change due to resource depletion, environmental degradation and increasing conflict as the productive capacity of the world's land and water and the capacity of providing ecological services on which both biological production and waste processing depend are both finite. Required is a move from trying to produce maximum biomass of farmed species to holistic management of natural resources and agro-ecosystems (Pullin *et al.*, in press).

As scientists and technologists we need to become increasingly responsible for the social and environmental consequences of the development and dissemination of technology, regarding other people, especially the poor as well as the environment or nature. As recently noted: "we are learning the fundamental principle that ethics is everything" (Wilson, 1999). Clearly philosophy is crucial in our aim to achieve sustainable development.

Philosophical basis of development

Philosophy is intellectually important because it influences our world view, way of thinking and behaviour. It is particularly important that we adopt an appropriate philosophy for aquaculture to fulfil its potential to alleviate poverty as well as to better utilise nutrients at the small-scale farmer household level, the themes of this workshop. The ethical basis of sustainability is reflected in the title of FAOs Code of Conduct for Responsible Fisheries.

The philosophy of Popper (1979) is relevant for our quest for sustainable development because of its systematic unity i.e., it covers evolution of life, the development of human knowledge and science, and social and political issues that influence development. All organisms engage in problem solving. Their genetic make-up determines both the problems that they experience due to unfulfilled expectations; and the flexibility of their response. Survival is the primary problem, and problem solving is the primary activity of all organisms, including humans. Humans additionally are opportunity seeking as we seek to improve our welfare through development. Popper's philosophy helps us to understand the nature of problems, how to identify them, and how to solve them. His fundamental problem-solving schema can be reduced to three steps: analyse the problem or opportunity; propose a hypothesis to solve the problem or exploit the opportunity; and test the hypothesis in relation to the problem or opportunity (Figure 6.1; Edwards and Demaine, 1998).

Problems with their background or context, the problem situation, are usually complex, especially those relating to small-scale farming households. From a research and development point of view this implies (CAB International, 2006a):

- Adequate understanding of problem situations, including those of the poor can only be achieved by working with target communities.
- Identification of, as well as working on, problems leads to ownership and increased commitment i.e., partnerships are required between farmers, service providers, and research and development specialists rather than top-down assistance.
- Changes can usually only be made incrementally i.e., by building upon an existing system and knowledge base rather than by introducing a completely new system as in top-down transfer of technology to farmers.
- Subject matter specialism is self defeating, i.e. there is a need for an interdisciplinarity team of natural and social scientists to adequately define problems of small-scale farming households.

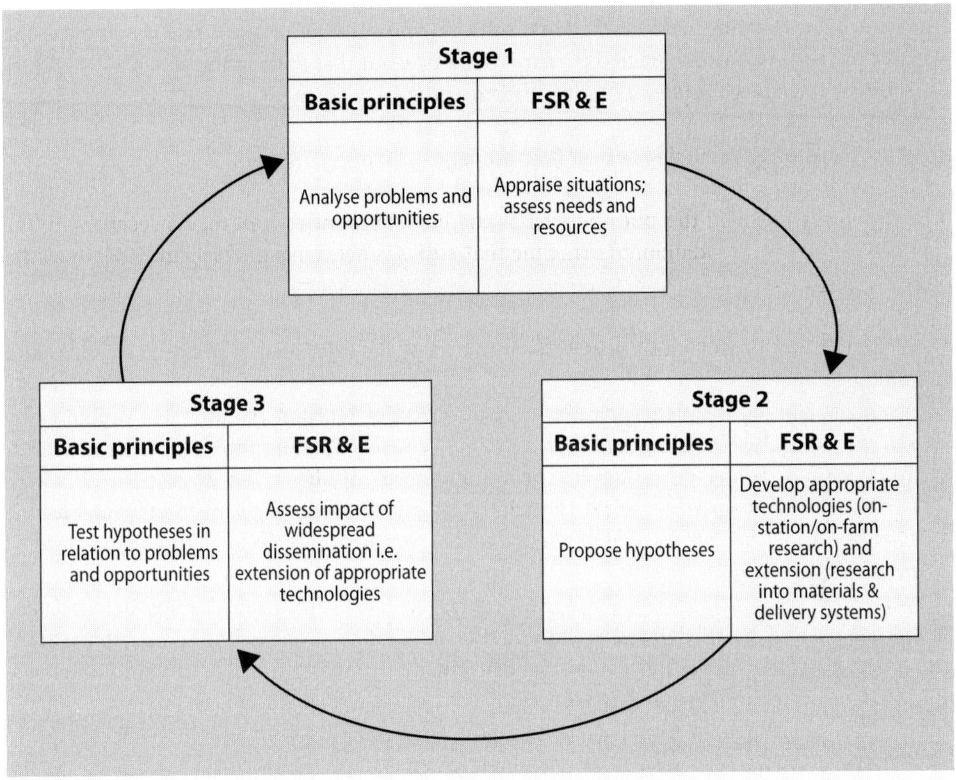

Figure 6.1. Three main stages in the identification of problems: basic principles according to the philosopher Popper (1979) and corresponding FSR&E (Edwards and Demaine, 1998).

- There is essentially a break in the research-development continuum, or problem-solving cycle, which requires redressing on a widespread scale.

Science and technology

Scientific disciplines

The real world is so complex that scientists study it by breaking it into components, parts or pieces so the scientific method of experimentation is reductionist although scientists are also involved in the crucial tasks of synthesis and integration. Scientific thinking is almost synonymous with analytical thinking, with analysis or detailed examination of the elements or structure of a substance, organism or problem, involving breaking down a system. Although science has lead to major improvements in human welfare, reductionism has limitations in understanding systems because a system is the sum of its components which interact. As soon as a system is separated into the parts or pieces, it is destroyed because the parts cease to interact.

Our knowledge is divided or classified into subjects or disciplines as the world is complex. The primary purpose of basic sciences is to explain nature, but that of the applied sciences such as aquaculture is to develop technology to improve people's welfare.

Relationships between science and technology

It is commonly believed that there is a one-way flow of ideas between the basic sciences, the applied sciences and technology. In fact there are two-way flows or reverse flows between the three.

There are two implications of a reverse flow model of the relationship between sciences and technology:

- Scientists can learn much, including identification of relevant researchable problems by getting out of the ivory tower of the university or research station to study the real world.
- To develop or adapt technology to the specific needs of farmers in diverse social as well as ecological and agro-ecological contexts (problem situations), it is necessary to combine scientific and other organised knowledge such as indigenous technical knowledge or local people's knowledge.

Technology is thus better defined broadly as the application of scientific and other organised knowledge to practical tasks, because scientific knowledge is not the only kind of knowledge used to develop technology, as discussed above.

Systems

Need for a systems approach

Traditional science has been highly successful in improving human welfare, especially for people in developed countries, but its limitations became apparent during the 20th century. Most of the major issues facing humanity cannot be solved without the integration of knowledge from natural and social sciences (Wilson, 1999). In aquaculture, science has been effective in production technology, leading to increased yields of fish or other aquatic organisms, but has been less successful in addressing social and environmental problems, such as equitable distribution of benefits and minimising pollution. From a systems point of view the only viable solutions are those that are sustainable in terms of both human equitability and the environment.

Disciplines in science are growing progressively narrower with increasing specialisation, causing a crisis of limited communication between people of different disciplines: such persons cannot state a problem in terms that the other can understand; and cannot answer in any form that the other person can understand. We have a limited ability to see the whole system or picture and generally do not appreciate the need for inputs outside our own area of expertise to solve a problem.

As scientists become increasingly specialised it is difficult for them to fully appreciate the system to which their knowledge applies. The more science breaks up into specialised subjects the less communication there is among disciplines:

> "It is quite remarkably difficult for specialists from one discipline to understand the concepts and language of another" (Checkland, 1982).

Poor communication between specialists from different subjects has been called "specialised deafness" due to a lack of "generalised ears" (Boulding, 1991). The major development problems are systems problems which cannot be solved by a disciplinary approach. The more that science becomes divided into specialised disciplines, the more important it becomes to find unifying principles. The systems approach has evolved in response to the limitations of traditional science which it complements. The basic philosophy is the concept of holism based on wholes and their properties with the systems approach seeking to include all major variables that affect the system to be understood.

Systems characteristics and properties

The relatively new discipline 'systems' is based on a concept of organised complexity in which:

- there is a hierarchy of levels of organisation;
- each level is more complex than the one below it;
- each level is characterised by emergent properties which do not exist at the lower level. This property of systems is synergism.

Systems thus represent a novel paradigm or shift in scientific thinking and have pervaded all fields of science as well as popular thinking, jargon and the mass media.

It is important to distinguish between open systems and closed systems. Agricultural systems are open systems with import of materials from, and export of materials to, their environment. Organisms can achieve a steady state or homeostasis which depends upon continuous exchanges with the environment. Most engineers are interested in closed systems. In machines where the components are not exchanged, but move in a state of equilibrium.

There is cause and effect in systems as in reductionist science, but rather than only linear causality, with a single observable effect being due to only a single cause as in reductionist science, in systems there is more likely to be multiple causality, with numerous variables influencing each observable phenomenon in positive or negative feedback loops chains, or continual repeated cause-and-effect relationships.

Agricultural (aquacultural) systems

The operational units of agriculture (or aquaculture) are agricultural (or aquaculture) systems and are arranged in a hierarchy:

- individual organism;
- enterprise or subsystem such as a crop, livestock or fish;

- farm or plantation;
- regional agriculture or economy;
- national agriculture or political economy;
- international agriculture or international system.

To fit in with ecological terminology and to consider humanity as part of nature, the term agro-ecosystem is best used. Agricultural and aquaculture systems are human-made agro ecological systems to produce mainly food. They have much lower diversity of species than ecological systems and are dominated by human goals and consequences of human social and economic actions.

The behaviour of a system in the hierarchy cannot be determined by simply studying higher and lower systems. Each level in the agro-ecosystem has to be analysed in its own right and in relationship to the other levels above and below it. Farming systems need to be analysed in relation to the specific farming system and its physical, socio-cultural and policy/institutional environments or contexts. To reflect this, alternative and broader categories than agriculture are given for the last three higher level systems in which non-agricultural factors play an increasingly important role. This totality of understanding should be used as the basis for development to improve people's welfare.

Scientific disciplines and the systems approach

Generalists or specialists?

There is a need to add a systems approach to conventional research and development because of the complexity of many problems, but classical science (reductionist) and systems science (holistic) should be viewed as complementary aspects of modern science. Both generalist and specialists skills are needed: a systems generalist to characterise the system and identify the major problem or unexploited opportunity that constrains the system, and a subject specialist to conduct in-depth study on ways to remove the constraint and to develop or improve specific technologies. A specialist may identify a problem but it may require a generalist to better define it or to identify more important problems that may constrain a given system

Reductionist scientific research on fish and their immediate environment (such as pond nutrient recycling, a major topic of the workshop) carried out by most aquaculture scientists is as valid as long as the problem is generated from the same system to which the solution from research will apply. In fact, the initial problem identified by the systems generalist may not be resolvable without specialist scientific research. Research, however, should be increasingly directed to solve real world problems in specific locations such as nutrient limitations on small-scale farms in the Mekong River Delta, and to disseminate the results of the research so that as many people as possible benefit. More research should aim to directly benefit society and less should be 'ivory tower' to primarily benefit researchers' interests.

Interdisciplinarity

As there is a limit to what one person can study, it is difficult to become a specialist in more than one subject. However, as most of us have been trained as specialists we need at least to:

- Step out of our specialised subject and try to understand the linkages and interactions between all factors affecting aquaculture.
- Be exposed to a range of subjects so that we develop broad minds and an appreciation that solutions to problems may lie outside our subject area.

To study the complexity of most biosystems, an interdisciplinary mix of at least the following are needed: biology (production technology); social sciences (people); economics (money), and environment. Interdisciplinarity runs against the traditional structures of tertiary educational institutes, but it needs to be promoted. Some interdisciplinary subjects have, however, evolved to address complexity: e.g. ecology, biochemistry and biophysics.

One way to provide a systems approach to identifying opportunities, properly identifying problems and resolving them is to set up an interdisciplinary team. Members of such a team need to be a specialist in a required discipline; possess a basic understanding of the disciplines of other team members and how they think, communicate, and act; and develop a habit of working together. However, teams are difficult to establish and manage. Considerable emotional and social intelligence are required to work with other people in a successful team.

Sustainable technology

A systems framework serves to indicate what is required in general for a given technology to be sustainable for poor farming households (Figure 6.2):

- Production technology: A technology needs to be sufficiently productive for aquaculture to be an attractive option to possible alternative and/or competing uses of resources such as labour and nutrients.
- Social and economic aspects: Low-unit cost input systems may be most appropriate for the limited resource base of most poor farming households; and low production costs mean that fish can be sold at a relatively low market price and be affordable to poor consumers.
- Environmental aspects: A technology needs to fit into the limited resource base of the poor, not use resources that may be used more productively in other ways, and be environmentally friendly.

It is essential that aquaculture not be considered only in its narrow, technical aspect in isolation from crucial social and economic and environmental contexts. The latter invariably determine sustainability. Technologies need to be characterised with respect to their relevance for contributing to improved welfare of the poor in each and every local situation or context in which they may have potential. Most of the poor reside in unfavourable or resource-poor areas which are mainly rainfed, with undulating terrain and problem soils. The marginal type of agriculture into which aquaculture has to be integrated has been called CDR or complex, diverse and risk-prone (Chambers *et al.*, 1989).

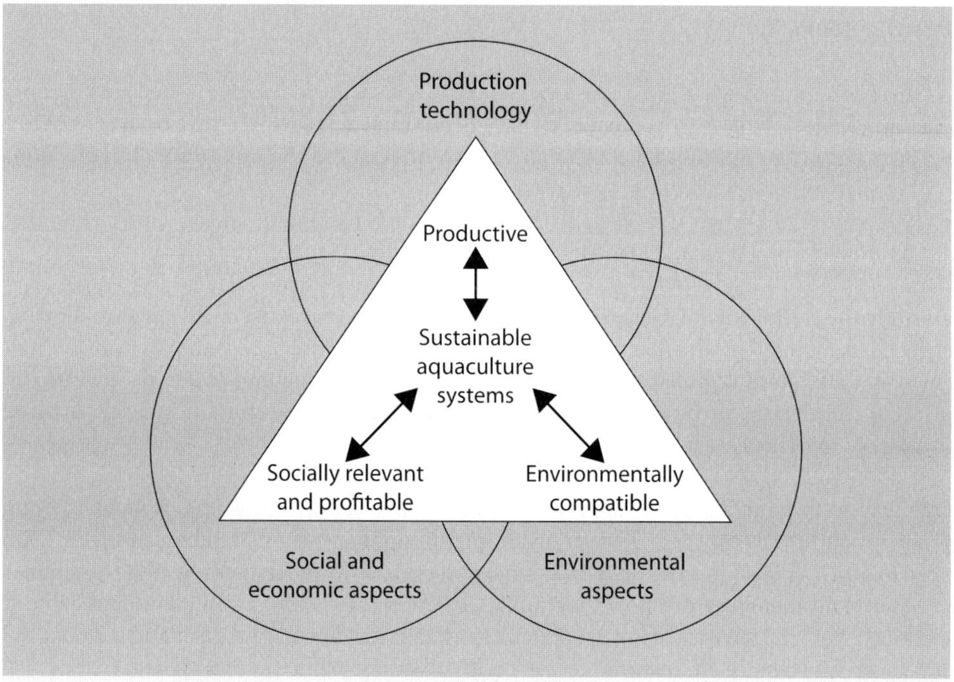

Figure 6.2. The three inter-related aspects of the sustainability of an aquaculture system (AIT, 1994).

There is a dichotomy regarding sources and amounts of nutrients in aquaculture in many areas, including the Mekong River Delta, where the workshop was held:

- Large-scale farmers with usually stand-alone fish farms with better access to inputs and markets have a nutrient disposal problem from using large amounts of trash fish, feedlot livestock manure, agro-industrial by-products or formulated pelleted feed.
- Small-scale farmers in which aquaculture is one component on a crop-dominated farm usually have limited nutrients for aquaculture.

The challenge is to make both types of farming systems more nutrient efficient, the former by reducing nutrient discharge to the environment, the latter by increasing the efficiency of the on-farm nutrient pool and by sustainably increasing the nutrient flow from judicious use of off-farm nutrients.

Systems methodologies

A general systems methodology

A useful systems methodology for aquaculture involves the following seven steps:

1. define the problem (or opportunity);
2. identify the system to which the problem (or opportunity) applies;

3. characterise the system;
4. analyse the problem (or opportunity);
5. hypothesise a solution;
6. test the solution;
7. implement the solution to resolve the original problem (or exploit the identified opportunity).

The crucial first step is to identify the primary problem that stands in the way of improving the system but this is seldom easy. As Einstein once said: "the definition of the problem is more important than the solution". The system within which the problem lies may be at one or more levels in the hierarchy of aquaculture systems. In classifying and describing the system with the problem, the following should be established: its boundary; all relevant factors, elements and variables; all relevant processes that influence the system; its initial state; and how it operates. Only factors essential to the operation of the system should be included to avoid unnecessary complication of the analysis of the system. The problem should be analysed in relation to system purpose. Following hypothesising a solution, it can be tested in one or more ways: 'guess and go', physical experimentation, or abstract experimentation. Much farming technology has been developed by farmers in a more or less *ad hoc* way without science. Physical experimentation involves testing the hypothesis in a controlled scientific experiment whereas abstract experimentation involved using models. The final step, implementation of the solution, is rarely adequately addressed, leading to a break in the research - development continuum as outlined above.

Farming systems research and extension

There is a long standing imbalance between strategic research to develop new knowledge of general significance, and adaptive research to assist farmers to adapt and adopt new or improved technology (Edwards and Demaine, 1997). Increasingly a farming systems research and extension (FSR & E) approach is being followed by projects which involves an interdisciplinary team of natural and social scientists working in partnership with farmers and service providers to:

- assess needs and resources of poor farming households in each target area;
- identify and field test appropriate technologies;
- analyse and monitor extension options for both extension materials and extension systems to deliver appropriate technologies to the target group.

Farming systems approaches such as farming systems research and extension (FSR & E) and rapid rural appraisal (RRA) were developed in the 1980s in response to limited success of the 'transfer of technology' model for CDR agricultural systems. Limited impact of the technical innovations developed by scientists on research stations with little to no input from farmers was due to the development of technology inappropriate for the target group rather than lying with the farmers themselves or the extensionists (CAB International, 2006b)

From the 1990s there has been the development of a series of approaches and methods, initially referred to as participatory rural appraisal (PRA). In RRA information is more elicited and extracted by outsiders such as scientists, whereas in PRA it is more shared and owned by local people such as farmers (Table 6.1). Participatory methods include mapping and modelling, analytical diagramming, seasonal calendars, wealth-ranking and well-being, and trend and change analysis.

To address the diversity of CDR farming systems, the concept of 'recommendation domain' was introduced, a group of farms with similar socio-economic and agro-ecological characteristics (CAB International, 2006b). Research conducted on, or for, a representative sample of farms in a given recommendation domain might be expected to be relevant for others in the group.

On-farm trials are fundamental to the farming systems approach but there is no standard methodology. The on-farm trial spectrum ranges from 'researcher-managed' trials to 'farmer-managed' trials with the degree of control of researchers decreasing and that of farmers increasing along the spectrum. Researcher-managed trials are usually to establish the feasibility of a technology which may have been developed on-station and tend to be relatively few and could be subsidised to minimise farmer risk. Farmer-managed trials are usually numerous and are aimed at establishing if a proven technology could be adopted by farmers and disseminated throughout a specific recommendation domain.

Participatory research is likely to be most effective if it combines both indigenous or local knowledge and scientific knowledge. Although local knowledge of aquaculture is limited in many Asian countries, as aquaculture is relatively new practice, new technology will be more likely to be adopted if it fits in with the local resource base and builds on local knowledge. This will also increase the farmers' natural tendency to experiment and possibly adapt the innovation prior to adoption. The role of scientists is likely to be of particular importance where appropriate technologies may not be readily identified through participatory methods, where there is limited farmer knowledge, or where aquaculture technologies are unfamiliar to farmers (Martin and Sherington, 1997).

Sustainable rural livelihoods

A useful conceptual framework to analyse the potential contribution of aquaculture technology to alleviate poverty is that of sustainable rural livelihoods (Carney, 1998). The heart of the framework are five categories of 'capital assets' of rural farming households: financial, human,

Table 6.1. The RRA-PRA continuum (Chambers, 1994).

Nature of process	RRA	PRA
Mode	Extractive elicitive sharing	Empowering
Outsider's role	Investigator	Facilitator
Information owned, analysed and used by:	Outsiders	Local people
Methods used	Mainly RRA plus sometimes PRA	Mainly PRA plus sometimes RRA

natural, physical and social (Figure 6.3). These are called assets to indicate that, although we are referring to the poor, we should build upon the 'positives' that they may already have rather than starting from an analysis of their needs. The pentagon also forces us to think in a systems or holistic way. The framework contains a 'vulnerability context' which includes culture, to reflect long term trends and short term shocks which the poor may be particularly affected by. Capital assets are improved by various 'transforming structures and processes' to generate various livelihood strategies, including one that is natural resources (NR)-based such as aquaculture and livelihood outcomes. Feedback loops connect transforming structures and processes to vulnerability context; and livelihood outcomes to capital assets.

Most small-scale farmers will have a diverse livelihood strategy, the main component of which is likely to be agriculture rather than aquaculture, there is a need to see aquaculture as one aspect of rural development rather than as an isolated technology.

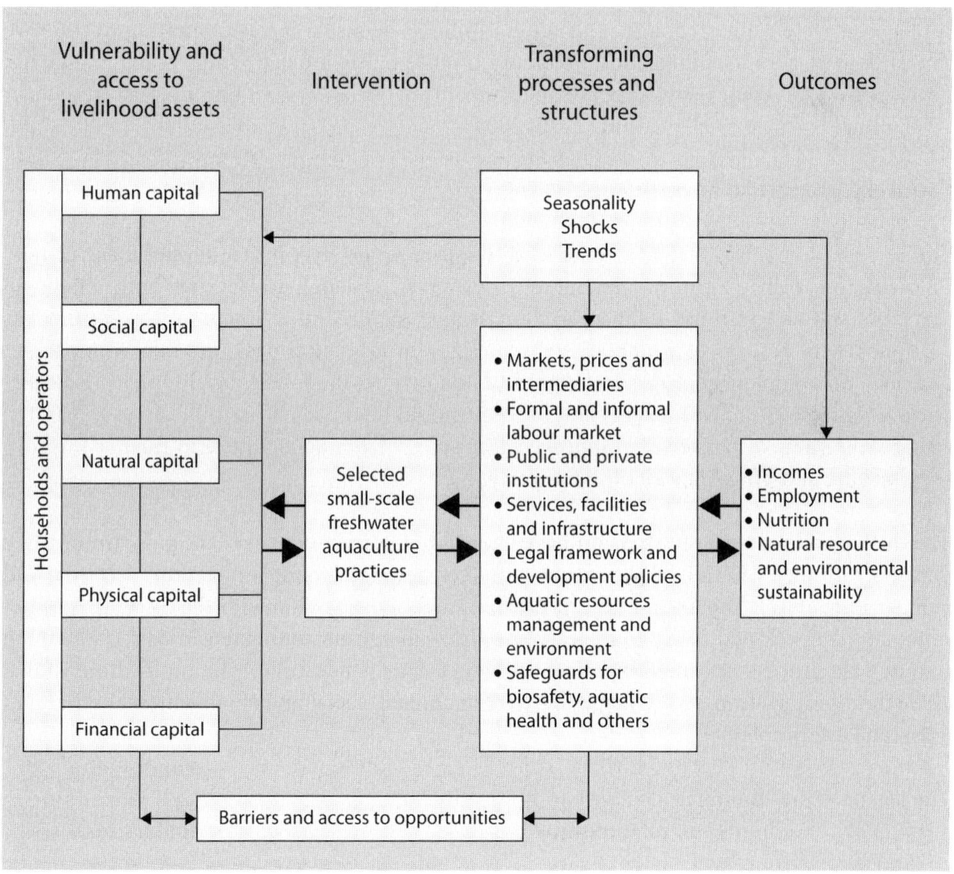

Figure 6.3. A conceptual systems framework for analysing the channel of effects of aquaculture on poverty (ADB, 2005).

Systems hierarchy

There is a need to consider the hierarchy of systems that influences rural aquaculture if its potential is to be fully realised at household level (Edwards and Demaine, 1997). There is a range of systems from the individual farmed animal or plant, through aquaculture technology contributing to a livelihood, to larger macro levels at community, regional, national and international levels. Aquaculture is a subsystem on most land-based small-scale farms in which agriculture predominates but is usually a stand-alone farming system on large-scale land-based farms and on water-based farms in inland and coastal areas.

Aquaculture is usually ignored by most agriculturalists, both researchers and development agencies, but its potential role to contribute to poverty alleviation and sustainable farming systems should be assessed. Major issues at macro level are market and input supply, government policy, and institutional issues. The sustainable rural livelihoods framework discussed above also recognises the importance of institutions and organisations to improving livelihood. It has been widely adopted, although often modified by researchers, attempting to improve the welfare of poor farming households. e.g. Figure 6.3 is a recent conceptual framework developed for an Asian Development bank study to analyse and assess the effects of freshwater aquaculture on poverty alleviation (ADB, 2005).

From extension to communication

The 'E' of FSR & E was originally intended to represent extension as a development activity following research to produce technology. However, research to develop both extension materials and an extension delivery system, is also required as a precursor to extension as a development activity to widely disseminate proven extension packages to a wider target audience on completion of the first round of adaptive research and development (Edwards and Demaine, 1997, 1998). Ideally the process should be repeated periodically to satisfy the rising aspirations of farmers for improved technology from aquaculture to further improve their welfare and lift them well out of poverty.

The theory and practice of extension have changed fundamentally over the past two or three decades with the evolution of "extension" into "communication for innovation" (Leeuwis and van der Ban, 2004). The linear and top-down model of innovation with the dissemination of innovations developed solely by researchers through agricultural extensionists is changing to a flexible process informed by all stakeholders. This is because of the multi-dimensional character of innovation with linkages between technical, social and environmental aspects of a particular technology.

The key processes to induce change and innovation are:

- network building as new relationships are needed between various stakeholders, technology and the environment;
- social or experiential learning through 'learning by doing' or 'discovery learning' is required to adequately identify and resolve problems regarding technology;

- negotiation as change is likely to generate conflicts of interest between stakeholders that need to be resolved to make change possible.

Therefore, there is a need to change the mission of agricultural extension organisations from transfer of technical knowledge, to increase agricultural productivity and providing technical and economic advise to process management to stimulate innovation (Leeuwis and van der Ban, 2004). Stakeholders should not be only extensionists, farmers and researchers but, depending on context, could include consumers, agro-industry, agri-business, traders and policy makers.

The systems approach and aquaculture

There is a well recognised discipline of systems within the field of agriculture with a dedicated journal, Agricultural Systems (see Collinson, 2000 for an historical account of the use of the systems approach in agriculture). There has been less explicit use of the term 'systems' in aquaculture, although its principles, approach and methods are being increasingly used to address a range of issues, such as environmental management, best management practice and codes of practice as well as small-scale aquaculture practice (Phillips *et al.*, 2001).

Following on from experience in agriculture, a FSR&E approach has been applied to promote small-scale aquaculture for poor farming households in Africa (Van der Mheen, 1998; Brummett and Costa-Pierce, 2002) and South and South East Asia (see examples of case studies in Edwards *et al.*, 2002). The first attempt to synthesise the lessons learned from use of RRA/PRA in aquaculture projects was a workshop held in 2000 (FAO, 2000). The involvement of most of the researchers in the nine case studies prepared for the workshop was essentially a learning exercise, as they had limited prior familiarity with participatory approaches. While such learning processes by researchers need to continue, it was recognised that the real potential of participatory approaches lies in capacity building of the end-users of aquaculture, the farmers, so that they will be able to assess its potential role as a component of their livelihoods.

Concerns have been raised about the likely impact of participatory approaches as usually practiced. Typically facilitation in participatory development has taken place at the grassroots level, involving people who are involved in or likely to be directly affected by decision making, such as introducing a new or improved aquaculture technology. People at higher levels in the systems hierarchy, such as bankers and those involved in input supply and marketing are usually not involved, even though they would possibly be able to provide financial and wider institutional support for inputs and sale of produce. Absence of higher level players may be an important reason for the problems frequently experienced in trying to scale-up or widely disseminate a technical innovation developed through participatory research with a few farmers throughout its likely recommendation domain, involving possibly thousands of potential beneficiaries (Groot and Maarleveld, 2000).

Some macroeconomists have also commented that it is naïve to expect that an aquaculture innovation developed only through a local level participatory approach would spread organically, without more active intervention through higher levels in the systems hierarchy, especially marketing (see Ruben, 2007; Niek Koning, personal communication). Clearly, a participatory approach for the development of innovation is more likely to have a widespread impact if all supporting conditions are favourable. This may require the involvement of a wide range of stakeholders from farmers at the bottom of the systems hierarchy through higher levels in the systems hierarchy: traders and local development agencies, national level line agencies and policy makers, and even international level players if produce is to be exported.

References

ADB, 2005. An evaluation of small-scale freshwater rural aquaculture for poverty reduction. Operations Evaluation Department, Asia Development Bank, Manila, The Philippines, 163 pp. www.adb.org/Documents/Reports/ Evaluation/sst-reg-2004-07/default.asp?p=opereval.

AIT, 1994. Partners in development, the promotion of sustainable aquaculture. Asian Institute of Technology, Bangkok, Thailand, 98 pp.

Boulding, K.E., 1991. General systems theory – the skeleton of science. In: G.J. Klir (ed.), Facets of systems science. Plenum Press, New York, U.S.A. & London, U.K., pp. 239-248.

Brummett, R.E. and B.A. Costa-Pierce, 2002. Village-based aquaculture ecosystems as a model for sustainable aquaculture development in Sub-Saharan Africa. In: B.A. Costa-Pierce (ed.), Ecological aquaculture, the evolution of the blue revolution. Blackwell Science, Oxford, UK. pp. 145-160.

CAB International, 2006a. Topic on Aquaculture Systems [original text by P. Edwards]. In: Aquaculture CompendiumCAB International, Wallingford, U.K. www.cabicompendium.org/ac.

CAB International, 2006b. Topic on Research and Development [original text by H. Demaine]. In: Aquaculture Compendium, CAB International, Wallingford, U.K. www.cabicompendium.org/ac.

Carney, D., 1998. Implementing the sustainable rural livelihoods approach. In: D. Carney (ed.), Sustainable Rural Livelihoods, What Contributions Can We Make? Department for International Development, London, UK. pp. 3-23.

Chambers, R., 1994. The origins and practice of participatory rural appraisal. World Development 22, 953-969.

Chambers, R., A. Pacey and L.A. Thrupp (eds.), 1989. Farmer first, farmer innovation and agricultural research. Intermediate Technology Publications, London, U.K. 218 pp.

Checkland, P., 1982. Systems thinking, systems practice. John Wiley & Sons, Chichester, U.K.

Collinson, M., 2000. A history of farming systems research. FAO, Rome, Italy and CABI Publishing, Wallingford, UK. 432 pp.

Edwards, P. and H. Demaine, 1997. Rural aquaculture: overview and framework for country reviews. RAP Publ. 1997/36, RAP/FAO, Bangkok, Thailand, 61 pp.

Edwards, P. and H. Demaine, 1998. Completing the problem-solving cycle in aquaculture: the AIT Aqua Outreach. Aquaculture Asia 3, 10-17.

Edwards, P., D.C. Little and H. Demaine, 2002. Rural aquaculture. CABI Publishing, Wallingford, UK. 358 pp.

FAO, 2000. Report of the Workshop on Participatory Approaches in Aquaculture, Bangkok, Thailand, 28 February – 1 March 2000. FAO Fisheries Report No. 630, FAO, Rome, Italy. 41 pp.

Groot, A. and M. Maarleveld, 2000. Demystifying facilitation in participatory development. Gatekeeper Series No.89. International Institute for Environment and Development, London, UK. 19 pp.

Leeuwis, C. and A. Van den Ban, 2004. Communication for rural innovation, rethinking agricultural extension. Blackwell Science, Oxford, U.K., Third edition, 412 pp.

Martin, A. and J. Sherington, 1997. Participatory research methods – implementation, effectiveness and institutional context. Agricultural Systems 55, 195-216.

Phillips, M., C. Boyd and P. Edwards, 2001. Systems approaches to aquaculture management. In: R.P. Subasinghe, P. Bueno, M.J. Phillips, C. Hough, S.E. McGladdery and J.R. Arthur (eds.), Aquaculture in the third millennium. Technical Proceedings of the Conference on Aquaculture in the Third Millenium, Bangkok, 20-25 February 2000, NACA, Bangkok, Thailand and FAO, Rome, Italy, pp. 239-247.

Popper K.R., 1979. Objective knowledge, an evolutionary approach. Clarendon Press, Oxford, U.K., 395 pp.

Pretty, J.N., 1995. Participatory learning for sustainable agriculture. World Development 23, 1247-1263.

Pullin R.S.V., R. Forese and D. Pauly, in press. Indicators for the sustainability of aquaculture. In: T.M. Bert (ed.), Ecological and genetic implications of aquaculture. Kluwer Academic Publishers, Dordrecht, The Netherlands.

Ruben, R., 2007. Participatory approaches in the Mekong Delta, Vietnam. In: A.J. van der Zijpp, J.A.J. Verreth, Le Quang Tri, M.E.F. van Mensvoort, R.H. Bosma and M.C.M. Beveridge (eds.), Fishponds in farming systems. Wageningen Academic Publishers, Wageningen, The Netherlands, pp. 281-292.

Van der Mheen, H., 1998. Achievements of smallholder aquaculture development in Southern Africa, experiences from the ALCOM Aquaculture Programme. ALCOM GCP/INT/SSS/SWE. 37 pp.

WCED, 1987. Our common future. Oxford University Press, Oxford, U.K., 400 pp.

Wilson, E.O., 1999. Consilience, the unity of knowledge. Vintage Books, New York, U.S.A., 367 pp.

Participatory approaches in the Mekong delta, Vietnam

Le Quang Tri[1] and Martinus E.F. van Mensvoort[2]
[1]Soil Sciences & Land Management Dept., College of Agriculture & Applied Biology, CanTho University, Can Tho City, Vietnam
[2]Department of Environmental Sciences, Wageningen UR, The Netherlands

Abstract

This article describes our experiences in the use of participatory methods in research as applied in the Mekong delta. Studies on Bottom-up Research (starting from farmers' knowledge), Participatory Rural Appraisal (PRA) and Participatory Community Appraisal (PCA) show that the degree of participation and their success depends on the goals, intentions and interests of the institutions which initiate a PRA/PCA, as well as the local participants. The formation, training, attitude and long-term involvement of team members are very important as are project resources and the setting for the PRA/PCA. The PLUP (Participatory Land Use Planning) approach gives farmers the opportunity to present their knowledge on the land, how to use it and express their needs/priorities. Such PLUP studies provide researchers and local land use planners, with valuable information about the farmers' perspectives. Although in Vietnam a number of PRA studies were carried out over the years, they mainly extracted information from the rural people. Recent PCA practices were useful for field staff to understand the situation in villages and communities, their problems and constraints, and opportunities for future development. They failed, however, to achieve participation in terms of empowerment of the people from whom the information was collected. The paper ends with a discussion on how participation in research and development projects can keep up better with recent developments in the field.

Keywords: Participatory Rural Appraisal (PRA), Participatory Community Appraisal (PCA), participatory research methods

Introduction

In 1980 FAO launched the People's Participation Program (PPP). Since then, PPP has been implemented in many projects throughout the developing world in an attempt to give local people a voice in development schemes (FAO, 1990). At the time of the PPP launch, Vietnam was virtually impenetrable for such new ideas. The country was governed by means of a strictly centralised political and economic system, right down to the grassroots. The central, provincial and district governments had agriculture departments whose main task was to control farmers' production quota and the state run rice distribution and export. Parallel to the economic liberalisation of the late 1980s, the issue of more formal methods of participation became the subject of debate in agricultural research and rural development interventions. Our effort focused on the application of Participatory Rural Appraisals (PRA) as a methodology to demonstrate local knowledge and capacities for problem-resolution, and also to provide a voice for different people in a community (Feitsma *et al.*, 2000).

Over the past decade the Mekong delta seems flooded with participatory activities. Barzman *et al.* (1996) claimed a participatory approach when studying the role of the weaver ant in citrus growth. Binh *et al.* (1997) used a participatory approach to investigate the existing mangrove – shrimp systems in Ngoc Hien district, the extreme South-West of the delta. International research institutions such as IRRI now claim to do almost only participatory research. The Mekong Delta Development Institute in Can Tho, established in 2005 but formerly known as the Farming Systems Research Institute, says in its mission statement, to intend to carry out training courses in 'Techniques in PRA, PPD, PAP, and PME'. All P's stand for participatory, even in participatory plant breeding and variety selection and participatory technology development (PTD). The Participatory approaches are not limited to agriculture. UNDP (2003) carried out a Participatory Poverty Assessment; the Mekong delta Development Institute has a project, together with Belgian experts, to improve the participatory interaction between extension workers and farmers.

While all these activities were going on in Vietnam, participation for development and particularly PRA came under increasing criticism. Angarwal (2001) indicated the exclusion of significant sections, in particular women, in PRA's in India and Nepal. Gladwin et al (2002) tried to deal with the criticism that participatory methods in research do not lead to science of quality and discussed the trade-offs between microlevel research/extension through participatory methods versus rigorous ways to obtain statistically sound data. The criticism on participation culminated in it being called a 'tyranny' (Cooke and Kothari, 2001). The core of the criticism was that participation (be it for research or for development in more general contexts) has failed to live up to its promises of empowerment and staying away from issues of power and politics. PRA was depicted as a technical approach, depoliticising the situation.

The criticism concentrated on PRA, but participation is much wider than just this methodology. It stands for a wide range of functions, purposes and projects. Recent publications try to do justice to participation and think about better ways of definition and application. Hickey and Mohan (2004) mention ideology, politics, level of engagement, citizenship and development theory as important clarifying aspects of participation. Contemporary approaches should be based on a more coherent ideological and theoretical stance, while participation should be reframed as multi-scale citizenship. They see development as an underlying process rather than a series of technical interventions. They aim at transformation (instead of abrupt change) in power relations, take into account temporal aspects (slow processes versus the short-term timeframes of development projects), see spatial aspects in terms of situated practices (e.g. sites of resistance and dynamic political fields) and, finally, they emphasise strongly the need to consider better proper representation.

The objective of this paper is to show successes and failures of participatory approaches to research, and the pitfalls that can be expected when applied in the rural Mekong delta. These are experiences of staff from the College of Agriculture and Applied Biology of Can Tho University. It describes how in the 1980s research approaches developed from top-down to bottom-up by taking farmer knowledge as a starting point, followed by the above mentioned PRA (Feitsma *et al.*, 2002). There are recent applications by Trung (2006) who developed a Participatory Land Use Planning approach and by Le Thanh Phong (Department of Crop Sciences, College of

Agriculture and Biotechnology, University of Can Tho, personal communication), who carried out a Participatory Community Appraisal. We end the paper with a discussion on the methods used and their participatory character (or lack thereof) and some ideas for improvement.

Initial experiences

Our initial research work in the 1980s dealt with finding better methods to use acid sulphate soils, problem soils hampering agricultural production in about half the Mekong delta. Our initial approach was top-down, common practice in agronomic research in those days. This resulted in an unbalanced exchange of knowledge between farmers, experts and specialists. Figure 6.4 shows how in the field of acid sulphate soils research most knowledge was exchanged between specialists, and very little was suitable for exchange with farmers. The arrows indicate that the knowledge is generated by the group the arrow starts from and can be used by the group the arrow points at. The figure indicates the number of publications found in each category in proceedings of acid sulphate soils conferences. In our work a different approach emerged when field surveys revealed methods of improved land use developed by the farmers. We took this farmer knowledge as starting point for research in farmer fields and combined with our own knowledge and findings from other areas, aimed at optimising the farmer methods.

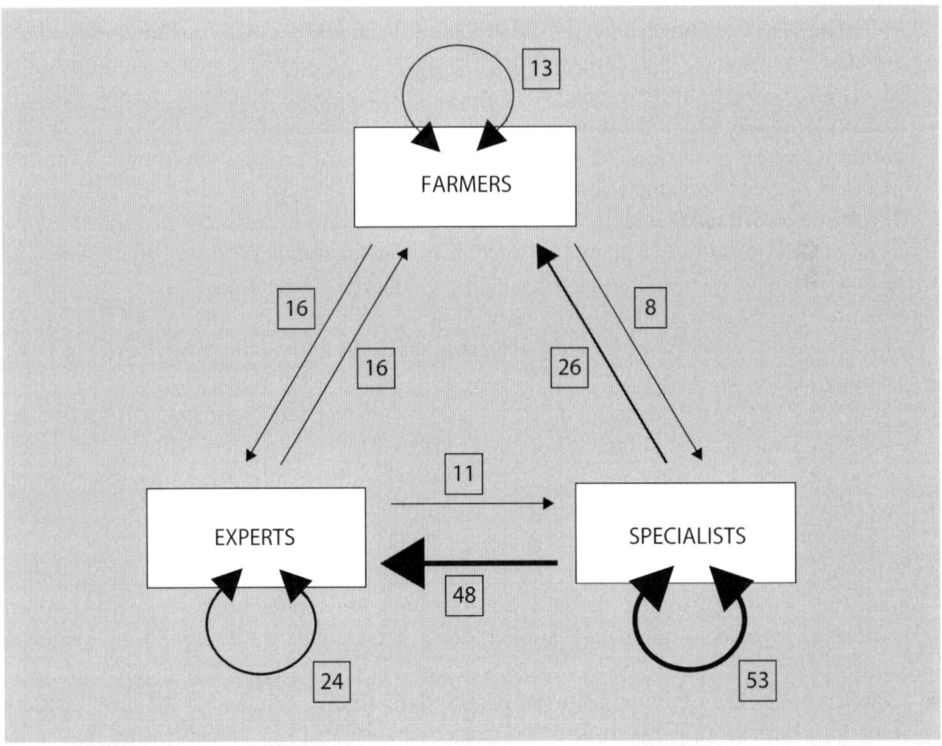

Figure 6.4. Exchange of knowledge between farmers, experts and specialists involved with acid sulphate soils (van Mensvoort, 1996).

Experience in PRA in the Mekong delta (Feitsma et al., 2002)

Much of the discussion of participatory rural appraisal (PRA) has been about its methodologies, its execution and claims to be a 'better' form of research than other methods of social learning, and a means of collaborative technology design that can relate with farmers needs. There are advantages and limitations to PRA, and many authors stress how it should be part of a range of data collection methodologies. The additional issue is that it should be relevant to farmers needs, and not just provide data for researchers, with information always flowing back to farmers, aiming for empowerment. Initial discussions centred on the how and why of collecting farmers' information. It also figures prominently in the analysis of how the wider ideals of PRA were not achieved.

The forms of participation are also relevant. To use the typology of Pretty (1994), most of the activities were passive and consultative participation where farmers listened to discussions and answered questions. The possibilities and stresses for interactive participation and self-mobilisation were not properly discussed in project design, where people should participate in joint analysis and planning, and even self-mobilisation. It was impossible to explore in advance what was really feasible in a village system still controlled by communist hierarchies. The following issues describe the initial difficulties in applying our first PRA that may still be relevant for similar future work in the Mekong delta.

- Work began without much reflection on problems that might occur in situations where there is little experience in PRA, and a political history that discouraged open communication. Moris and Copestake (1993) suggest this can result in disillusionment among villagers and practitioners, enhance difficulties in divided communities, and bring unanticipated communication problems. We experienced exactly these problems, and urge all future planners to give them ample consideration.
- The presence of a village elder, or village leader, can be very disturbing when interviewing farmers. They tend to fall silent and wait for the leader to talk on their behalf. After the first experiences this problem was solved by giving the village leader (often a person in a political position) a special task, away from where the farmer interviews took place. Since the leaders are often very knowledgeable about the entire village they performed the transect walks.
- No or very little staff was trained in PRA and, whenever they performed these studies, rarely had time to do it in multi-disciplinary teams. This situation has definitely improved with more such activities going on, but it is still very helpful to offer good, intensive training and proper planning sessions in advance. Attention on organising follow-up activities after an initial PRA and constructive discussion of challenges involved is also vital.
- The starting situation with little secondary information made it difficult to define topics to be treated, and to define groups to work. This situation has much improved with more experience and better defined PRA's (Trung, 2006; Le Thanh Phong, personal communication).
- With the large village communities and the scattered setting, common in the rural Mekong delta, it is difficult to perform flexible research open to changes and evolution during its performance, especially when rigorous plans for group contact have to be made in advance.

Time limitations and inflexibility made it difficult to adjust to topics and ideas as they came up. They, therefore, need to be thought about in advance.
- The presence of foreign experts may be a strongly contra-productive factor in participatory research activities in the Mekong delta.
- The large heterogeneity in production activities among farmers of similar economic status in villages caused some particular stresses in PRA work, in the form of boredom among participants and superficial discussions. Heterogeneity within the community can hamper the PRA methods in themselves. With very divergent activities and polarised economic status among people, it is unrealistic to expect structured discussions, because people present will be focusing on totally different subjects. Workable group size, groups with similar focus and how to move through contact with different groups still needs reflection.
- The PRA team felt that it had arrived in a local community to "do a quick PRA". This abrupt and exploitative approach is all too common in project-based PRAs where there is a deadline to meet, or in scheduled training courses.
- Local expectations can easily be raised. If nothing tangible emerges, local communities may come to see the process as a transient external development phenomenon, and be disappointed in the outcome.
- Threats. The empowerment implications of PRA, and the power of its social analysis, can create threats to local vested interests.
- Rich/poor poverty criteria based on food security, proved too general for definition of research needs in an area of rapid agricultural change. Working with well-defined groups focused on specific issues will better provide in-depth information on problems and priorities, and give a clearer expression of the dilemmas in which the villagers find themselves. Such focused groups can emerge from preliminary PRA contacts.
- Reserved behaviour by farmers in the Mekong delta has been further shaped by communist styles of communication. If one wants to perform a 'real' PRA, in which people are brought together to discuss, plan and execute a project together, there is a need for sufficient time for work and for sustained contact.

Participatory land use planning (PLUP)

This participatory LUP method was part of a broader study that applied three kinds of LUP methods to the same study area (Trung, 2006). Core problems of the area were land use conflicts, in particular between shrimp and rice farmers, environmental problems caused by land use and the desire of the local government to maintain a protective mangrove belt along the coast.

Description of the PLUP

The PLUP approach focuses on the capacities and needs of local users. It is based on the assumption that sustainable resources management can be achieved only if natural resources are managed by the local population. LUP is done both by and for the actual land users, with minimal involvement of official or professional land use authorities. Consequently, LUP focuses on the village or traditional community boundaries. This is in contrast to planning for large 'functional' areas such as watersheds or larger administrative units. Christ (1999) describes the objective of PLUP as "to create the framework for sustainable land use that is socially acceptable, environmentally sound, politically desired and economically viable". He pursues this objective by assisting local stakeholders in planning the use of locally available resources and to strengthen their capacities for managing the resources in a sustainable way. Local land users need to agree with the results of the planning process, as they will have to live with it. LUP resulting in regulations and prohibitions that are trying to prevent local people from carrying out land use activities which they practice for sound reasons, are bound to fail.

Participatory land use consists, in fact, of four central questions (Christ, 1999): (i) What is the present land use situation? (ii) What needs to be changed, what does not? (iii) How can the changes be made and what are the best options? (iv) How, when and by whom can the changes be implemented? Researchers streamlined the discussions with hints in order to obtain useable answers to these questions.

Application of PLUP

Trung *et al.* (2005) used Christ's questions and the toolbox designed by Ticheler *et al.* (2000) and tried to avoid difficulties regarding communication, lack of secondary data, large and scattered hamlets and limited time, as reported by Feitsma *et al.* (2002). In this approach, groups of about 10 key informants (experienced farmers) were formed in each hamlet. In total 26 of these groups (about 20% of all households per hamlet) were interviewed in the two villages of the study area. The villagers themselves determined who would be present. The PLUP was repeated three times, in 2002, 2003 and 2004. Two weeks per village was needed for each survey, the work was done during the summer time (August-September). To have thorough perspectives, agriculture farmers and aquaculture farmers were grouped separately. In each group, farmers participated in reviewing the hamlet's land-use history, described their land conditions and production systems, explained the reasons for land-use change, defined the socioeconomic factors that affect the change decisions, drew a sketch map showing the land use and land constraints of their hamlets, and proposed the preferred future land use. Transect walks were also conducted to verify the farmers' resource map. During the transect walk, farmers were asked for information on the land and the land-use types they practiced. To facilitate the discussion, in each group two researchers were involved. The first one initiating the debate by hinting questions, he/she also helps the villagers in drawing the resource sketch map, graphs or tables. The second person is responsible for taking notes. One day was needed for completing the land use plan of one hamlet, and another day in the subsequent years.

Initially farmers got involved with enthusiasm in the PLUP, but this attitude often receded if the discussion was too long (Fagerstrom *et al.*, 2003; Feitsma *et al.*, 2002). In our experience, the discussions should remain shorter than 3 hours. The presence of a hamlet leader during discussions and during farmer interviews usually makes villagers hesitant to give their ideas that differ from the government target (Feitsma *et al.*, 2002). This was solved by inviting the hamlet leader for a transect walk during the villagers discussion. Separating the aquaculture and agriculture villagers into focus groups can make discussions more specific, reduce boredom among participants and avoid superficial discussions (Moris and Copestake, 1993).

From the land use preference proposed by different villager groups, potential land use conflicts can be derived. This information is very valuable for land use managers, planners and decision makers.

Trung repeated the PLUP three times (2002, 2003 and 2004) which enabled him, not only to see exact land use changes over a year, but also in how far the farmers had realised the changes they had in mind (Figure 6.5). It was possible to compare the land use changes and preferred land use of three consecutive years, and enquire why changes had not been executed. Figure 6.5 shows that about 60% of the study area changed land use, while most changes that the farmers planned, had also been executed. Plans that had not been executed usually comprised mixed cropping systems. Figure 6.6 shows where the areas of potential land use conflicts are, i.e. where farmers plan things that do not go well together, e.g. a farmer plans rice while his neighbours plan shrimps (using salt water).

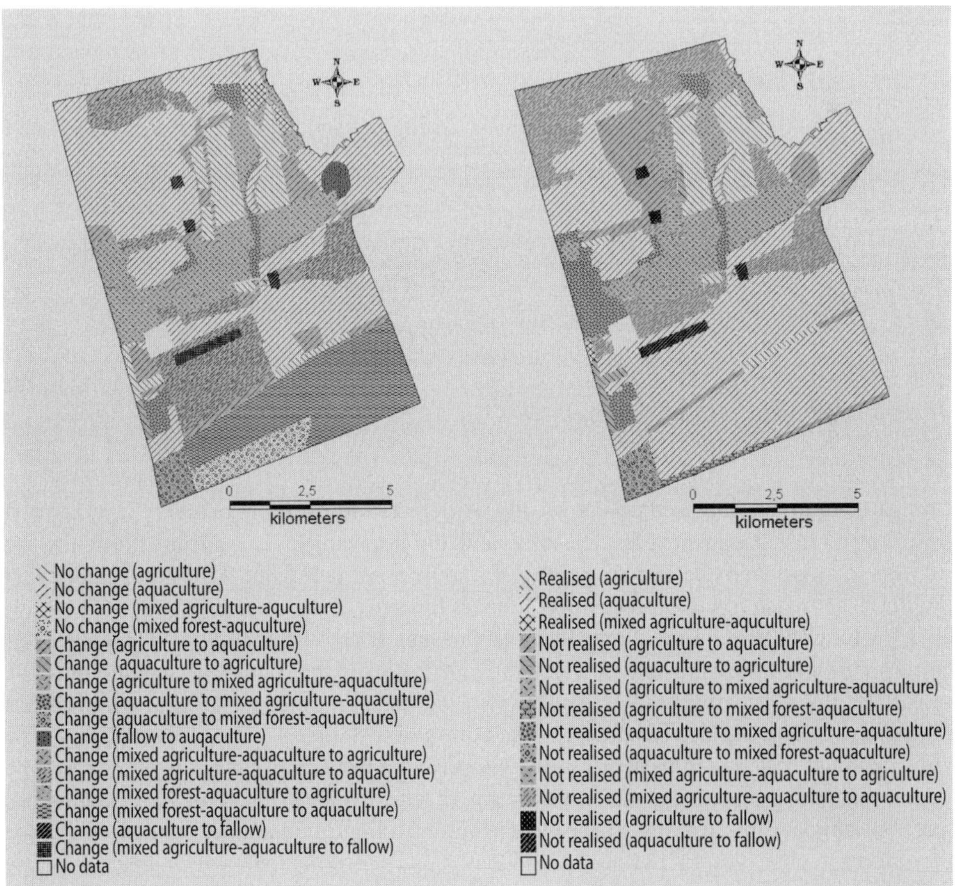

Figure 6.5. Land-use change 2002-2003 and realisation of the farmers' preferences (Trung et al.*, 2005).*

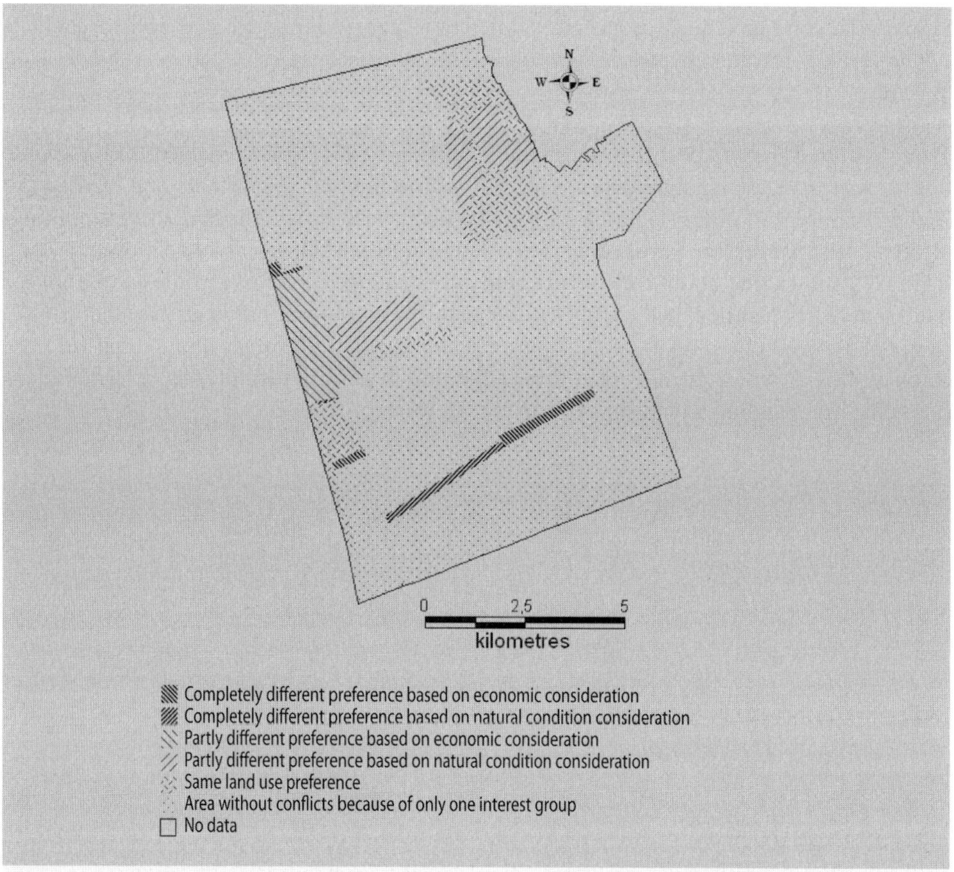

Figure 6.6. Preference conflict (Trung et al.*, 2005).*

Compared to other LUP methodologies, PLUP predicted much more precisely land use that was going on in subsequent years. The interviews did not really include social information on subjects such as ownership and the politics of resource use. Trung *et al.* (2005) concluded that the PLUP approach is a good tool to get farmers involved in land use planning as they have an opportunity to present their knowledge on the land, express what they need and give their opinions on how to use the land. It should be said, however, that this is the standpoint of the researcher who gets better results from this method than other methods. The actual decisions in land use change are driven mainly by economic considerations. The only area where the government is in charge in the planning is the mangrove protection zone along the coast. The largest change that occurs is from rice to shrimp, an almost irreversible change that sometimes forces neighbours to tag along. There is no actual planning in operation, farmers decide for themselves.

This PLUP, therefore, is not a truly participatory development plan, it is a research activity that included farmer knowledge in the study and found that PLUP was the best method. The plan was put before various stakeholders (researchers, extension workers, local government) but not to farmers.

Participatory Community Appraisal (PCA)

This PCA is part of a larger study that tries to find out the sustainability of farming systems that include fishponds (Le Thanh Phong, personal communication). Upon completion of that study, feedback sessions with the farmers are planned. The study forms part of a large project that includes breeding of fish for low input systems and optimising nutrient management in the ponds. All these activities have feedbacks with farmers in their programmes.

What is a PCA?

The participatory community appraisal (PCA) aims to (1) familiarise field researchers with participatory approaches and share experiences among them, and (2) field test and modify a detailed protocol for a PCA that will be used at each site of the same project. The results of the PCA were discussed with groups of farmers. The basic attitude of PCA is to respect the farmers' rhythm and form of dialogue; work with methods that empowers the villagers to express their views; visually shared and verify the information gathered with those concerned; and clearly outline the objectives of the PCA.

Application of PCA to understand IAA systems

Le Thanh Phong, personal communication) used a Participatory Rural Appraisal (PRA) at the beginning of 2002 to analyse the trends in the development of Integrated Agriculture-Aquaculture (IAA) farming systems. The survey areas are described by location, soil and water conditions, cropping income, fishpond activity, animal husbandry and ranking of farm components. With regard to levels of fish culture intensification, the three districts are classified into high input (O Mon), medium (Tam Binh) and low input (Cai Be) system.

Phong also used a Participatory Community Appraisal (PCA) as a tool to collect some general information about communes and hamlets, such as the status of agricultural production systems, the life of people, and the wealth/socio-economic ranking of households. Subsequently, a monthly schedule for monitoring was set up and data were collected by book-keeping and by interviews with families operating farming systems integrating fishponds. Data were collected on a large number of items, e.g. farm size, family composition, cropping systems, land use history, economic parameters on incomes (on-farm and off-farm). Two hamlets, administrative units just lower than village, were chosen in each district. For each hamlet a well-being ranking was done by three knowledgeable farmers who know the inhabitants well. These people were indicated by the local extension workers. By using PCA, Le Thanh Phong (personal communication) have classified the well-being of 743, 693 and 773 farm households in O Mon, Tam Binh and Cai Be, respectively, and selected a commune of 13 farmers in O Mon, 17 in Tam Binh and 20 in Cai Be for the detailed PCA. This included timelines, transect maps, bio-resource flows and production activities. Data on timelines were collected from 1972 onwards. After the PCA, a baseline survey was held in December 2002 on 30, 29 and

31 randomly selected farms in O Mon, Tam Binh and Cai Be districts, respectively, using a structured questionnaire. The questionnaire involved farming system patterns, land use and production characteristics. The survey was repeated, with the same households, in December 2004. In the second survey only 80 households were visited, 28, 23 and 29 farms in O Mon, Tam Binh and Cai Be, respectively, and were asked for the farm activities. The ten missing farmers had left agriculture or the village because of urbanisation or land sale.

The timeline of the high input system of O Mon is shown in Figure 6.7. It shows important developments over time such as the introduction of double rice systems, political changes, and natural catastrophes. Figure 6.8 shows farm transects reflecting common patterns of farm settlements in the Delta. Farmers usually live nearby rivers or creeks to have access to water for transportation. An orchard is established around the house, or behind the house on raised beds with a well-drained soil. Animals are kept in the farmyard surrounding the house or in the orchard. The pond is constructed nearby the house or fish is raised in ditches, if present, in the orchard. Behind the orchard land is used for rice fields (low and wet), sometimes with acidity problems. The orchards with high input (a) are located close to the main river where soil fertility is higher. The medium input systems (b) lie between the two main branches of the Mekong and have a low land surface, often acid. The low input systems (c) are found north of the rivers, partly on low land with slight acidity problems but also on higher fertile soils, close to the river.

The components in the IAA systems are related through bio-resource flows. Traditionally the rice field is the main source of food and cash income for the family and feed for the animals. Rice straw can be used to mulch beds of vegetable, orchard or to produce mushrooms. Pig manure can be used to supply feed for fish or fertilisers for the orchard (e.g. in low input system). Free-range chickens and ducks may be fed rice or rice bran. Their manure is a source of organic fertiliser for fruit trees or for the fishpond when farmers make the poultry pen above the fishpond. Other feed sources for pigs, poultry and fish can come from the orchard: weeds

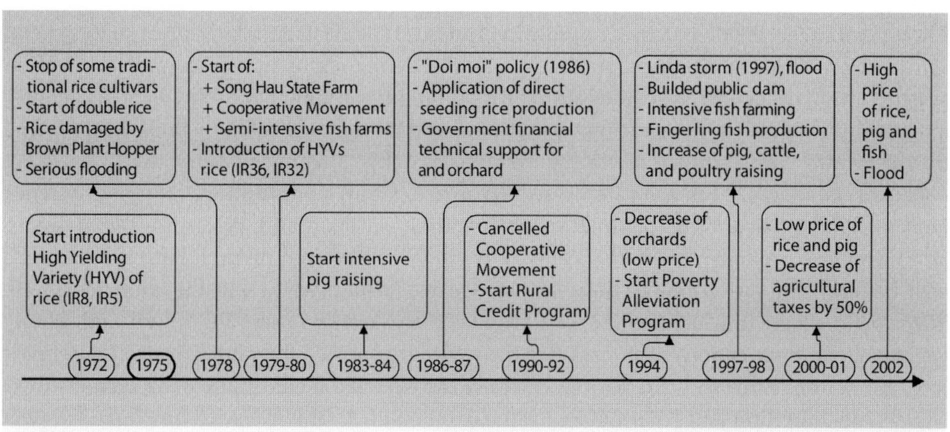

Figure 6.7. Timeline in O Mon district, Can Tho city, Vietnam (Le Thanh Phong, personal communication).

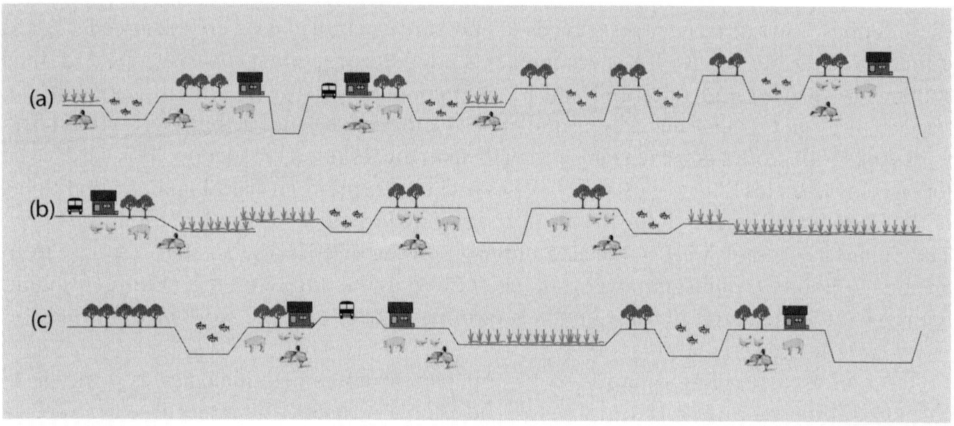

Figure 6.8. Transect maps of high input (a), medium (b) and low input (c) systems (Le Thanh Phong, personal communication).

and wastes from vegetables and fruits. The enriched mud from the bottom of the pond, or decomposed rice straw after mushroom harvesting, is placed under the trees in the orchard. The pond can be used as a latrine to supply human excreta to fish (e.g. catfish). In addition, the fishpond can supply water for the fruit trees and feed for pigs and poultry, such as water spinach, snails, or crabs. Almost all wastes and excreta are thus recycled on the farm.

Discussion and conclusions

The work described above was all in the framework of research projects. None of those had objectives with obligations towards application of the new knowledge. In our initial experiences in acid sulphate soils the experiments were carried out on farmer fields. They were frequently visited by farmer groups and some of the results were presented on the local TV. But how many farmers accepted the newly developed technology was never surveyed. Feitsma *et al.* (2002) placed the P between brackets in her original report, showing clear doubts about the participatory character of her work. The PRA provided 'social learning' for the team, and for some local support staff, but not for the larger R&D community or policy makers, and it never became integrated into a larger-scale initiative that could specifically target low-income groups in the region.

Trung's (2006) work is very recent. The PLUP approach is a good tool to get farmers involved into the land use planning. They have an opportunity to present their knowledge on the land, express what they need and their opinions on how to use the land. But Trung only presented his work in a conference and to some stakeholders in the district he studied, but not to the farmers. His study was mainly geared towards local decision makers who, so far, have only relied on top-down planning studies that did not involve farmers.

Phong's PCA was useful for field staff to understand clearly the present situation of a village or a community, problems and constraints, and opportunities for future development. However, the PCA has some disadvantages such as (1) no hard quantified data are produced; (2) it is difficult to replicate - each PRA is unique; (3) if not done properly it can easily yield falsified statements, distrust and rejection; and (4) it encourages intensive communication which triggers too high expectations. Phong and his Vietnamese colleagues in the POND project held workshops with groups of farmers informing them of the results of his surveys. This needs to be followed up when his PhD work is ready. We have to be careful when claiming direct usefulness of the participatory research for the farmers compared to other forms of research. However, some modest successes were in the acid sulphate soils experience, the positive response from farmers to Phong's results and expectations of more useful advice on fish culture, fishpond management and the role of these ponds in farming systems once the work is completed.

In Vietnam the prevailing desire is to present participatory methodologies as a means to harmonious interaction between local people and external advisors. This desire obscures a debate about overall approaches to technology development, poverty alleviation and empowerment. In the future, research programmes involving Natural Resources Management might benefit from the processes of participatory monitoring and evaluation (methods of information collection, frequency of feedback, kind of meetings and contents) as described by Vernooij *et al.* (2006). Such an approach would eliminate the feeling that all stakeholders are only involved in the stage of data collection but only a few of them are involved in further steps. Surely Can Tho University cannot be charged with the management of large development projects using participatory methods to involve the local people. But Can Tho could be instrumental and catalysing in the use of proper methods. It should not simply apply PRA as a methodology, referred to as tyranny by Cook and Kothari (2001), but incorporate the recent thoughts of how participation should be included properly in projects. One chance to do this will be the next Dutch financed project, RESCOPAR, that studies livelihoods in the coastal area and offshore in the delta.

References

Angarwal, B., 2001. Participatory Exclusions, Community Forestry and Gender: an Analysis for South Asia and a Conceptual Framework. World Development 29, 1623-1648.

Barzman, M.S., N.J. Mills and Nguyen Thi Thu Cuc, 1996. Traditional knowledge and rationale for weaver ant husbandry in the Mekong delta of Vietnam. Agriculture and Human Values 13, 2-9.

Binh C.T., M.J. Phillips and H. Demaine, 1997. Integrated shrimp-mangrove farming systems in the Mekong delta of Vietnam. Aquaculture Research 28, 599-610.

Christ, H., 1999. Participatory Land Use Planning in Cambodia. Proceedings of the Introductory Workshop14 - 15 September 1999, Department of Forestry and Wildlife, Phnom Penh. www.mekonginfo.org/mrc_en/doclib.nsf/0/A6470DC37BEFBE5AC72567FA002E057D/$FILE/MAIN-REPORT.htm

Cooke B. and U. Kothari (eds.), 2001. Participation: The New Tyranny? Zed Books, London.

FAO, 1990. Participation in Practice: Lessons from the FAO People's Participation Programme, Rome.

Fagerstrom, M.H.H., I. Messing and Z.M. Wen, 2003. A participatory approach for integrated conservation planning in a small catchment in Loess Plateau, China - Part I. Approach and methods. Catena 54, 255-269.

Feitsma, M., L. Vincent, and N.D. Can, 2002. Understanding the challenges of PRA for farming systems research: lessons from coastal Vietnam. In: M.E.F. Van Mensvoort and. L.Q. Tri. (eds.), Selected papers of the workshop on integrated management of coastal resources in the Mekong delta, Vietnam. C.T. de Wit Graduate School for Production Ecology & Resource Conservation (PE&RC), Wageningen.

Gladwin, C.H., J.S. Peterson and A.C. Mwale, 2002. The Quality of Science in Participatory Research: A Case Study from Eastern Zambia. World Development 30, 523-543.

Hickey, S. and G. Mohan, 2004. From Tyranny to Transformation? Exploring new approaches to Participation in Development. Zed Books, London.

Mensvoort, M.E.F., 1996. Soil knowledge for farmers, farmer knowledge for soil scientists: the case of acid sulphate soils in the Mekong delta, Viet Nam. PhD thesis, Agricultural Wageningen University, The Netherlands.

Moris, J. and J. Copestake, 1993. Qualitative Enquiry for Rural Development. Intermediate Technology Publications on behalf of the Overseas Development Institute, London.

Pretty, J.N., 1994. Alternative Systems of Inquiry for a Sustainable Agriculture. IDS Bulletin Vol. 25 No. 2. IDS, Brighton.

Ticheler, J., T. Defoer, and L. Kater, 2000. ResourceKit for participatory learning and action research. Detailed field tools for PLARUser's guide for the ResourceKIT, p. 1 CD-ROM. Managing soil fertility in the Tropics. KIT, Amsterdam.

Trung, N.H., 2006. Comparing land use planning approaches in the Mekong Delta, Vietnam. PhD thesis, Wageningen UR, The Netherlands.

Trung N.H., L.Q. Tri, M.E.F. van Mensvoort and A. Bregt, 2005. Participatory land use planning (PLUP). proceedings "2004 AFITA/WCCA. The 4th International Conference of the Asian Federation of Information Technology in Agriculture and the 2nd World Congress of Computers in Agriculture and Natural Resources. August 9-12, 2004. Bankok, Thailand". Published by Hydro and Agro Informatic Institute, National Science and Technology Development Agency, Thailand.

UNDP, 2003. http://www.undp.org.vn/undp/docs/2003/ppa/mekongppae.pdf. accessed March 2006.

Vernooij, R., Sun Qiu and Xu Jianchu, 2006. The power of participatory monitoring and evaluation: insights from south-west China. Development in practice 16, 400-411.

Banging the bells: communicating hope for rural environments

Lorna C. Malicsi
Knowledge Management Unit, Southeast Asian Regional Center for Graduate Study and Research in Agriculture (SEARCA), Philippines; lcm@agri.searca.org

Abstract

This paper describes several communication methods used by development organisations to help improve the quality of life of people in rural environments around Laguna Lake. Four of these include the Community-Based Coastal Resource Management (CBCRM), Farmer Participatory Extension (FPE), Conservation Farming Village (CFV) and Information, Education, and Communication (IEC). These methods relied heavily on (1) heavy community organising and immersion work with non-government organisations; (2) high visibility of local government units in supporting/sponsoring CFV, and (3) a strong Information, Education, and Communication (IEC) component. In particular, IEC was used to create a strong public information awareness on the worsening situation of the lake, and at the same time to elicit commitments from stakeholders to rehabilitate the Laguna Lake of the Philippines. The IEC campaign culminated into a multi-sectoral dialogue that gathered, for the first time, 600 participants representing different sectors. Fifty participants were surveyed for their impressions on the conduct of the multi-stakeholder dialogue. Seventeen wanted to know specific programs to save the lake and ten participants recommended that the next dialogue will report the accomplishments after the first dialogue.

Keywords: communication, participatory, co-management, sustainable, stakeholder

Introduction

The good news with today's information system is that rural people communicate better and faster than they used to in the past years. This information system has lately made meaningful changes in the quality of the life of rural people. But the story does not stop from here. Information technology is never meant as a stand-alone. Development planners need to bring in other ingredients in order to reduce poverty in rural areas.

In the ancient days, the church would 'bang the bells' during times of festivities or emergencies. Community members, as soon as they heard the bells ringing, would immediately converge in the church vicinity to find out what matters were to be discussed and resolved. People understood perfectly what these ring tones meant. If the bells rang with a sad note, it meant somebody receiving his/her last rites on earth. If the bells rang joyously, it meant a couple being finally tied in marriage. If the bells rang with a note of urgency, it meant an impending disaster. Today, in many parts of the world, we hear another kind of ringing. We hear ring tones emanating from phones, computers, and other gadgets which have meanings perfectly understood by respective individual users. However, even with this very efficient transmission of messages, many questions remain floating in the air: Why is poverty still persistently creeping like weeds in the rural areas? Why are forest areas disappearing? Why is the environment silently wasting?

This paper gives a bird's-eye view of various communication and extension strategies, but discusses in detail the use of Information, Education, and Communication (IEC) in promoting participatory management of Laguna Lake in the Philippines.

Communication methods for rural environments

Except for a few cases, rural environments are generally poor. Increase in population, lack of education, capital, and disappearing forests because of lack of alternative means to earn a living other than farming, and many more – these all contribute to a highly marginalised society. This is why, even if the peripheral cities or countries for that matter are bursting with economic zest, rural areas are moving with lethargic energy towards the economic ladder of independence.

Despite this lethargy, there is hope as development-oriented organisations use approaches that empower the rural poor. They use various communication and extension approaches, like:

1. Community development communication, development journalism, and village drama. The International Center for Agricultural Research in Dry Areas (ICARDA) uses these methodologies in disseminating new technologies to farmer communities. Through Knowledge Management and Dissemination (KMD) type of research programs, ICARDA engages in a co-learning approach that aims to change culture and behavior of various stakeholders (ICARDA, 2004).
2. One-Stop Aquaculture Shop and Information System. The Network of Aquaculture Centres in Asia-Pacific (NACA) arranges sessions for shrimp farmers to meet and exchange information in Vietnam. Various agencies get to meet these farmers and monitor/track information particularly for problem areas in raising shrimps (NACA, 2004).
3. Information and Communications Technology for market-related information. The Centro International de Agricultura Tropical (CIAT) in Columbia uses ICT to support the creation of local content. In doing this, CIAT explores the use of participatory approach by which community-based stakeholder groups can be tapped. The approach embraces the use of conventional media (print, radio, printed bulletins, local drama).
4. Community-Based (Coastal) Resource Management (CBCRM). Pioneered by Silliman University to save coral reefs and other coastal resources in Apo Island, Dumaguete City, Philippines. This approach uses community organising in educating coastal pilot communities on the importance of maintaining sanctuaries. At present, the community maintains a fish sanctuary in its coastal area.
5. Farmer Participatory Extension (FPE). The ASIALAND Management of Sloping Lands Network finds FPE effective in empowering resource-poor farmers in Southeast Asia. FPE follows nine stages: training research and extension workers, site survey and characterisation, participatory client selection, participatory planning, enabling farmer volunteers to conduct on-farm trials, participatory monitoring and evaluation, training farmer trainers, disseminating knowledge and technology by farmers, and establishing farmer networks. FPE uses the Farmers' Field School Approach, but concentrates more on

training individual farmer-leaders who teach other farmers by showcasing their own farms as demonstration sites of soil conservation technologies.

6. Conservation Farming Village (CFV). The Philippine Council for Agriculture, Forestry and Natural Resources Research and Development (PCARRD) and Care Philippines collaborated in a CFV pilot addressing selected upland communities. CFV taps local government units' involvement to formally declare some villages as CFVs (Malicsi, 2005).
7. Information, Education, and Communication (IEC). Studies of the IEC approach, such as those conducted by FAO (FAO, 199X) suggest that the use of a comprehensive IEC support for development projects unites a community and motivates this community to actively participate in resolving issues that concern their lives. Lessons learned from these studies include:
   a. Knowledge of existing community structures (e.g. who are the people who are highly credible and what formal or informal groups are they members of) and making communication campaigns flow through these structures.
   b. Rather emphasise positive behavior change than the negative consequences of current behavior to obtain development.
   c. People learn new behaviors best when they are learning something they feel is useful, when they can put into practice what they are learning, and when they receive feedback and are rewarded for doing well. Showing specific examples or demonstration sites is often the best way to teach complex behaviours.

These innovative approaches are supported by knowledge exchange and dissemination through institutes like e.g. the Southeast Asian Regional Center for Graduate Study and Research in Agriculture (SEARCA). SEARCA and other institutes build knowledge assets basically from its four offerings: graduate scholarship, short-term training, research and development, and consulting to advance agriculture and rural development. Dissemination and exchange of this information are through the print, web, face-to-face dialogues, roundtable discussions, conferences, seminars, etc. Clients are policy makers, development planners, executives, and managers who can influence policies and decisions on agriculture and natural resource management that might improve rural economies in South East Asian countries.

Using IEC campaign to save the Laguna Lake

An IEC campaign[5] to save the Laguna Lake in the Philippines was done by the AsiaDev Consulting Group[6] from June 2005 to February 2006. The IEC campaign consisted of four major activities, namely: (1) Multi-sectoral Dialogue, (2) Launching of the Integrated Watershed Management Learning and Resource Center, and (3) Multi-sectoral Participatory IEC Planning-Workshop; and (4) the Capacity Building of volunteer IEC representatives from various sectors. This paper discusses only the first major activity: the Multi-sectoral Dialogue. The objective of the dialogue was to gather all groups which had a stake or an issue on saving of Laguna Lake. The Laguna Lake Development Authority (LLDA) led this activity. LLDA is

[5] Funded by the World Bank and the Philippine Government.

[6] The author is one of the IEC Specialists working on the project.

an agency which coordinates and leads government initiatives in sustainable managing the lake and its resources.

Laguna de Bay

The Laguna de Bay provides life-support services to more than 10 million people in various ways. It is the largest and one of the most vital inland bodies of water in the Philippines. Its waters cover an area of approximately 900 square kilometres. The lake can hold more than 2,000 million cubic meters of water. Laguna de Bay is used for many purposes: supplying drinking water, irrigation water, fish and other foodstuffs, trading, recreation, and many others.

Today, it serves as a sink to most of the industries and households surrounding it. Some tributaries of Laguna Lake have water quality that is deteriorating because of high nutrient loading. Specifically a yearly load increase of nitrogen by 400% and phosphorus by 600% in San Pedro and Morong Rivers, have been observed within a ten-year period. Household wastes contribute about 69% of the organic wastes discharged into the lake; the rest come mainly from industry (19%) and agriculture (12%).

Biodiversity, on the other hand in Laguna Lake, has declined in the last ten years. Brackish water species, *Scatophagus argus*, for example no longer exists in the lake. Migratory species have been affected by government's policy to divert rivers for irrigation and dam construction. These decisions, somehow, have changed the habitat (drying of riverbeds in some areas) and the flora and fauna thriving in affected areas.

Ravaged through the years by abusive anthropocentric activities, Laguna de Bay is a centrepiece of campaign for resuscitation. The process of reviving the lake and its resources is multifaceted and will need the involvement of all stakeholders. Basically, IEC was used as a tool to mobilise and educate people on how to protect, preserve, and prepare the Lake for the next generations.

The Laguna Lake Development Authority (LLDA)

The Philippine government established LLDA in 1966 to lead, promote, and accelerate the development and balanced growth of the Laguna de Bay basin within the context of national and regional plans and policies for social and economic development. Primarily, it carries programs and activities that prevent undue ecological disturbances, deterioration and pollution of Laguna de Bay. LLDA's programs aim to influence stakeholders to become co-managers of the lake and its resources (http://www.llda.gov.ph/llda.htm). Because of the intensity of degradation that is taking place, there is a need for a multi-sectoral approach in attacking the problem of rehabilitating the lake. Thus, the IEC campaign.

Methodology

The AsiaDev Consultants used the IEC Conceptual Framework (Figure 6.9) in implementing the campaign for the Multi-sectoral Dialogue:

1. Consulting. AsiaDev consulted the different stakeholders of Laguna Lake about their perceptions and preferences on participating in the First Multi-sectoral Dialogue. AsiaDev consultants used Focus Group Discussions (FGDs) and Key Informant Interviews (KIIs) in getting information from stakeholders.
2. Analysing Content. After the FGDs and KIIs, IEC implementers analysed the contents and designed a theme as the rallying point by tying together all messages into one coherent IEC campaign theme.
3. Designing strategy/IEC theme and support activity. Contents woven into a thematic presentation of messages were packaged into different communication support materials: posters, radio scripts, straight news, video documentary, etc.
4. Approving Content and Budget. Lead organisation, in this example, the Laguna Lake Development Authority evaluated the content vis-à-vis technical content accuracy and budget and approved the actions, including the corresponding IEC strategy materials.
5. Pre-testing. This validated the intended stakeholders' perceptions on the types of communication activities and their corresponding IEC materials. It assessed the stakeholders' understanding and acceptance of the message being given to them.
6. Implementing IEC strategy. The Multi-sectoral Dialogue was implemented and support IEC materials were distributed.
7. Evaluating. The Multi-sectoral Dialogue and the IEC support materials were evaluated by letting the participants answer a simple questionnaire.

The Multi-sectoral Dialogue

The implementation of the Multi-sectoral Dialogue started with the creation of public awareness on the existing state of the Lake. Twenty-one press releases about the existing condition of the lake and the need for stakeholders to meet and dialogue were published in the Philippine National Dailies and Community Newspapers before the event. Posters and brochures were pre-tested with actual stakeholders before these were strategically posted and disseminated. Institutional radio announcements for the schedule of Multi-sectoral Dialogue were made. Some Laguna Lake sectoral representatives were interviewed live in the national and community radio and TV stations. The first among the four major IEC events in working towards a participatory lake management and resources sustainability, the Multi-sectoral

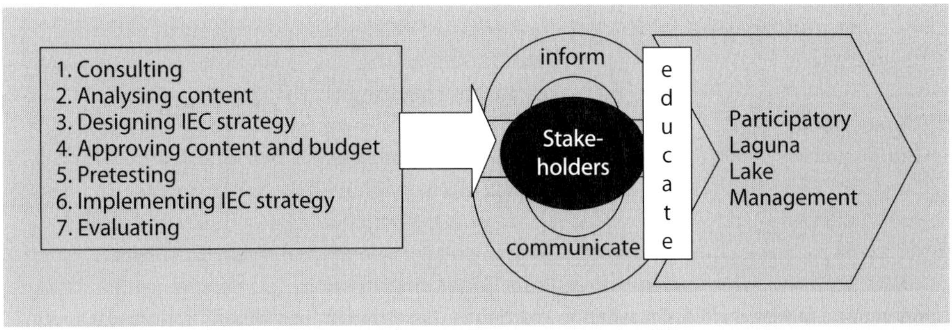

Figure 6.9. The conceptual framework of IEC (Information, Education and Communication.

Dialogue, or the Laguna de Bay Summit was held last August 19, 2005, at the Ayala Greenfield Estate Clubhouse in Maunong, Calamba City, Philippines. The activity was well accepted by the stakeholders. Four hundred stakeholders from different sectors were originally invited, but the turn-out was greater than expected: more than 600 participants during the day of the event.

The Multi-sectoral Dialogue had the following contents (gleaned from the Consulting phase):

- Programs and Accomplishments of Laguna Lake Development Authority in Saving Laguna Lake in the Philippines.
- Testimonies/best practices of other sectors in Saving the Lake.
- Open Forum after presentations.
- On-the-spot poster-drawing contest among high school students. Theme covered a united effort in saving the lake.
- Evaluation of the Multi-sectoral Dialogue, including the Exhibit materials during the day.

From this event, various issues were sieved to LLDA (Table 6.2). As a result of the multi-stakeholder dialogue, LLDA was able to build up its database on existing groups/key persons involved in Laguna Lake Rehabilitation. This database includes 600 key participants from the 5 provinces and parts of Metro Manila.

Evaluation of the Multi-sectoral Dialogue from the participants recorded the following:

1. Majority found the Multi-sectoral Dialogue as highly educational, useful and productive. It helped them become more aware of the mandates and functions of LLDA.

Table 6.2. Issues raised by stakeholders.

Stakeholders	Issues
Business Group	Need for Clear Policy Guidelines on the Environmental User Fund (EUF). The group feels that the domestic households where majority of the wastes come from, do not pay EUF; while they who are following the rules by using solid waste management technologies are penalised.
Fishermen	The lake waters smell bad, the fishes and other resources which were plentiful in the past, have significantly dwindled.
Federation of River Basin Council	They are tired of the lake clean-up drives and feel nothing is being accomplished. They feel the local government units are not implementing strictly the laws on solid waste management.
Youth (Students)	They want to see the local government units to apprehend people/groups who are breaking the law, particularly on open waste dumping disposal.
Local Government Units	They are happy with the lake rehabilitation projects that they are currently doing.

2. High school students said that the activity widened their understanding on the importance of the lake. They were impressed that government through LLDA is doing something good in protecting the lake.
3. Participants from various groups highly welcomed focus group discussions as an avenue where they can air their issues and concerns. Participants made recommendations on what issues to discuss in the next Multi-sectoral Dialogue.

Discussion and conclusion

Laguna Lake

Preliminary results show various stakeholders appreciating government's move to involve them in rehabilitating the lake. The IEC approach for saving the lake has gathered overwhelming responses from its stakeholders, particularly on the use of the multi-sectoral dialogue as a major IEC strategy. This strategy is unanimously recommended by stakeholders to be done on a yearly basis. One policy that can be implemented is for government to earmark at least 15% of its total budget for IEC purposes in saving the lake. And part of this expense would be used for implementers and stakeholders alike to constantly dialogue; with the former mobilising stakeholders to get involved in lake rehabilitation.

During the Multi-sectoral Dialogue, stakeholders learned of the existing condition of the lake and the factors and practices contributing to its deterioration. The presentation of results of studies conducted by the academe served as an eye-opener to the participants during the day. The dialogue (question-and-answer) portion further clarified issues concerning the management of the lake. Written responses of 50 participants for the topics they wanted to learn more is in Table 6.3. An exhibit about the Lake was featured to complement the presentations.

Lessons learned

At the organisational level, development agencies can efficiently make use of knowledge exchange through print and web; development communication; one-stop-information shop; and information communication technology because they have control over their human resources, budgets, facilities, and equipment. Externally, these development agencies are well connected. They are able to tap expertise, database, and new funding sources from partner organisations with similar concerns. The IEC approach of spreading information to raise public awareness was very successful: the number of participants at the multi-stakeholders dialogue turned out much greater than was expected. IEC is one of the strategies that can be used to create a large participation of stakeholders.

On the other hand, at the community level, CBCRM, FPE, and CFV can be used by development agencies as extension approaches. They are very similar in goals and approaches. Once anchored in the lifestyle of community people, these approaches redound to sustainable ecosystem resource management in upland and coastal areas.

Table 6.3. What stakeholders want to know about the lake.

Information wanted	No. of respondents
1. About Laguna lake- plans/ programs/implementation	17
- Hydrology of the lake including its tributaries	
- Specific programs to save the lake	
2. Lake improvement after the first dialogue	10
3. Youth-oriented projects/programs	6
- Special project for the kids in elementary grades	
- How can children help to save the environment	
4. Stakeholder workshop	5
- Stakeholders' commitment to make the lake clean	
- Commitment of industries and households	
5. Fishermen participation	4
- Livelihood program for displaced fishermen	
- The testimonials of the fishermen and the youth	
6. Others	8
- The improvement of the Tanay River Basin	
- Funding the different projects of the river councils	

The success of the CFV approach hinges on six factors contributing to successful adoption of sustainable farming practices among farmers (Manzanilla *et al.*, 2004). Three of these had something to do with interpersonal communication or face-to-face contact: (1) heavy community organising and immersion work with non-government organisations; (2) high visibility of local government units in supporting/sponsoring CFV, and (3) a strong Information, Education, and Communication (IEC) component that aggressively promotes soil conservation technologies.

Concluding, development agencies can use IEC as a complement to extension innovations like 'Community-Based Resource Management' and 'Farmer Participatory Extension' to promote new paradigms and technologies. IEC can help development and extension planners to engender participation and behavioural changes among stakeholders, and to increase awareness and knowledge of communities. IEC would be a major approach that planners can use in banging the bells to communicate hope and healing for rural environments.

References

Centro International de Agricultura Tropical (CIAT). http://www.ciat.cgiar.org/.

FAO, 199X. Development Support Communication Studies in the Philippines. Philippine Council for Agriculture, Forestry, and Natural Resources Research and Development (PCARDD). Los Banos, Laguna; Philippines.

International Center for Agricultural Research in Dry Areas (ICARDA), 2004. Annual Report 2004. pp. 64-65. http://www.icarda.org/Publications/AnnualReport/2004/AR_04.html.

Laguna Lake Development Authority (LLDA). http://www.llda.gov.ph/llda.htm.

Malicsi, L., 2005. Factors Contributing to Successful Sustainable Farming Adoption Among Selected Farmers in the Philippines. In: European Forum. International Conference on Agricultural Research for Development: European Responses to Changing Global Needs. Abstracts. 27-29 April. Swiss Federal Institute of Technology Zurich, Switzerland.

Manzanilla, D., L. Malicsi, E. Paga and G. Bordado, 2004. Sustainability and Effectiveness of Conservation Farming Village (CFV) as a Technology Transfer Modality under the AsiaLand Project in the Philippines. In: Conference Book SWM International Conference on Innovative Practices for Sustainable Sloping Lands and Watershed Management. September 5-9, 2004, Chang Mai Hill Hotel, Chang Mai, Thailand.

Network of Aquaculture Centres in Asia-Pacific (NACA), 2004. STREAM Agreement with the Ministry of Fisheries, Vietnam on the Development of a Communications Strategy. http://www.enaca.org/modules/news/article.php?storyid=282.

Southeast Asian Regional Center for Graduate Study and Research in Agriculture (SEARCA). www.searca.org

Extension services and market integration of small farmers: Case studies of INVE and VN-Dutch Lady in Vietnam

Roel H. Bosma[1], Rudy H.R. Bijnens[2], Dick Harting[3] and Tran Thanh Bê[4]
[1]Aquaculture and Fisheries Group, Department of Animal Sciences, Wageningen University, The Netherlands; roel.bosma@wur.nl
[2]Director Vietnam of INVE (Thailand) Ltd., Ho Chi Minh City, Vietnam
[3]Technical advisor Dutch Lady Vietnam Food & Beverage Co. Ltd., Ho Chi Minh City, Vietnam
[4]Director of the Mekong Delta Research Development Institute, Can Tho City, Vietnam

Abstract

Extension services play an important role in the poverty alleviation of small farmers in developing countries. In the past most of the extension work has been government and project driven. Two cases, one in dairy and one in aquaculture, demonstrate that for specific sectors the profit-oriented companies are complementary to government extension service, if next to inputs they provide e.g. training and are active in problem-oriented research. However transferring technology alone is not enough to improve livelihoods of the rural people; the extension services also need to equip the end-users with information, knowledge and skills on various issues relating to rural economy. On the other hand agricultural government agencies should not seek to completely confine their extension services to areas with public good characteristics, they are also called to develop initiatives stimulating market integration.

Keywords: extension, private, public, aquaculture, dairy

Introduction

The development of small farmers is commonly considered a tool for the reduction of poverty in developing countries. In development practice most of the extension work has been government and project driven, while in the industrialised countries e.g. the animal feed industries have partially replaced government extension services (Smith and Thompson, 1991). Here we present two cases in Vietnam, one related to aquaculture and the other to dairy processing, and subsequently discuss the opportunities and threats for industry driven extension and innovation of farming systems, compare them to government extension systems, and make suggestions on the tasks of public extension services.

The market development approach of INVE Aquaculture Health in Vietnam[7]

The Vietnamese shrimp sector is concentrated in five coastal provinces in the centre and the south of the country. The shrimp production is a steady growing industry; with a production of close to 250,000 ton year^{-1} Vietnam is ranked fourth after China and Thailand but is catching up

[7] Based on the contribution of Dr Rudy Bijnens that was presented by Prof. Patrick Sorgeloos.

with India/Bangladesh. Vietnam is taking over the global market share of the *Pineaus monodon* production from Thailand. Vietnam has an estimated 4,200 shrimp hatchery operators and over 450,000 ha of shrimp grow-out ponds involving over 700,000 families. Annually the hatcheries produce over 18 billion post larvae. Over 90% of the hatcheries are back-yard type producing between two and 10 million post larvae per annum. They are financially weak and do not buy a stock of feeds and chemicals. Most shrimp ponds are managed by a household and a typical shrimp farmer has one pond of 4,000 m^2 and limited financial means. Those small family farms have limited access to loans from the bank who consider shrimp culture a risky enterprise.

Commercial infrastructure

In the five coastal provinces, 55 out of the approximate 100 retail shops for hatcheries are supported by INVE, a Belgium based international enterprise for animal feed solutions. Other shops specialise in products for grow out. Most products are sold on credit and the farmers' purchases are based on their daily need. The shops are mostly small family businesses with turnovers between 20 million and 6 billion VND $year^{-1}$ (US$ 1,250 to 375,000). About 50 'big' shops have a turnover above 500 million VND from around 100 customers each having an average credit of over 5 million VND. The total credit in the five provinces to hatcheries amounts to over 0,5 billion VND. About 50 'small' shops have a turnover below 500 million VND from around 50 customers each having an average credit of over 3 million VND.

The total value of goods in credit to hatcheries amounted to 0.65 billion VND (over 40,000 US$). The hatchery owners only pay back their debt to the shop after harvest and profit has been made, which is a significant risk for the distribution channel. The shops have to decide whether to continue credit to a hatchery after a failed cycle. INVE trains the shopkeepers, provides the shop with decoration and advertisement tools, and minimises the commercial risks for shops, keeping their stock value low through frequent provision.

Client oriented approach

The core of the INVE approach is to provide solutions to the problems the end-users of the products experience by getting as close as possible to these clients. INVE applies a three stages approach: define the customers' needs; develop a solution and deliver the solution through the shops and fairs. To define the customer needs 15 first line agents visit shops, hatcheries and farms to collect information, to give after sales service, to check the targeted performance, and to train the technicians and extension workers. This allows these agents to identify, measure and analyse the needs of the customers and to support INVE's market intelligence. They assist INVE with describing the solution that might meet those needs and, in collaboration with local institutes involved in aquaculture and fisheries, create the technology and offer products or advice to meet those needs. Furthermore they support farmers' training sessions, seminars and workshops.

The partnership of Dutch Lady Vietnam with the local dairy farming community

Dutch Lady Vietnam Food and Beverage Company is a joint venture between the Dutch company Royal Friesland Food, organised under the People's Committee of Binh Duong. The company commenced its operation in 1996 under the name Vietnam Foremost Dairy Company, the name changed in June 2002. The Dutch participation is 70% of the capital shares and the Vietnamese partner 30%. Dutch Lady Vietnam's main products are: ready to drink milk, full cream, fresh milk, and sweetened condensed milk and a variety of yoghurts. All products are distributed nation-wide under the brand names 'Dutch Lady', 'Yomost' and 'Completa'. Dutch Lady Vietnam is involved in a Dairy Development Programme which is focussing on assistance to small dairy farmers. To date, the total number of farmers participating in the programme is 2,500. By end of 2005 they supplied 80 tons of fresh milk per day and the quantity is increasing steadily.

The company designed a transparent milk payment method to stimulate dairy farmers to produce high quality fresh milk. The purchase price of the milk by the company varies between 3,900 VND kg^{-1} and higher, depending on the additional milk quality. Dutch Lady Vietnam pays a premium in addition to the base milk-price. If farmer's milk quality surpasses the standard quality (12.0% Total Solids, 3.5% fat, bacterial contamination Resazurin 4.0; no additives), a premium is paid that is in accordance with the surplus quality. Payments are made directly to the farmers once every two weeks. To further stimulate the farmers to produce and deliver good quality fresh milk the company introduced a milk-price-bonus of 20 to 100 VND kg^{-1} (15,900 VND = 1 US$). The farmers can earn this bonus if their farm and milking utensils meet set hygiene standards and sound production methods.

The dairy development support programme

The programme aims to build a sustainable relationship with local dairy farmers supplying fresh milk to the company. The initial programme started in September 1995 for a period of ten years involving a total investment of 6.6 million US$, it was continued beyond 2005 with an additional funding of 6 million US$. The company has organised 40 milk collection points where farmers can sell their fresh milk directly to the company. These points are scattered over the provinces Binh Duong, Tay Ninh, Long An and Ho Chi Minh City. In addition three Milk Chilling Centres were established with a total investment of 1.7 million US$. All milk collected from the farmers is chilled at these locations before it is transported to the processing plant.

Other components of the projects are the organisation of a support service, farmers' training and a credit programme. Dutch Lady Vietnam technical staff, as well as especially employed workers, provide farmers, free of charge, with support service such as artificial insemination and veterinary service, as well as medicines at very low prices. The staff is always available to answer the farmers' questions. A special telephone help desk is made available solely for this purpose. The main objectives of the training, which is of mutual interest, are to raise awareness among farmers for a better quality of fresh milk, and to strengthen the farmers' knowledge and skills for improving the quality of this fresh milk. Not only are farmers and Dutch Lady

Vietnam concerned about income, but also about consumers' health. The subjects of training are veterinary healthcare, hygiene, the technique of milking, good design of breeding facilities, and how to look after cows during pregnancy and calving. So the training does not only focus on teaching farmers to produce milk of a high food safety standard. The training follows the approach of learning by doing and observing. Farmers are provided with learning materials and attend training at the company's office and at a small demonstration farm. From 1998 the project enabled farmers to get credit from Dutch Lady Vietnam via a Vietnamese bank in order to purchase one or two heifers. In 2003 Dutch Lady Vietnam imported 216 pregnant F1 dairy heifers from New Zealand and distributed the heifers among 96 dairy farmers in the dairy community in the Ben Cat district of the Binh Duong province.

To develop the relationship the company staff approaches farmers at home and farmers are invited to the companies' extension centre where they are informed about the support programme. When farmers agree to produce milk, they will be required to attend training sessions and the staff visits their husbandry facilities, monitors their progress and checks the condition of the milk cows. If the fresh milk of a farmer lacks quality, the staff will visit the farm and examine the hygiene of the husbandry-facilities, take samples to analyse the milk and give recommendations for improving its quality. To obtain credit for cattle, the farmers are required to bring their fresh milk to a collection point, to provide fresh milk according to the quality standards of Dutch Lady Vietnam, and to maintain the standards for good animal care.

Benefits of the project for the partners

The programme provides an efficient model of transfer of technology for small farmers who can rely on assistance from the company's technicians whenever they encounter problems. The farmers learned new techniques to raise cows and to improve the quality of the milk, and confirm that the risks with regard to raising cows have been reduced. As the farmers escape from intermediate traders, they get favourable prices when selling milk directly to the company. In addition, the company is creating a milk market for the farmers. Most of the farmers have increased their herd, and raising cows has become their main economic activity. The income of the farmers' households has increased 30 to 50 percent. Some households have even increased their income 2.5 times, from one to 2.5 billion VND month^{-1} (i.e. from 65 to 160 US\$ month^{-1}).

To date, the quantity of fresh milk supplied by the farmers represents only between 15-20% of the total milk demand of the company; the rest still comes from imported dry skimmed milk. Average cost of the investment in the dairy development programme was 3,600 VND per kg fresh milk collected. Dutch Lady expects long-term benefits, because the quantity of fresh milk will continue to increase in the future. The project has contributed to consolidate a better image of a foreign partner in a joint venture, which is dynamically involved in social activities of local people. The project enhances the local economy by creating jobs for the rural labour force and by strengthening the livestock sector and thus reducing dependence on crop farming. It employs 11 full-time extension agents, 36 people for the milk collection, and another 6 for the milk payment and administration. Many of the farmers like to shift from crops to raising cows as grassland is tax exempt. In 2005, the gross margin at a survey farm was

0.72 US$ day^{-1} cow^{-1}. Counting a gross margin of 0.5 US$ day^{-1} cow^{-1}, the average earnings from the dairy component of each of the 2,300 farm households delivering milk from 8,500 dairy cows, can be estimated at 1.9 US$ day^{-1}, i.e. close to 700 US$ yr^{-1}, The total gross margin of the participating farmers can be estimated at 1.5 million US$, being higher than the annual invested amount to date. After 10 years in operation the Dairy Development Programme of Dutch Lady Vietnam is most successful. The combination of fresh milk collection together with subsidised farm services, training and know-how transfer can be set as an example of dairy development.

Discussion and conclusion

Both cases show that extension of technology and improvement of production can be driven by profit oriented commercial firms. Extension can be organised more efficiently and is less bureaucratic by firms than by government (van den Ban and Hawkins, 1985). The firms may be able to hire professional agents and to attract good candidates by paying higher salaries. Private firms are more efficient in dealing with obstacles, in considering opportunities, and in identifying the required partners. Private firms are more flexible and learn faster, especially if their projects start small and incorporate good discussion platforms with all hierarchical levels.

INVE participates in the process of technological innovations and demonstrates that it efficiently addresses the technical problems of shrimp production. In the past, government extension services often worked in isolation that could lead to growers being seen and treated as ignorant and in need of modernisation (Carney, 1998). INVE also provides financial support to the intermediate traders and the end-user in the shrimp production sector. However, the link of finance to a specific operation may also create dependency and have perverting effects; project linked credit operation of government often lead to low reimbursement rates. The solution used by Dutch Lady Vietnam: an independent bank for specific loans, seems more appropriate.

However, the cases confirm that profit oriented companies provide mostly mono-culture extension, i.e. directed at only one product or activity (Smith and Thompson, 1991). Moreover, industry and in some cases also government based extension, did not address the poor farmers efficiently (Alston and Pardey, 1996). The public-good nature of agricultural information often leads to so-called market failure, but also the private extension services are subject to market failure (Fulton *et al.*, 2003). The government extension services should address partial service delivery and could withdraw from areas that are covered by private extension services. If agricultural extension is to become dependent on commercial priorities only, then the directions pursued may not be economically efficient from the point of view of society as a whole, or may be contrary to other goals related to social welfare or the environment (Marsh and Pannell, 1999).

In Vietnam and many other countries, agricultural extension means transfer of farming technologies (research findings) to farmers (to apply in their own agricultural production). In Vietnam, agricultural extension is translated into "*khuyến nông*" and commonly understood as "*chuyền giao kỹ thuật nông nghiệp*" (agricultural technology transfer). Every Vietnamese knows "*khuyến*": to promote, to encourage, to stimulate, to recommend, to advise. But "*nông*" may mean differently; that is "*nông nghiệp*" (agriculture), or "*nông dân*" (farmers), or "*nông thôn*" (rural). The task of the government is to provide "*khuyến nông là phát triển nông thôn*": extension is rural development and supports all activities aiming to improve livelihoods of the rural people. To achieve this goal transferring technology alone is not enough, the extension services also need to equip the end-users with knowledge and skills in household and community resources management, to provide information on prices and marketing issues, on development policy, on health, and on socio-cultural aspects. This requires collaborative participation of several individuals and organisations, depending on context. Therefore in Vietnam extension, in a broad meaning, is also the mission of e.g. Women's Union, Farmers' Association and Youth Union. Around the world, cutting costs within organisations has been achieved by getting rid of employees. But cost cutting without regard for 'effectiveness of services' is counterproductive (Carney, 1998). The two cases show that it is often more appropriate to build partnership with other organisations, rather than reducing the governments' workforce (if efficient). The potential costs associated with making agricultural information less accessible should be assessed against the benefits of cost recovery policies (Fulton *et al.*, 2003).

The cases demonstrate the strong interdependence of agricultural extension with research and marketing in achieving positive change for agricultural communities. Carney (1998) developed a framework for classifying goods and services to determine which were appropriate for public provision, private or mixed provision. Her examination of different types of research demonstrated that the stage of the technology (pre-technology; prototype technology or useable technology), and the type of technology being developed (mechanical, chemical, biological or agronomic), influenced what type of research and extension services were required, and who was best positioned to provide the service. The end-users should have the right to participate in decision processes of any research / development project relating to them. In the Vietnamese context that fits government policy and the national campaign to promote democracy at grass-root levels. A participatory approach should not be restricted to gathering information from stakeholders for analysis, synthesis and design of a project. It should be an approach empowering stakeholders. The challenging task of the extension services is to facilitate active participation to ensure successful and sustainable rural development.

For specific sectors the profit oriented companies are complementary to government extension service, especially if next to inputs they provide training, credit (if independent from the company), and support, and are active in problem oriented research. On the other hand agricultural government agencies should not seek to completely confine their extension services to areas with public good characteristics such as land planning or environmental protection. Government based extension is also called to develop initiatives stimulating market integration.

References

Alston, J.M. and P.G. Pardey, 1996. Making science pay: the economics of agricultural R&D policy. AEI Press, Washington, D.C.

Carney, D., 1998. Changing public and private roles in agricultural service provision. Overseas Development Institute, London.

Fulton, A., D. Fulton, T. Tabart, P. Ball, S. Champion, J. Weatherley and D. Heinjus, 2003. Agricultural Extension, Learning and Change. A report for the Rural Industries Research and Development Corporation. RIRDC Publication No 03/032 http://www.rirdc.gov.au/pu, 81 pp.

Marsh, S. and D. Pannell, 2000. Agricultural extension. A decade of change. Rural Industries Research and Development Corporation, Canberra. 20 pp.

Smith, L.D. and A.M. Thompson, 1991. The role of public and private agents in the food and agriculture sectors of developing countries. FAO, Rome.

Van den Ban, A.W. and H.S. Hawkins, 1985. Agricultural Extension. New York: John Wiley and Sons.

Assessing farmers' motives for livelihood diversification in the Mekong delta: household life cycle, virtual farm size, and index of integration

Roel H. Bosma[2], Cao Quoc Nam[2], Henk M.J. Udo[1], Johan A.J. Verreth[1]
[1]Wageningen University, ASG/AFI, P.O. Box 338, 6700AH Wageningen; roel.bosma@wur.nl
[2]Can Tho University, College of Agriculture; 3/2 street, Can Tho, Vietnam; cqnam@ctu.edu.vn

Abstract

Contrary to the global trend of specialisation in agriculture, in Vietnam mixed farming systems have emerged. An important motive was the wish to improve livelihoods and diet of the nuclear families with a household lifecycle of five phases. Off-farm diversification was especially important for a young married couple preparing the creation of their own household. During expansion the farmers increased virtual farm size by keeping more livestock; during accumulation they invested in land or education, and during consolidation old couples adjusted farm activities to their labour capacity. Livestock, including fish, was essential for livelihood. Technological innovations of rice, fruit and fish were drivers for change. The improved food security and reduced cash income from rice after the 1986 reforms, pushed farmers to take risks. The farm area and number of components providing cash determined the level of cash income from agriculture. Farms with at least 4 flows of biomass between components earned more, demonstrating that real integration improved profits. Know-how and a minimum area of land at, or close to, the homestead are required for an effective integration.

Keywords: agriculture, aquaculture, household, lifecycle, Vietnam

Introduction

Although specialisation is a global trend in agriculture, in Vietnam farming systems integrating aquaculture and agriculture (IAAS) have emerged during the past 30 years (Luu, 1992; Prein, 2002). Specialisation is often considered equal to intensification, i.e. a higher efficiency of labour, land and capital use. Intensification is needed to feed the growing global human population. However, there is a concern for ecological, economic and social sustainability of specialised farming. Mixed crop-livestock and livestock-fish-crop systems might have a higher potential to maintain eco-systems' functions and to absorb shocks of the natural and economic environment (Holling, 1995; Luu, 1999). Inversing the trend might be feasible if integration in itself gives higher profits. The identification of the factors in farmers' decision-making in the Mekong delta (MD) could be a first step to strategies for diversification in other regions.

After a short description of the sampling we analyse the drives and motives for the integration of farm components using the livelihood capital asset framework considering 5 capitals:

natural, social, human, physical and financial (Ellis, 2000). In a third section we assess the contribution of diversification and integration to the livelihoods.

Sampling and methods

We sampled in the two agro-ecological zones where IAAS can be found: the fresh water alluvial zone (delta) and the zone with hills and uplands (hills). In the delta, we retained three hamlets of an existing sample representing the agro-ecological variation and selected in districts having a land-use policy supporting the development of IAAS (Le Thanh Phong, personal communication). The hill zone is mainly located in the districts Tri Ton and Tinh Bien at the border with Cambodia, where we retained three hamlets that had dominantly rain-fed agriculture (Bosma *et al.*, 2006). We interviewed 144 farmers in 6 hamlets; in each hamlet 24 were selected through stratified random sampling based on participatory wealth rankings of poor, intermediate and well-off households. During on-farm interviews with the head of the household or his wife, we composed a resource flow diagram, collected some standard farm characteristics, and information on changes since the establishment of their farm. The household was defined as the persons living and eating at the farm, and distinguished non-working old and young, active young, adult and old people.

The extent of the integration of various components was quantified by assigning the value 1 to each flow between two components; the cumulated values represent an Indicator for the Integration of Components (IIC). For analytical purposes we also derived the number of Components contributing to Cash Income (CCI). The income was distinguished in net cash from: on-farm activities (CAF), off-farm and non-farm activities (CON). We did not ask farmers to quantify the home-consumed products. Statistical analysis was done in SPSS (version 12.0).

Livelihoods assets framework

Natural capital

Notwithstanding the relatively recent occupation of the MD the average farm size in the delta was less then 1 ha and in the hills it was slightly more than 2 ha. In the delta, the land use is mainly determined by the flooding period, the diurnal tides, by the possibilities of water management and the risk of acidification. To construct dry land for a homestead, people excavate soil, thereby creating a pond. In the hill zone, the land quality varies considerably, especially as some farmers had also access to fields in the flooded zone. Some upland farms had loamy soil and easy access to artesian water welling from the soil after removal of the upper loamy layer, while most had only shallow sandy soils and needed concrete water reservoirs.

Since 1992 the farmers could obtain either an owner or a user certificate, giving land a collateral value. Holding a user certificate required farmers to bring land use in line with regulations; e.g. in the MD land more than 30 m above sea-level had to be planted with perennial crops like trees for timber or fruit to prevent erosion. For land with certificates a market developed and prices increased. Sudden increases of land prices were due to improved water management, policy changes for export, or new technologies.

Social capital

The availability of credit, training and services for the farmers were related to the district land-use policy. The land-use policy in the delta supported the creation of IAAS of rice, aquaculture, orchards and livestock, while in the upland it focussed on mixed cattle-crop-orchard systems. After 1994 the state developed a policy to support the poor; households classified as poor had, for example, access to credits with a low interest rate and free access to schools.

Human capital

Farmers repeatedly mentioned the following motives for innovation: improving income, diversifying the diet, and the well-being of children. Households in the Vietnamese MD were mainly nuclear, i.e. consisted of father, mother and children. Mean farm household size varied somewhat between hamlets but the averages of the delta and the hill settlements were equal. Three sets of factors affected the valuation of human capital: the formal education and product know-how, the household life cycle, and the importance given to the rice-field for food security. In the hills the possibilities of making a livelihood from non-farm activities were limited; a demand for off-farm labour only existed for 3 months and was high for 21 days during rice harvest. According to farmers the availability of family labour was important in the decision to start or to stop raising livestock.

In the household life cycle five phases could be distinguished: preparation, separation, expansion, accumulation, and consolidation. During the preparation phase a newly married couple stayed in the household of the husbands' parents and saved and invested for their future by developing off-farm or non-farm activities. When the next son wanted to get married, the first married couple separated from the family and created its' own household. If the parents or the parents-in-law had accumulated enough fields they could give land to the young couple in order for them to create a new household of their own. After the separation most young families exploited the available capitals through optimal diversification of activities: non-farm, off-farm and on-farm activities requiring land or not. At the onset of the expansion phase of the farm household lifecycle the new mothers mostly changed off-farm for homebound activities like raising chicken, pigs or fish (Figure 6.10). During the expansion phase farmers virtually increased the farm size by keeping more livestock to employ available family labour, if land area was limited. The choice of the species was motivated by knowledge and experience and by the market opportunities. In the accumulation phase they invested in land or formal education for their children. During consolidation, especially if all children had left agriculture, the old couples adjusted the farm activities to their labour capacity and started or increased areas of fruit orchards and fishponds. This confirms that the availability of labour, as well as the most mentioned motives for change, are related to the household live cycle (Zoomers, 1999; Perz, 2001).

For people living in the MD, rice was, and still is, the main staple crop. Especially in the delta the cash income derived from fruit was lower when the importance given to the food-security

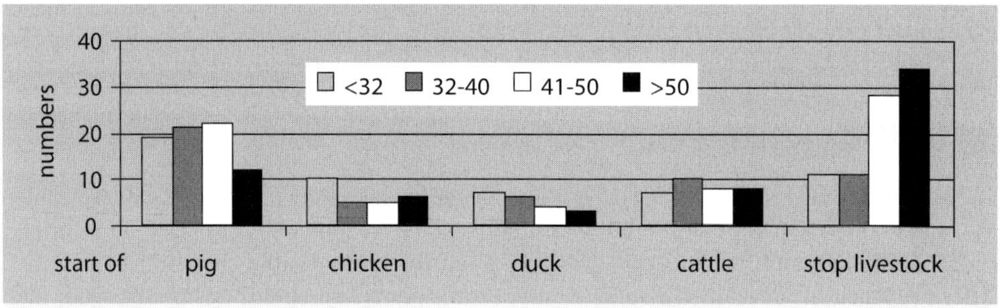

Figure 6.10. Frequency distribution of the age of the household head at starting or stopping a livestock activity (the age categories were adjusted to contain approximately an equal number of total on-farm changes each).

of rice was higher (rho = - 0.28, P<0.05). The continuous availability of rice on the market and its' relatively stable prices since 1989, diminished the necessity to produce rice on-farm for the family's food security.

Physical capital

The rank of well-being was related to the family farms' total land area (rho=0.43; (P<0.01). The importance of the distance between field and road or waterway was reflected in, for example, the prices of orchards in O Mon: the orchards on the border of a waterway or at the roadside having a higher market value than the in-field orchards. According to our sample, neither the distances between the fields and the homestead or the road, nor the distance to the market, affected the income, but for the delta, differences in management of IAAS farms were significantly related to their distance to main markets (Phong *et al.*, 2004, 2005), and farmers ranked the presence of outlet markets second of 7 factors driving the adoption of aquaculture in the delta (Nhan *et al.*, 2007).

The possibility of cultivating three crops a year in the delta was relatively recent in Tam Binh (1997) compared to O Mon (1983) and Cai Be (1986). Consequently, rice still occupied 70% of the cropland in Tam Binh while it was about 50% in Cai Be and O Mon (Le Thanh Phong, personal communication). Rice production became less attractive for at least 3 reasons: the double and triple rice technologies improved family food security, the gross margins of triple rice were lower than double rice due the high costs of inputs (Nhan *et al.*, 2003), and when market prices of rice were low due to overproduction the margins became negative (IFPRI, 1995, Linh, 2001). As the irrigated fields needed to be cropped to maintain their value[8], farmers looked for alternatives. At first they replaced the third rice crop with vegetables, but finally they

[8] If they do not use the irrigated field once, it will be too difficult to get rid of the weeds for a next crop; thus, needing food anyway, they plant rice.

replaced all 3 rice crops. To produce fruit and fish species with interesting market prices they transformed part of their rice-fields into pond, orchard or ditch-dyke systems[9].

Financial capital

A good market was a prerequisite for the development of new activities such as fish or fruit farming, but not a major condition for cultivating rice on the irrigated fields, nor for keeping ducks or chickens in small numbers, neither for raising fish in latrine ponds. In the emerging market economy, the cropping of only rice was disadvantageous compared to IAAS with fruits, vegetables and livestock besides rice. Rice made cash available only 1 to 3 times a year, while cash was needed more regularly. Orchards planted with several species provided a more regular cash income and had mostly a higher gross margin.

Before 1976 the farmers traded only pigs through middlemen. During the period of enforced village cooperatives and a centralised economy, between 1976 and 1986, farmers depended on the middlemen to cover administrative procedures for access to legal markets outside the village. Gradually the middlemen penetrated into most commercial chains favouring the integration of the small producers in the global market.

Most institutional loans for farmers were available for activities related to the land-use policy or to the poverty alleviation programme. Short run credits for inputs were available from banks, input providers, traders, and product chains. Private money lenders asked at least triple the bank' interest rates. Lending among relatives and friends was frequent and mostly without interest, and required no collateral. Empirical data showed that for poverty alleviation the distribution of pigs and goats by credit, sharing arrangements, or 'passing on the gift' was far more effective than cattle.

Diversification and integration

Contribution of farm components to livelihood

Most important contribution to cash income for close to 70% of farmers in the hills came from fruits, while it came from rice for almost 80% in the delta (Figure 6.11). Less than half of the farmers keeping chickens made cash income from chickens, but for most the level was low. The level of cash income derived from ruminants and ducks demonstrated an effect of specialisation, and while almost half of the interviewed farmers raised pigs for 8% only, it was a substantial cash income (>15 kVND d^{-1}). On average, the cash income from livestock, including fish, composed close to 1/3 of total cash income; it contributed for 64% and 32% to CAF, respectively in the delta and the hills. For the average household of 5 capita it represented 1 to 2 US\$ day^{-1}.

[9] Raised beds or embankments that gradually developed into ditch-dyke systems; the ditches provided water for irrigation and contained fish, either naturally retained or stocked (Sanh *et al.*, 1998).

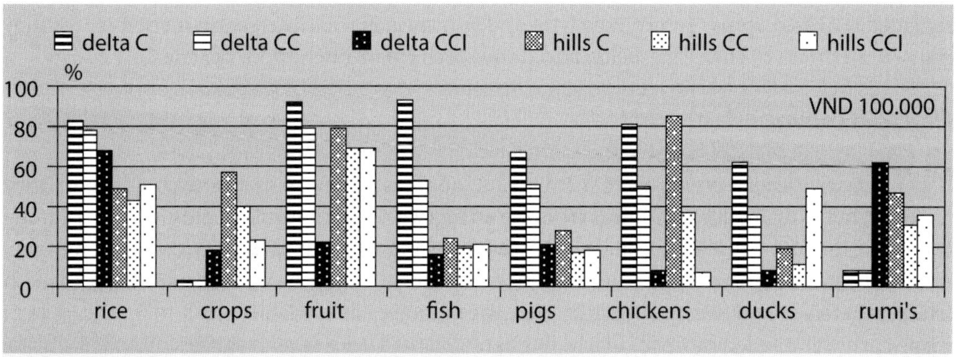

Figure 6.11. The percentage of farmers doing the activity (C), making cash from this component (CC), and the income the farmers earn from it (CCI, in 100,000 VND), in the delta and the hills of the MD (rumi's = ruminants).

Farm diversification: cumulate or integrate

The CCI was positively correlated to the rank of wealth (r=0.4; P<0.01) in both delta and hills, and to the NC and the IIC in the delta only (r=0.3, P<0.01). In the delta both IIC and CCI correlated significantly to the CAF (r=0.3; P<0.01), but in the hills the CAF was correlated significantly to the CCI only (r=0.3; P<0.05). In the hills the farm size and the distance between homestead and fields dominated the effect of the IIC; the correlations were r=0.3 (P<0.01) and r=-0.3 (P<0.05), respectively. In the delta some specialised farmers earned a high income; however those farmers need more cash for livelihood. The NC and CCI determined 2 distinct clusters of CAF: 15.2 ± 1.6 and 10.0 ± 1.2×10^6 VND year^{-1}. In general farms having at least 3 CCI performed better (Figure 6.12).

The fluctuating market prices and the insecure production potential, drives farmers to diversify, i.e. to keep multiple options to secure sources for food and cash. Farmers keep fish and small livestock for this purpose: one can either eat it, or sell it when the need for cash is

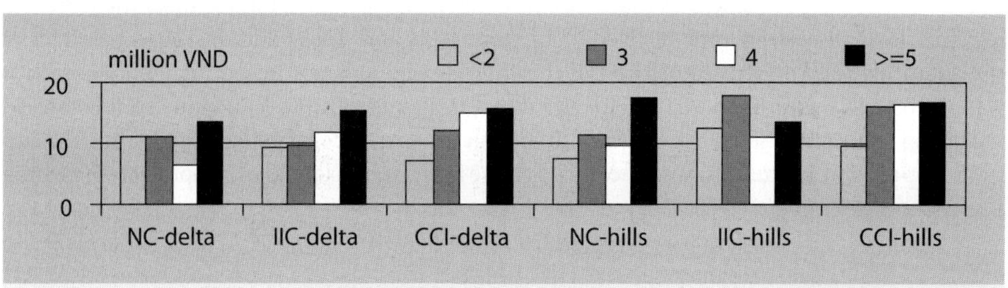

Figure 6.12. The average total net cash income of agriculture (million VND.yr^{-1}) as a function of the Number of Components (NC), the Indicator for Integration of Components (IIC) and of the number of Components contributing to Cash Income (CCI), in the two zones of the Mekong Delta.

higher. Farmers exploited all possible means to cumulate and integrate farm components; e.g. the houses of ducks and chickens were constructed above the fishpond where the manure was recycled, but also vegetable could be grown in or above the pond. Farmers stated that they increased the number of components for an optimal use of the small available area of land. They recycled farm residues as much as possible considering the labour availability; e.g. the rice straw could be used to feed ruminants but also to cultivate mushrooms. Most occurring integration in the delta was the recycling of human waste: to prevent spreading through flood and tides, the use of latrine ponds with catfish was widely practised. As food safety is an issue in this respect, research is needed to secure this source of income and sustainable waste management (Bunting, 2004).

Diversification in income or in farm composition

The number of farmers producing rice or fruit either for home-consumption alone or for both marketing and home-consumption has hardly changed over the past 10 years (Figure 6.13). However, about half of all farmers intensified one or more activities over the past 30 years. The frequency of intensification was higher since 1995 compared to the period between 1975 and 1994 (Table 6.4). Over the past 30 years 40% of the farmers replaced part of the rice-field with an orchard. The number of farmers raising fish increased only slightly, but close to 40% of the farmers intensified fish production either through using artificial brood-stock or increasing the total pond-area. Though 8% of the farmers have intensified their pig production in the past 10 years this has lead only to a temporary increase in the number of farmers raising pigs (Figure 6.13). The intensification of the chicken production was entirely realised after 1995 (Table 6.4), and is also reflected in an incremental increase in the number of practising farmers (Figure 6.13).

Discussion and conclusions

The marketing systems of middlemen enabled the integration of small farmers in the global market of fruit and fish. The distance to markets has a two-sided effect: proximity of a market drives the development of components to earn cash, while relative isolation pushes households to increase the number of components for food autonomy.

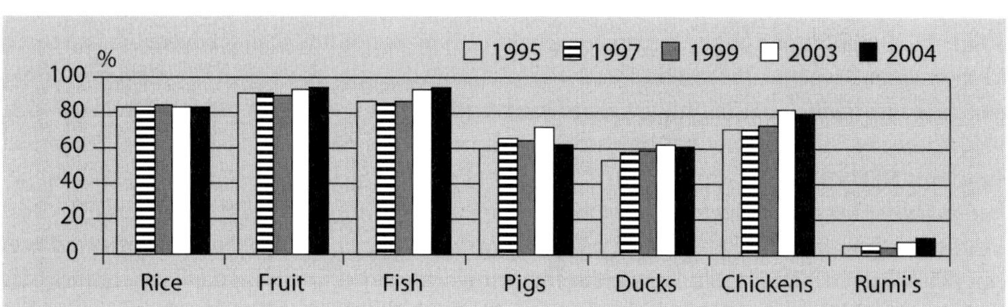

Figure 6.13. The change in the percentage of farmers in the delta practising a specific component between 1995 and 2004 (rumi's = ruminants).

Table 6.4. The percentage of farmers that intensified or increased one of the mentioned on-farm activities since 1976 and 1995.

Period	Pigs	Chickens	Brood-fish	Fish-pond	Orchard
1976-2004	13	11	17	38	32
1995-2004	8	11	13	26	18

The technological innovations of rice, fruit and fish were essential for the development of IAAS in the MD, partly contradicting Netting (1993) who stated that scientific and technological innovations were not the crucial causal factors in agriculture intensification. After the economic reform in 1986, diversification was triggered by a low market price of rice and the improved food-security for rice, the developing markets for products and for off-farm labour, and the free choice of farmers to initiate activities. The improved rice food security changed the farmers' reference frame and pushed them to take risks and to engage in activities in which they had less experience and knowledge. The MD farmers intensified activities to increase food production and cash income from their limited area against the lowest possible cost. The farmers claimed that IAAS used resources more efficiently than mono-cultures, but very often the integration was not effective but a mere accumulation of components. The effective integration of cumulated components could be hampered by the distance to the fields or by a small homestead, but also by lack of know-how.

Most important motivation for the on-farm innovations was the will to improve family livelihood, while the household cycle of the nuclear families determined the labour and subsequently, the capital availability to engage in more market-oriented activities. Between 2003 and 2005, in the delta less than 20% of the average farm income came from off-farm activities, and the non-cash returns (home-consumption) represented close to 20% of the total farm turn over (Le Thanh Phong, personal communication). This last figure demonstrates the high level of market integration of the MD farmers.

Livestock allowed farmers to expand the farm turnover and was essential for poverty alleviation. The diversification of livelihood to non-agriculture was important to accumulate capital for on-farm investments and especially important during preparation for the young couple's separation and in the expansion phase of the farm household lifecycle. As the focus of the farm activities changes during the household lifecycle, the extension services should consider this when formulating the recommendations.

Acknowledgements

The first author acknowledges Prof. Leontine Visser and Mr. Dang Kieu Nhan for their advice, Mr. Pham Minh Duc for his assistance in the fieldwork and the farmers and village authorities for their availability and hospitality. The authors are grateful for the support of the partners at Can Tho University: Dr. Tri, Dr Be, and 'teacher' Le Thanh Phong. This study was funded through the INREF-POND programme of the Wageningen University.

References

Bosma, R.H., Cao Quoc Nam, H.M.J. Udo and J.A.J. Verreth, 2006. Farmers' motives for agriculture diversification and its' contribution to livelihoods in the Mekong Delta. Asian Journal of Agriculture and Development (accepted).

Bunting, S.W., 2004. Wastewater aquaculture: perpetuating vulnerability or opportunity to enhance poor livelihoods? Aquatic Resources, Culture and Development 1, 51-75.

Ellis, F., 1998. Household Strategies and Rural Livelihood Diversification. The Journal of Development Studies 35, 1-38.

Holling, C., 1995. Sustainability: The Cross-scale Dimension. In: M. Munasinghe and W. Shearer (Eds.), Defining and Measuring Sustainability. IBRD/WB, Washington, D.C., pp. 65-76.

IFPRI, 1995. Rice Market and Monitoring and Policy Options Study. Statististical Publishing House, Hanoi.

Linh, N.V., 2001. Agricultural Innovation: Multiple grounds for technology policies in the Red River Delta of Vietnam, WUR, Wageningen, 200 pp.

Luu, L.T., 1992. The VAC system in northern Vietnam, Farmer-proven Integrated Agriculture Aquaculture: a technology information kit. IIRR & ICLARM.

Luu, L.T., 1999. Small-scale aquaculture in rural development - trends and constraints. In: FAO (Ed.), Report of the Asia-Pacific Fishery Commission Ad hoc Working Group of Experts in Rural Aquaculture. FAO, Bangkok, Thailand.

Netting, R.M., 1993. The Technology and Knowledge of Intensive Farm Practices. In: Netting (Ed.), Smallholders, Householders: Farm Families and the Ecology of Intensive, Sustainable Agriculture. Blackwell, Oxford, pp. 28-57.

Nhan, D.K., N.T. Duong, M. Verdegem and L. Duong, 2003. Towards Agricultural Diversification and Rural Poverty Alleviation - Development of Agri-Aquaculture farming in the Mekong Delta, Vietnam. In: C. Wollny, *et al.* (Eds.), Technological and Institutional Innovations for Sustainable Rural Development, Deutscher Tropentag 2003. Klartext GmbH, Gottingen.

Nhan, D.K., L.T. Phong, M. Verdegem, L.T. Duong, R.H. Bosma and D.C Little, 2007. Integrated Freshwater Aquaculture, Crop and Animal Farming Systems in the Mekong Delta, Vietnam: determinants and role of the pond. Agricultural Systems (in press).

Perz, S.G., 2001. Household Demographic Factors as Life Cycle Determinants of Land Use in the Amazon. Population Research and Policy Review 20, 159-186.

Phong, L.T., H.M. Udo, M.E.F. van Mensvoort, L.Q. Tri, R.H. Bosma and D.K. Nhan, 2004. Understanding integrated agriculture-aquaculture systems in the Mekong Delta, Vietnam. In: H.K. Wong, *et al.* (Eds.), The 11th Animal Science Congress, Asian-Australian Association of Animal Production Societies, Kuala Lumpur, Malaysia.

Phong L.T., M. Udo, M.E.F. van Mensvoort, L.Q. Tri and D.K. Nhan, 2005. Dynamics of Integrated Agriculture-Aqauculture Systems in the Vietnamese Mekong Delta. Ìn: Rowlinson et al (ed). Integrating Livestock-Crop Systems to Meet the Challenges of Globalisation, Vol. 2 Proc. of AHAT/BSAS conferennce Nov 14-18, 2005. Khon Kaen, Thailand.

Prein, M., 2002. Integration of aquaculture into crop-animal systems in Asia. Agricultural Systems 71, 127-146.

Sanh, N.V., V.T. Xuan and T.A. Phong, 1998. History and future of farming systems in the Mekong Delta. In: V.T. Xuan and S.E. Matsui (Eds.), Development of farming systems in the Mekong Delta of Vietnam, Ho Chi Minh City Publishing House, pp. 16-80.

Zoomers, E.B.A., 1999. Linking livelihood strategies to development. Amsterdam: KIT-Press, 89 pp.

Livestock intensification in mixed farms: benefits and trade-offs

Henk M.J. Udo, B.O. Bebe, Tashi Samdup, Sutresniwati, I. Gede S. Budisatria, Aklilu H. Asgedom and Le Thanh Phong
Animal Production Systems Group, Wageningen Institute of Animal Sciences, Wageningen University, P.O. Box 338, 6700 AH Wageningen, The Netherlands

Abstract

This paper discusses benefits and trade-offs of livestock intensification for mixed farming systems on basis of case-studies in the African and Asian context. The majority of rural households are smallholder crop-livestock farmers. Smallholder dairying can significantly contribute to household income on these farms, provided the necessary resources and market infrastructure are available. Small animals can help to improve the livelihoods of poor households, but will not become main income earners for rural households. They are an appreciated secondary activity. Intensification of livestock implies that the smallholder mixed farmers have to produce more for the market and that they can pay less attention to other livelihood functions of animals. Intensification can increase cash income of the farming households and it provides food to the urban population, however it can have a negative effect on the resilience of smallholder systems. It can also have a negative impact on the environment and local genetic resources are one of the big losers of intensification. For some smallholders the predicted doubling in demand could well become a key means for integration into a market framework, for others the livestock revolution will lead to increased competition with other smallholders and with large-scale operators. This will result in a reduction of the number of smallholder livestock farmers. Sustainable development strategies have to address securing the traditional assets of livestock keeping, as well as increasing productivity and improving market participation, all in relation to the context of the farming systems concerned.

Keywords: intensification, cattle, small ruminants, poultry, fish, livelihoods

Introduction

Human population increases, urbanisation, economic progress, and changing consumer preferences are predicted to double the demand for livestock products in the first two decennia of the 21st century (Delgado *et al.*, 2001). This 'Livestock Revolution' is expected to occur in the same areas where increased demand is expressed, that is in the developing countries and in the new emerging economies. At present, the demand for animal food is already growing at a rate of 3% a year. This increased production mainly occurs outside the traditional rural sector (FAO, 2005a). The majority of the rural households are smallholder crop-livestock farmers. At their farms, livestock can be complementary to crop production through the use of manure, draught power and crop residues, or can be competing for instance for cash inputs and labour use. An important motive for keeping livestock is their function as a capital asset, in particular in areas lacking formal insurance and developed financial markets. Livestock development policies generally focus on marketed production, and thereby neglect the multiple functions livestock has at farm level (Udo and Cornelissen, 1998; Moll, 2005). According to Delgado *et al.* (2001), the livestock revolution could well become a key means of integrating smallholders into a market framework. There is often a trade-off between the benefits from marketed production and the intermediate and intangible benefits.

At present, arguments and justification for livestock development seem to centre on the demand for meat and milk by the wealthier consumers. At the same time, about 800 million people are starving, and close to 3 billion people are living on less than US$ 2 per day. About three-quarters of the rural poor keep livestock. The type of livestock kept varies according to agro-ecological zone and region of the world, and is related to tradition, resources, age category, and wealth status of the households concerned. Poultry and small ruminants are closely linked to the poorest people. In a study involving 1700 households in Kenya, Kristjanson *et al.* (2004) found that once households can acquire cattle they are no longer regarded poor within a community and when they can acquire dairy cattle they are considered relatively well-off.

Development policies promote intensification of livestock production, both to meet the increasing demands and to contribute to development of rural households. Intensification can be defined as the increased use of external inputs and services to increase the output quantity and/or value per unit input (Bebe *et al.*, 2002). In practice, it is often understood as applying more advanced production styles. Crossbreeding is an important driving force for livestock intensification. Governmental institutions in developing countries often support smallholder dairying, in response to the increasing demands and to reduce imports of dairy commodities. It is, however, often stated that small animals, such as sheep, goats and poultry, can make a contribution to better livelihoods of the really poor by encouraging them to engage in economic activities (Dolberg, 2001; Devendra and Chantalakhana, 2002). This paper discusses benefits and trade-offs of livestock intensification in mixed farming systems on the basis of case-studies in the African and Asian context.

Smallholder dairying in Kenya

Kenya is prominent among developing countries for integrating dairy into smallholder farming systems, particularly in the highlands. Major determinants of this success in the highlands are colonial history, its favourable agro-ecology, supportive agricultural policies and the traditional value of milk in the diet (Bebe *et al.*, 2002). A case-study on the consequences of intensification of smallholder dairying in the Kenya highlands involved a cross-sectional survey of 1755 households, with additional information from longitudinal semi-structured interviews with 50 households (Bebe, 2003). In the highlands about 60% of the rural households have integrated dairy in their mixed farming systems. A typical crop-dairy farm is one hectare with a herd of one to three cows. Households with dairy cattle have, on average, twice as much land as households without dairy cattle. Overall, the size of farms in the highlands has decreased by more than half over the past two decades, mainly because of subdivision through family inheritance. This shrinking of land holdings is a major concern. Milk production levels were not different (4 kg per day of calving interval) amongst the different dairy breeds (Friesian, Ayrshire, Jersey, Guernsey). Availability of good quality feeds was a major constraint. Research and development recommend the use of the smaller dairy breeds, farmers, however, prefer

Friesians because of their larger body size and higher market values, and the perception that Friesians can produce more milk (Bebe, 2003).

Crop-dairy farmers intensify their farming practices by shifting from free-grazing to semi-zero or zero-grazing (stall feeding). The semi-zero and zero-grazing farms already comprise over three-quarters of all smallholder dairy farms. The zero-grazing farms produce more marketable milk per unit of land, which is a major objective of policy makers. The individual free-grazing farms, however, show considerable higher economic returns, based on cash income, income in kind (home consumption, draught power, manure) and capital asset function than the zero-grazing and semi-zero-grazing farm types (Bebe *et al.*, 2002). The free-grazing farms have to produce the replacement stock needed for the other farms, as semi-zero and zero-grazing farms are unable to produce sufficient heifers for replacement. Shortage of replacement stock is thus a major obstacle to the continuity of intensification of smallholder dairying in the medium to long term. Intensification can only continue when there is a minimum proportion of free-grazing farms (Bebe, 2003) that can produce replacement animals for the semi-zero and zero-grazing farms. However, with human population densities continuing to rise and landholding shrinking, it is inevitable that the number of free-grazing farms will continue to decrease.

In Kenya, as in many other developing countries, the informal milk market has about 85% of the market share (Omore *et al.*, 2001). Consumers generally prefer raw milk, even those who can afford pasteurised milk. This represents trade-offs between milk safety and economics in milk marketing.

Crossbreeding for dairy development

Kenyan dairy farmers mainly use European dairy breeds, whereas farmers in many other developing countries have started dairying through crossbreeding local breeds with European dairy stock. The impact of crossbreeding goes much further than increasing milk yields. Samdup (1997) studied the impact of Brown Swiss x local Siri crossbreeding on mixed farming in a district with a temperate climate in Bhutan. Input, output and internal flows were quantified on an annual basis for 73 households with or without crossbred cattle. The introduction of crossbred animals resulted in an increase in the density of agricultural throughputs. Farms with crossbred cows used 1.7 times more external inputs and had 2.5 times higher total agricultural outputs than farms without crossbred cows (Samdup, 1997). Crossbreeding changed the production system from a crop-oriented system to a more livestock-oriented system. In households without crossbreds, 18% of the labour was used for livestock, which also contributed 18% to farm gross margins. In households with crossbreds, 28% of the labour employed in agriculture was used for livestock-related activities and livestock contributed 62% to farm gross margins. The data from the Bhutan case-study refer to a survey done in 1996 (Samdup, 1997). Recent information from 2000 and 2004 indicates that most households in this district have crossbred cattle on their farms. Local cattle are still dominant in remote areas with seasonal forest grazing systems and low market access. Milk from these animals is processed into cheese and butter, whereas milk from crossbreds is usually brought to milk collection centres.

The consequences of crossbreeding have to be evaluated at system level. Crossbreeding for dairy production has definitely brought economic gains for the households involved. In India, crossbred cattle provide employment opportunities for grossly underemployed household members, in particular in landless farming systems (Patil and Udo, 1997). The long-term sustainability of crossbreeding needs to be addressed. In Bhutan, reproductive rates were low and calf mortality was high in higher level Brown Swiss crossbreds. In smallholder dairying, replacement stock (purebred or crossbred) is generally in short supply, because of low calving rates and high calf mortalities (de Jong, 1996). Crossbreeding is often done haphazardly and it contributes to the genetic erosion of well-adapted local breeds (Thomas *et al.*, 2002). The local genetic resources are the big losers of intensification.

Crossbreeding as hype

Crossbreeding has become a standard approach to increase livestock productivity in almost all farm animal species. Indonesia is a country in Asia were crossbreeding is widely promoted to increase livestock production. In this paper a case of cattle crossbreeding involving the Ongole is used. During the colonial period the Dutch administration introduced Ongole cattle from India to replace local Javanese cattle (Barwegen, 2005). Ongole were considered better suitable for carrying the heavy loads for the sugar industry. On the main island Java, the Ongole has completely replaced the Javanese cattle and it is now considered as a local breed in Indonesia. At present, the Ongole is replaced by Simmental crossbreds. In Central Java a study was done on perceptions and perspectives of different stakeholders on cattle crossbreeding. Group discussions were followed by individual interviews of different stakeholders, varying from farmers (n=122), veterinarians (n=2), inseminators (n=14) to food sellers (n=35) (Sutresniwati, 2006). Crossbred cattle farmers were wealthier and better educated than local cattle farmers (Sutresniwati, 2006). Farmers keeping local cattle had often insufficient capital resources to acquire crossbreds. Veterinarians and inseminators mentioned some health and reproduction problems for crossbreds, however, farmers invariably indicated that they wanted to keep crossbreds, whether they had sufficient capital and feed resources or not. Inseminators received a higher fee from farmers when they inseminated their cows with semen form exotic bulls. The Simmental cross was the favourite breed because of their red shiny coat, yellow lips and their unselective feeding habits. The main motive for keeping cattle (crossbreds or local) was their capital asset function. Manure was considered to be the second priority in keeping cattle. Prices of crossbred cattle exceeded the meat value of animals. It seems that crossbreeding with Simmental has become a 'hype'. Only sellers of *bakso*, a local meat dish, preferred meat from local cattle, because of the texture of the meat and the number of meat balls that could be prepared per unit of meat (Sutresniwati, 2006).

Small ruminants

Often, it is said that small ruminants can make a contribution to better livelihoods of the rural poor in developing countries (Morand-Fehr and Boyazoglu, 1999; Sinn *et al.*, 1999; Devendra and Chantalakhana, 2002; Lebbie, 2004; Peacock, 2005). This is mainly based on the premise that sheep and goat keepers are among the poorer groups in society (Morand-Fehr and Boyazoglu, 1999; Kristjanson *et al.*, 2004).

Budisatria (2006) studied small ruminant production systems in Central Java, an area in Indonesia known for the quality of its small ruminants. He collected the data through interviewing farmers (150 small ruminant farmers and 71 neighbouring households) and key persons (n=30), group meetings, field observations, and through monitoring feeding practices, animal performances and marketing strategies. Added to this, quantitative data were collected from 10 small ruminant markets, 44 mosques and 42 roadside sellers. Small ruminant development programmes started about 80 years ago. Major changes have occurred since that time in the types of animals kept (farmers have replaced their Javanese thin-tailed sheep by fat-tailed sheep and the promotion of Etawah goats has resulted in a change over from local Kacang goats into Etawah-grade goats), in animal numbers (a strong increase) and in farmer's management (Budisatria, 2006). Driving forces for changes in small ruminant systems have acted at national, regional, agro-ecosystem and household levels. The intensification of land use has resulted in major changes in management. The majority of small ruminants are now kept in confinement, or in a combination of grazing and confinement. Resources at household level, i.e. family labour and capital availability, were major factors determining whether farmers keep small ruminants or not. The housing of small ruminants close to the family quarters resulted in very high levels of faecal bacteria contamination of drinking water sources (Budisatria, 2006).

The case-study farmers referred to their small ruminants as a saving that provides security and helps to accumulate capital. Manure was the second main reason for keeping small ruminants. The returns per unit of family labour from keeping sheep or goats were well below the minimum labour wage (Budisatria, 2006). The contribution from sheep and goats to income was very small. Innovation scenarios indicated that if farmers can start specialising in sheep fattening or goat breeding the technical and economic results can be improved compared to the present production systems. However, on a daily basis the returns remained low: 41-49% of the minimum labour wage for sheep fattening and 65-83% of the minimum labour wage for goat breeding. Farmers with sufficient household labour available will not consider the labour costs as real costs. Alternative employment opportunities are limited either because they are not available, particularly in areas with a poor infrastructure (uplands), or because household members are not competitive in the job market.

In Indonesia small ruminants, in particular sheep, play a key role in religious festivities since the majority of people are Moslem. The demand for and price of small ruminants increase 2-3 fold during the weeks before the Moslem feast of sacrifice (Budisatria, 2006). Farmers do not seem to be able to profit from this increased demand. They own 4-6 animals, which makes it too difficult to plan their sale of animals in relation to the period when the prices increase. Farmers rarely sell their animals directly on the small ruminant market or to the consumers. In the Central Java study, more than 80% of the small ruminants sold were sold to village collectors. The farmers in this study also complained that there was no marketing information available to them, the information they receive is via the village collectors. Urgent cash needs, e.g. for school fees and preparation of rice fields, were the main reason for selling small ruminants (Budisatria, 2006).

Budisatria (2006) concluded that it is unlikely that small ruminants will become a main income earner in rural households. It seems that keeping small ruminants is more a sign that farmers are poor, than that it can help them to move out of poverty. If households have sufficient family labour for the management of small ruminants, small ruminants are an appreciated secondary activity.

Poultry

Village poultry

In developing countries, many rural households keep poultry in their farmyard. Poultry provide food and a small cash income, and are used as presents to strengthen social relationships. There is considerable interest in traditional poultry systems. Projects to improve village poultry systems are considered to be a vehicle for poverty alleviation and to increase self-reliance, in particular for women (Mack *et al.*, 2005). There are many technical possibilities for improving free-range and backyard poultry keeping. Rural households, however, are not adopting these technologies widely (Udo, 1997). A model approach, based on literature on village poultry, showed that NCD (Newcastle Disease) vaccination, daytime housing, supplementary feeding, and control of broodiness each had a positive effect on bird offtake, egg production, egg offtake, and flock size (Udo *et al.*, 2006). Daytime housing showed the most positive response in flock output parameters. Housing facilities help protect animals from predators. Housing implies, however, that the animals have to be cared for; the farmer has to provide feed and water. This feed has to be balanced because the chickens can no longer look for feed to meet their requirements. Crossbreeding had a two-phase effect on output parameters. After the start of crossbreeding, egg production increased due to the earlier age at start of lay and the higher rate of egg production of crossbred hens compared to local hens. The simulations indicated that after nine months egg production of crossbreds dropped because of higher mortality rates and loss of broodiness in crossbred hens. Prospective interventions will be sustainable only if they fit the limited physical and economic resources of farming households. Cost-benefit calculations for the Tigray region in Ethiopia and village poultry research sites in Kenya indicated that NCD vaccinations were economically most effective (Udo *et al.*, 2006). Housing and crossbreeding had a highly negative impact on net returns. Vaccination, housing, crossbreeding, and supplementary feeding all require additional cash inputs. This means that farmers have to become more market-oriented. This change-over is very difficult for village poultry farmers.

The success of the Bangladesh poultry development model shows that access to credit with small and frequent repayments is essential for village poultry development (Dolberg, 2001). This 'model' is based on experiences from modern commercial poultry keeping, building on 10 hens of improved breeds per family. At village level the farmers have been divided into groups specialising as breeders, chick rearers, feed manufacturers or vaccinators. These groups have been given credits with small and frequent repayments (Dolberg, 2001). Experience has shown that ten laying hens are enough for one family to earn 100 US$ per year. Close to 2 million women are involved in smallholder poultry production in Bangladesh.

Small-scale commercial units

In poultry, commercial methods are relatively easy to copy. In every country you can find small-scale units keeping 50-1,000 birds on a commercial basis involving commercial hybrids, compound feeds, and commercial housing methods. These units have to produce for a market. In some countries in Asia, there are commercial poultry units of 1,000-5,000 chickens in villages, which produce for big commercial firms on contract basis. This is already the transitional stage into large, so-called integrated poultry farms. The international livestock firms are needed to establish poultry activities. The major problems of the small-scale commercial units are coping with the uncertainty of the supply of sufficient cheap raw materials for feed and the inconsistent demand for their products. This is why many of these farms fail in the long term (Udo, 1997). Large-scale commercial poultry keeping can rapidly respond to increasing demands. Industrial livestock systems are capital and energy intensive. They make high demands upon health, hygiene and management. Foreign exchange is needed for imported management, equipment, breeding stock, feed ingredients, and processing requisites (Udo, 1997). The commercial poultry units have suffered most from the Asian economic crisis in the late 1990's, because of their reliance on imported feeds and the decrease in buying power of the consumers (Ørskov, 1999).

Integrated agriculture aquaculture systems

In South East Asia, farming systems integrating agriculture with aquaculture are expected to make farmers' livelihoods more sustainable. The Vietnamese Mekong Delta is an example of a region where many farmers have diversified their production system into Integrated Agriculture Aquaculture Systems (IAAS), with many variations in terms of crops, orchards, fish, and livestock integration (Bosma *et al.*, 2006). Economic liberalisation in 1986, natural disasters and market demands were driving forces for this diversification (Phong *et al.*, 2005). The potential benefits of IAAS are: risk spreading, a more even distribution of cash-generating opportunities, and more efficient resource use. Manure from pigs and chickens and other wastes are used to fertilise the fishpond. The pond sediments are subsequently used to fertilise the crops of which the residues, in turn, can be used as feed for the livestock. A baseline survey of 80 farms in the Mekong Delta over the period 2002-2004 showed that most farms had four components. Farm gross margins were highest for rice, followed by fruits, fish, pigs, cash crops and were negative for poultry (Phong *et al.*, 2005). The negative returns for poultry were caused by the Avian Influenza outbreaks in 2003 and 2004. The AI outbreaks caused a considerable decrease in demand and prices for poultry products, whereas the demand and prices for pork and fish increased. The farmers responded with an intensification of the aquaculture component and the pig production. Thus, the IAAS farmers were able to respond very quickly to the changes in market demands. This shows the resilience of integrated farms. The AI outbreaks, however, very much affected farmers that had solely specialised in poultry production.

The intensification of pig production on the IAAS farms is done through the use of compound feeds and hybrid pigs. Fish production is also intensified through the increased use of compound feeds. Feeding feed pellets with large amounts of fish proteins to herbivorous fish species used in the IAAS farms does not seem to be sustainable. It forces farmers to use a whole intensification package including different species of fish and higher stocking densities

(Hans Komen, personal communication). Charo-Karisa (2006) showed that growth of Nile tilapia can be improved without requiring expensive supplementary feeds. He concluded that selection for improved fish growth under high input conditions often produces breeds which demand more resources than can be provided on traditional fish farms. Phong *et al.* (2007) evaluated ecological sustainability in terms of indicators representing diversity, efficiency and productivity for low, medium and high external input fish systems in the Mekong Delta. It was found that the medium input fish system was superior to the high and low input systems in terms of sustainability of resource use.

Discussion

The case-studies indicate that intensification can increase cash income of the farming households. In Kenya, dairy production has become the main income source for over 600,000 mixed farmers. Small animals, such as poultry and small ruminants, do not seem to have the potential to become main income earners for mixed farmers; they may act as starters in a development process (Dolberg, 2001). Intensification usually involves a whole package of innovations based on external inputs. This implies that farmers have to produce more for the market and they can pay less attention to the other functions of livestock for the livelihoods of the farming households. So, intensification could have a negative effect on the resilience of smallholder systems. High input farming is likely to cause pollution of water, air and soil, and resource degradation also occurs in low input farming, e.g. mining of resources (de Haan *et al.*, 1997).

Livestock production will intensify further. In all case-studies it was reported that intensification necessitates increased use of purchased inputs and services, such as feeds, replacement stock, breeding and health services and with these, a need for credits and an increase in knowledge. The integration of smallholders into a market framework could lead to increased competition with other smallholders and with large-scale operators. Smallholders have to compete in markets that are more demanding in terms of product quality and food safety. Globalisation and removal of barriers to trade could also imply that local producers have to compete with cheap imported products. These developments will result in a reduction of the number of smallholder livestock farmers. Sustainable development strategies have to address securing the traditional assets of livestock keeping, as well as increasing productivity and improving market participation, all in relation to the context of the farming systems concerned. This implies also to strategies to minimise the potential damage to environmental sustainability.

A recurring question is whether the livestock revolution will help or hinder the alleviation of poverty of rural livelihoods. FAO has estimated that large-scale commercial livestock operations are growing twice as fast as traditional mixed farming systems (FAO, 2005a). In Asia, where growth in livestock production has been most dramatic, large-scale industrial production accounts for roughly 80% of the total increases in livestock products since 1990 (FAO, 2005b). Without pro-poor policies, that deliberately promote poverty reduction and give voice to the poor, small producers are likely to be excluded.

References

Barwegen, M.W., 2005. De geschiedenis van de veehouderij op Java, 1850-2000. [History of livestock husbandry on Java, 1850-2000]. PhD thesis Animal Production Systems, Wageningen University, Wageningen, the Netherlands, 316 pp.

Bebe, B.O., 2003. Herd dynamics of smallholder dairy in the Kenya highlands. PhD thesis Animal Production Systems, Wageningen University, Wageningen, the Netherlands, 155 pp.

Bebe, B.O., Udo, H.M.J. and W. Thorpe, 2002. Development of smallholder dairy systems in the Kenya highlands. Outlook Agric. 31, 113-120.

Bosma, R.H., L.T. Phong, U. Kaymak, J. van den Berg, H.M.J. Udo, M.E.F. van Mensvoort and L.Q. Tri, 2006. Assessing and modelling farmers'decision-making on integrating aquaculture in the Mekong Delta. NJAS 53, 281-300.

Budisatria, I.G.S., 2006. Dynamics of small ruminant development in Central Java, Indonesia. PhD thesis Animal Production Systems, Wageningen University, Wageningen, the Netherlands, 144 pp.

Charo-Karisa, H., 2006. Selection for growth of Nile tilapia (Oreochromis niloticus L.) in low-input environments. PhD thesis Animal Breeding and Genetics, Wageningen University, Wageningen, the Netherlands, 169 pp.

De Haan, C., H. Steinfeld and H. Blackburn, 1997. Livestock and the Environment. Report of a study of the Commission of the European Communities, FAO, and WorldBank, WRENmedia, UK.

De Jong, R., 1996. Dairy stock development and milk production with smallholders. PhD thesis Animal Production Systems, Wageningen University, Wageningen, the Netherlands, 308 pp.

Delgado, C., M. Rosegrant, H. Steinfeld, S. Ehui and C. Courbois, 2001. Livestock to 2020: the next food revoluition. Outlook Agr. 30, 27-29.

Devendra, C. and C. Chantalakhana, 2002. Animals, poor people and food insecurity: opportunities for improved livelihoods through efficient natural resource management. Outlook on Agr. 31, 161-175.

Dolberg, F., 2001. A livestock development approach that contributes to poverty alleviation and widespread of nutrition among the poor. Liv. Res. Rural Dev.13, 1-14.

FAO, 2005a. Responding to the "livestock revolution". Livestock Policy Brief 01, Livestock Information Sector Analysis and Policy Branch, Animal Production and Health Division, FAO, Rome, 8 pp.

FAO, 2005b. Pollution from industrialized livestock production. Livestock Policy Brief 02, Livestock Information Sector Analysis and Policy Branch, Animal Production and Health Division, FAO, Rome, 8 pp.

Kristjanson, P., A. Krishna, M. Radeny and W. Nindo, 2004. Pathways out of poverty in Western Kenya and the role of livestock. PPLPI Working Paper No 14, FAO, Rome, Italy, 23 pp.

Lebbie, S.H.B., 2004. Goats under household conditions. Small Rumin. Res. 51, 131-136.

Mack, S, D. Hoffmann and J. Otte, 2005. The contribution of poultry to rural development. World's Pltry Sc. J. 61, 7-14.

Moll, H.A.J., 2005. Costs and benefits of livestock systems and the role of market and nonmarket relationships. Agric. Ec. 32, 181-193.

Morand-Fehr, P. and J. Boyazoglu, 1999. Present state and future outlook of the small ruminant sector, Small Rum. Res. 34, 175-188.

Omore, A., S.J. Staal, L. Kurwijilla, E. Osafo, G. Aning, N. Mdoe and G. Nurah, 2001. Indigenous markets for dairy products in Africa: trade-offs between food safety and economics. In: J.H.A. de Gooijer (ed.), Dairy Development in the Tropics, Office for International Cooperation, Faculty of Veterinary Medicine, Utrecht, The Netherlands, pp 19-23.

Ørskov, E.R., 1999. New challenges for livestock research and production in Asia. Outlook Agric. 28, 179-185.

Patil, B.R. and H.M.J. Udo, 1997. The impact of crossbred cows at farm level in mixed farming systems in Gujarat, India. AJAS 10, 621-628.

Peacock, C., 2005. Goats-a pathway out of poverty. Small Rumin. Res. 60, 179-186.

Phong, L.T., H.M.J. Udo, M.E.F. van Mensvoort, L.Q. Tri and D.K. Nhan, 2005 Dynamics of integrated agriculture-aquaculture systems in the Vietnamese Mekong Delta. In: P. Rowlinson, C. Wachirapakorn, P. Pakdee and M. Wanapat (editors). Integrating Livestock-Crop Systems to meet the Challenges of Globalisation, Volume 2, Proceedings AHAT/BSAS International Conference, Khon Kaen, Thailand, British Society of Animal Science, T150.

Phong, L.T., H.M.J. Udo, M.E.F. van Mensvoort, A.A. van Dam, L.Q. Tri and A.J. van der Zijpp, 2007. Quantitative agro-ecological indicators and productive performance of Integrated Agriculture-Aquaculture Systems in the Mekong Delta. In: A.J. van der Zijpp, J.A.J. Verreth, Le Quang Tri, M.E.F. van Mensvoort, R.H. Bosma and M.C.M. Beveridge (eds.), Fishponds in farming systems. Wageningen Academic Publishers, Wageningen, The Netherlands, pp. 135-146,

Samdup, T., 1997. The performance of Brown Swiss crossbred cattle types and the impact of crossbreeding at farm level, Bumthang, Bhutan. MSc thesis Animal Production Systems, Wageningen University, Wageningen, the Netherlands, 109 pp.

Sinn, R., J. Ketzis and T. Chen, 1999. The role of woman in the sheep and goat sector. Small Rum. Res. 34, 259-269.

Sutresniwati, 2006. The invasion of crossbred cattle: stakeholders' perspectives in Central Java, Indonesia. MSc thesis Animal Production Systems, Wageningen University, Wageningen, the Netherlands, 66 pp.

Thomas, D., E. Zerbini, P.R. Rao and A. Vaidyanathan, 2002. Increasing animal productivity on small mixed farms in South Asia: a systems perspective. Agric. Syst. 71, 41-57.

Udo, H.M.J., 1997. Relevance of farmyard animals to rural development. Outlook Agric. 26: 25-28.

Udo, H.M.J. and T. Cornelissen, 1998. Livestock in resource-poor farming systems. Outlook Agric. 27, 219-224.

Udo, H.M.J., A.H. Asgedom and T.C. Viets, 2006. Modelling the impact of interventions on the dynamics in village poultry systems. Agr. Sys. 88, 255-269.

Farmers and fishponds in South East Asia: let the fish talk about the water[10]

Ruerd Ruben
Centre for International Development Issues Nijmegen (CIDIN), Radboud University Nijmegen, P.O. Box 9104, 6500 HE Nijmegen, The Netherlands; r.ruben@maw.ru.nl

Abstract

Fish plays an important role in the livelihood strategies of rural households in Southeast Asia, both as a source of protein and as a component of family income. In response to the reduced availability of natural capture fisheries, integration of fishponds into agriculture-based systems is frequently considered as a feasible alternative. Notwithstanding the increasing availability of knowledge regarding appropriate technological regimes for integrated agriculture-aquaculture management, farmers appear to be rather reluctant towards such resource use intensification. This is explained by the simultaneous adjustments in external market conditions and institutional regimes that make fishponds more suitable as a defensive risk management strategy, but far less attractive as a pathway towards rural modernisation and poverty alleviation.

Keywords: aquaculture, livelihood, diversification, integration, Vietnam

Introduction

Smallholder farmers living in rural areas throughout South East Asia are increasingly incorporating aquaculture as part of their production systems. Given the decreasing availability of wild fish from rivers and the sea, fish cultivation in pond systems represents a major strategy for getting access to animal proteins at relatively low costs (Edwards, 2000; Lewis, 1997; Lightfoot *et al.*, 1996). Different research efforts have addressed the wide range of options for further improving resource efficiency of fishpond systems, in particular by strengthening the integration with cropping and livestock regimes to ensure better feed utilisation, to improve water exchange and to reduce nutrient losses (Verdegem, 2005). While fish production systems based on high external input use can be technically efficient, in economic terms low input systems tend to be more profitable. Interestingly enough, many farmers seem to prefer rather simple aquaculture management options that consider seasonal labour demands and permit to stabilise household income (Rothuis, 1998; Verdegem, 2005).

Different pathways for the integration of aquaculture into smallholder farming systems and livelihoods can be envisaged, either based on diversification or specialisation of activities.

[10] The title of this presentation is inspired by Carla Risseeuw's book (1988) 'The Fish don't talk about the Water' referring to the rather limited (gender) role interpretation in rural villages in Sri Lanka due to scarce consideration of interactions between (male and female) agents. We paraphrased this statement to emphasise the necessity to evaluate different strategies for optimising aquaculture management based on considerations that explicitly consider the socio-economic context that makes their adoption more or less feasible.

Aquaculture appears to be mostly a livelihood strategy that enables poor and vulnerable farm households to diversify their activities and revenue streams, thus serving primarily objectives of smoothing household consumption and reducing risks. Further intensification of aquaculture tends to be an option that is only available to better-off farmers that have more access to inputs and are better linked to market outlets. It remains, however, highly questionable whether poorer households will, and can, be interested in investing their scarce resources and efforts in more intensive integrated aquaculture-agriculture systems.

The remainder of this paper is structured as follows: First we discuss the linkages between farming systems and household livelihood strategies to provide a better understanding of the potential roles and practical contributions of aquaculture. Hereafter, we outline two different strategies for incorporating aquaculture into smallholder farming systems, referring to 'passive' and 'active' pathways for activity diversification and their likely implications for returns to land and labour. This is followed by a discussion of appropriate technological options, market outlets and governance regimes that could enable a pro-poor oriented aquaculture development. Finally, we present some strategic policy options for better tailoring integrated aquaculture management towards sustainable poverty alleviation objectives.

Linking livelihoods and farming systems

Farming and fisheries represent two important strategies for rural households to satisfy their subsistence and survival objectives. The decision-making regarding their engagement in these activities finally depends on the available resources (land, labour, capital and knowledge) and the surrounding market- and institutional conditions (Ellis, 1998). This is illustrated in Figure 6.14, where the choice of activities and production technologies is guided by household preferences and takes place within the context of market exchange options and institutional access constraints. Farm-households can thus be characterised as units of production and consumption, where decisions regarding production activities (at the farm) are strongly co-determined by the (labour) resource endowments and the consumptive needs (within the household).

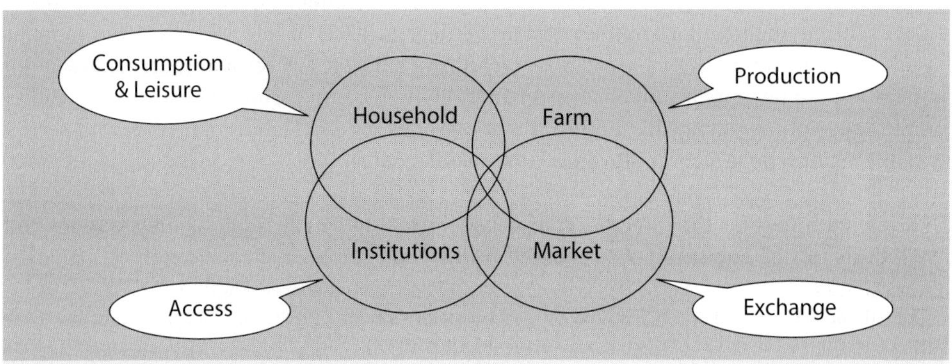

Figure 6.14. Integrated rural livelihoods.

The complex character of integrated decision-making within farm-households implies that adoption and/or adaptation of fishpond systems can only be expected to take place if household welfare can be improved. Higher yields or optimal production, as such, are necessary, but are certainly insufficient conditions for adoption. Recognising farm-household rationality regarding endogenous adjustments in farming systems requires a more detailed understanding of the diversity of objectives that poor rural households perceive. Empirical studies on the emerging rural differentiation in the Asian countryside (Sikor, 2001 on North Vietnam; Stokke *et al.*, 1991 on Sri Lanka; Jayaraman and Lanjouw, 1999 in Indian villages) indicate that agrarian households perceive different sets of objectives, depending on their resource base and access to alternative factor markets. In this respect, we can distinguish between five major farm household objectives and indicate their differential importance for three groups of typical rural farmers that are most relevant in the South East Asian countryside. Table 6.5 provides a concise overview of this matrix of objectives:

Different types of households perceive specific objectives, and the relative weight of these objectives determines which types of activities are most likely to be undertaken (Netting, 1993). The poorest farmers (usually landless farmers or sharecroppers) tend to give high priority to satisfaction of consumptive needs and risk reduction, given the uncertainty of access to resources and markets for food, land and labour. Poverty is thus partly derived from the fact that poor farmers select less risky - usually lower productivity - activities (Townsend, 1995). Consequently, production systems of chronically poor households are strongly diversified and highly intensive in family labour (thus sacrificing leisure). Transitory poor households devote somewhat more attention to balancing labour use and leisure objectives, but still derive a major share of their family wealth from home consumption. Only less poor rural households can afford to focus on profit objectives, and - given their surplus production - more attention is given to the reduction of transactions costs for market exchange.

This stratification of rural households and the related objective structure, enables us to better understand the different roles and functions that may be attributed to the integration of fishponds in arable farming systems. Chronically poor households perceive fishponds as an attractive device for activity diversification and expect major returns through increased fish consumption, particularly during the off-season. It is rather unlikely to expect that these farmers will be interested in further resource intensification or fishpond expansion, given the scarcity of land and labour and the limited market access (Ahmed and Lorica, 2002).

Table 6.5. Farm household objectives for different rural strata.

	Chronic Poor	Transitory Poor	Non Poor
Consumption	++	++	+
Profit	(+)	+	+++
Leisure	+	++	+
Risk management	+++	++	++
Reduction Transaction Costs	0	+	++

In a similar vein, fewer poor farmers may be willing to adopt improvements in fishpond systems that economise on their labour use. Technical options that increase total production, but decrease labour productivity (e.g. improved sediment recycling), are therefore less attractive. Otherwise, somewhat better-off farmers are expected to be more inclined to adopt productivity-enhancing technologies and management systems, given their better access to profitable market outlets.

Understanding the interlinkages between farming systems and household livelihood objectives can thus be a useful tool to identify feasible and potentially attractive (i.e. profitable) innovations for fishpond production and management. Empirical research indicates that mixed and integrated agriculture-aquaculture systems are indeed adopted if attractive pay-offs can be ensured (Bosma *et al.*, 2006). Interestingly enough, systems based on high external input use proved to be technically efficient (at the farm level), but economically low-input systems are more profitable (for the household). Therefore, policies and incentives should be identified that permit to overcome these trade-offs between farm and household objectives.

Pathways for rural development

Integration of aquaculture into agriculture-based farming systems is frequently mentioned as a strategy that enables farm households to improve their livelihoods and increase their welfare (Allison and Ellis, 2001). Although this might be a correct view when looking only at the start and the end point of farming systems diversification, in reality rural smallholders can rely on various sequential strategies to adjust their farming systems over time. Therefore, a more balanced insight into the different dynamic pathways for agricultural development is required.

In the Mekong delta of Southern Vietnam, integration of fish cultivation into arable farming systems usually starts with the construction of a small family pond. This pond is fed with human excreta and household residues while using residue water from the rice fields. Almost no external inputs are used and the fish grows under natural conditions, without requiring much labour effort. Transforming such a system into a semi-intensive fish production unit requires a substantial change in mentality. In fact, it is highly unlikely that smallholders who basically use the fish for home consumption, will be motivated to make such efforts and investments.

Pathways towards the intensification of mixed systems are usually of a non-linear nature. Farmers will only be inclined to make direct investments in farming systems intensification (e.g. devoting more labour and using external inputs) if their basic objectives of risk control are satisfied. This pathway is usually only available to wealthy and better-off farmers with stable access to markets and resources, who may be expected to engage directly in more specialised and intensive fish farming. As illustrated in Figure 6.15, the large majority of smallholder farmers are likely to start with gradual diversification and improvements of their traditional mixed systems in order to achieve internal resilience. The transition towards more input-intensive fishponds will only occur once farm-household food security and risk management objectives are guaranteed (Sadoulet and De Janvry, 1995).

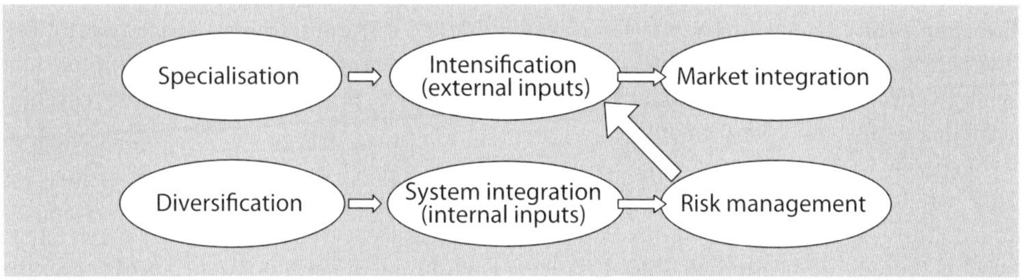

Figure 6.15. Pathways for agricultural development.

Aquaculture and fishponds may therefore play a decisive role in the transition from internal to external input use, and eventually enable farm-households to engage into further steps towards specialisation and input intensification. It is likely that such transition takes place in different phases, gradually improving farmers' willingness and abilities to invest. Diversification based on internal inputs thus tends to be a necessary stage before agricultural development really takes off. Similar results regarding development pathways are found in other fragile settings, where external support is, by preference, used for strengthening food production before any in-depth investments are made for improving yields (Ruben and Clercx, 2003). Assistance towards small-scale farmers will therefore initially benefit most from yield-stabilising technologies.

In subsequent stages, farming systems diversification through the integration of fishponds into arable cropping, can also provide interesting options for improving total factor productivity. Contrary to what is usually expected (Kurosaki, 2003), such diversification tends to reduce land productivity, albeit improving the returns to labour. Based on a comparative study of farming systems in three different countries (Figure 6.16), we found that farm households engaged in more different (farm and off/non-farm) activities are mainly motivated by the higher labour productivity. This is especially the case when local labour markets become more developed, thus offering alternative opportunities to (young) family workers. The only real incentive to remain involved in farming is when on-farm labour rewards keep pace with

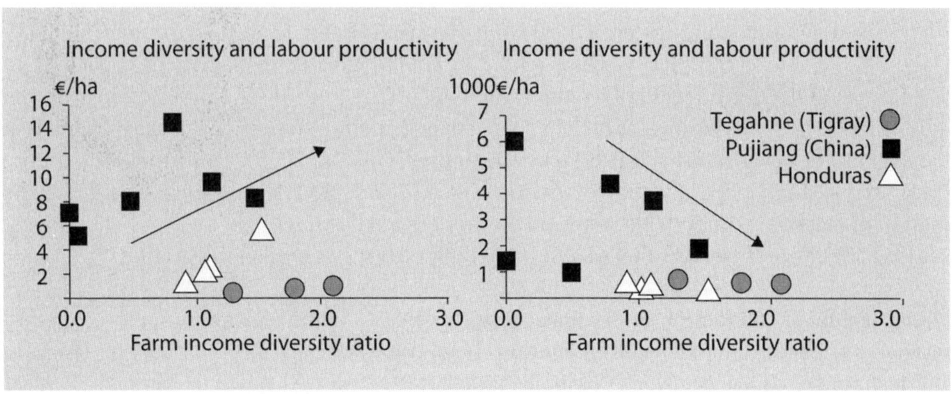

Figure 6.16. Impact of activity diversification on factor productivity (Langeberg et al.*, 2005).*

the opportunity costs of labour for off-farm activities. It should also be noticed that more diversification not necessarily means better activity integration; in fact fairly limited real synergy effects of mixed farming are registered in practice, probably because most recycling activities are just too labour intensive.

This brings us to another important point regarding feasible development pathways for aquaculture-agriculture integration. Poor farmers generally search for an activity pattern that permits them to optimise their labour returns. Even in the densely populated areas of South East Asia, farm labour is relatively scarce. This is even more the case when engagement in regional migration and off-farm opportunities are rapidly increasing. Labour-led intensification is therefore a highly unlikely development pathway out of poverty. This does not imply that some households (especially with older farmers) demonstrate their willingness to devote more labour time to fishpond intensification, but this is mainly due to life-cycle effects (see Bosma *et al.*, 2007). Since older people have less alternative employment options, the fishpond may basically represent their pension. While younger farmers may prefer some more commercially-oriented fisheries in order to permit capital accumulation.

While there are different reasons for farmers to engage in aquaculture production, not all mixed systems equally contribute to poverty alleviation. Even when aquaculture intensification is technically feasible, it will not always be economically attractive. Moreover, transition pathways towards more intensive production systems need to consider (1) their initial contribution towards farm household risk management objectives, and (2) the competitive position of family labour on the local and (extra)regional labour market. The existing heterogeneity in development pathways followed by poor and better-off farmers and/or by older and younger households can, therefore, be largely understood as their diverse response to internal and external resource and market constraints (Ruben and Pender, 2004).

Potential contributions of integrated aquaculture

The integration of aquaculture into cropping systems provides wide options for creating a number of technical complementarities in terms of nutrient recycling, re-use of water and waste management. The internal interaction involved in the input-output interfaces within such mixed systems point towards increasing complexities for farm management practices (Figure 6.17). More complex systems tend to demand greater management capacities of the farmer, but could eventually reduce their flexibility regarding responses to external adjustments. It is well-known that better risk management practices are usually adopted at the expense of lower returns to production activities (Dercon, 2005). This implies that more integrated systems perfectly suit the interests of resource-poor farmers that operate at the margin of markets, while a tendency towards more specialisation might be expected amongst somewhat better-off farmers that search for a higher degree of market integration.

While technical interaction at the farm-level is already highly complex, socio-economic interfaces with local markets and institutions have equal (or even more) impact on land use and technology choice decisions (Sadoulet & de Janvry, 1995). As noted in Figure 6.17, rural households maintain simultaneous exchange relations with several markets:

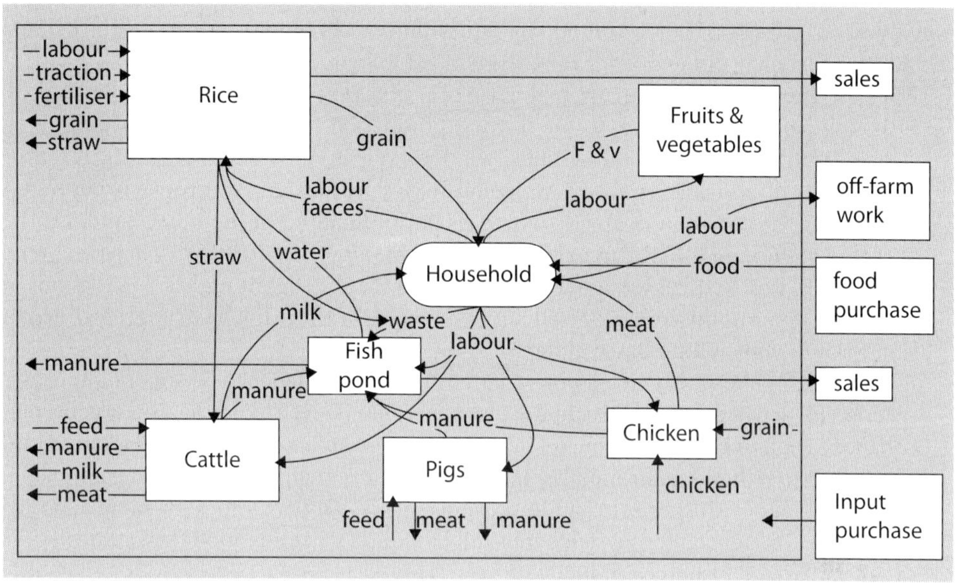

Figure 6.17. Internal and external flows in diversified farming systems.

1. with the labour market: hiring-in of labour for peak operations, hiring-out of family labour during slack periods, off-farm work and non-farm self-employment, seasonal or temporal migration to other areas (while sending remittances back to the household);
2. with the land market: renting land out or in, land lease contracts with other farmers, seasonal land rent, land mortgage transactions, and sharecropping arrangements;
3. with the capital market: borrowing money from banks, local moneylenders, traders, friends or relatives, savings and insurance arrangements at individual or group level, traditional savings practices (savings in kind), etc.
4. with the input market: purchase of fertilisers, seed, animal feed, and pesticides, sometimes with back-payments against the harvest;
5. with the output market: selling of rice, animals, fish and vegetables to local traders and retailers, or deliveries to the wholesale market and to supermarket outlets.

Most of these markets show a remarkable development in Southeast Asia. In Vietnam, off-farm income and remittances together already represent more than 30% of household income in the Mekong Delta (Adger *et al.*, 2002), and land lease transactions are increasing after the *Doi Moi* policy. Markets for credit and capital are rapidly emerging in other countries of the region, and both input and output markets have been privatised, giving rise to a strong dynamics in market exchange transactions.

These external market linkages determine, to a great extent, the available options for crop and technology choice. Whereas technical efficiency depends mainly on options for increasing input complementarities, economic efficiency is ultimately dependent on relative prices of inputs and outputs (Lee, 2005). 'Appropriate' aquaculture technology can therefore be defined considering three key dimensions (Kuyvenhoven and Ruben, 2002):

1. From an efficiency viewpoint, optimal *resource productivity* is perceived as the major goal, focussing particular attention to the scarcest production factors. Comparing traditional rice cultivation with fishpond systems, we observed in Vietnam a declining resource productivity of land, labour, as well as, capital, in rice production, mainly due to depressed prices after the abolition of state controls. Fish production, on the other hand, can be attractive to increase the returns to land, but performs almost equally to rice in terms of the returns to labour. Returns to capital from fishponds may be slightly higher. In situations of increasing opportunity costs of labour, farmers might therefore not be strongly inclined to shift towards aquaculture-based systems.
2. From the farm household viewpoint, *resource use flexibility* tends to be a more important criterion for activity choice. This implies that preference is given to production systems that enhance options for shifting resources in case of changing market conditions. Fishponds are considered to reduce labour mobility (since aquaculture requires permanent care), while diversification of fish species could improve resilience. Opportunities for more adaptive management of aquaculture systems thus might require more attention than strategies for increasing overall resource productivity.
3. From a long-term perspective, *sustainability* of the farming system and the reduction of externalities deserve specific attention. Strangely enough, most POND research focussed nutrient dynamics, while pest control receives only minor attention. The integration of rice and aquaculture systems usually requires, however, a strict control on pesticides use. Berg (2002) showed that IPM-based Rice-Fish systems clearly outperform non-IPM based systems in terms of net revenues and better satisfy farmers' expectations regarding reduced externalities.

Farm-household engagement in integrated rice-fish systems can thus be considered from multiple perspectives with sometimes contradicting goals and objectives. Trade-offs between technical and economic efficiency are likely to appear (i.e. improved resource productivity only attainable at high input costs), particularly for the less commercially-oriented farmers. Therefore, most *de facto* integration can be observed amongst family farms during their consolidation stage, especially in households that are close to retirement and therefore interested in fishponds as a strategy to optimise their returns to family labour. Otherwise, higher degrees of specialisation are likely to be found amongst farmers with more resources and better access to (input and output) markets.

The role of markets and institutions

The engagement of farm households with integrated fish farming activities does not take place in isolation. Local market conditions and institutional regimes exercise a decisive influence on the relative attractiveness of integrated systems, and could contribute to their intensification and/or integration. This is particularly important to understand, since the appraisal of supportive policies should be based on an adequate interpretation of farm-household responses to changing external conditions.

At the marketing level, aquaculture production is scarcely considered as a full substitute for rice cultivation, even when relative prices of the latter crop are strongly declining in most South East Asian countries. Molini (2002) reports that in Vietnam the rice-fish price ratio increased from 1:7.5 (in 1988) to 1:10 (in 1993), occasioning a strong decline in rice production. Similar tendencies are observed in other Southeast Asian countries. Shoichi *et al.* (1989) even suggest that rice is gradually becoming an inferior good.

Rice has become a less rewarding crop (Minot and Goletti, 2002), but its cultivation is rarely fully abandoned by smallholder producers, given the existing uncertainty regarding the access and availability of rice at the market. This implies that most farmers will consider rice unattractive as a producer, but still remain involved in rice production as a consumer. Their condition of net buyer of rice during the main part of the year (they might sell most of the rice directly after the harvest, usually at low prices) motivates them to continue rice production and thus refrain from full specialisation in aquaculture.

Another important reason for giving preference to mixed rice-fish systems is found in the seasonal complementarities. Fish sales provide a good source of income to cover consumptive expenses during the 'tough period' when the rice crop is still in the field and many expenses need to be made. In addition, the selling of the rice harvest can be slightly delayed (thus receiving better prices) if some fish revenues are available. This strategy allows farmers to avoid having to rely on distress sales and thus provides an important higher income to resource-poor farmers that are usually dependent on moneylenders to cover expenses in the pre-harvest period.

Finally, some aquaculture farmers might try to improve their marketing position by engaging in contract farming arrangements with local supermarkets. Given the increased rate of urbanisation, large-scale sales of high quality fish offer a promising new outlet. This is only feasible when due attention is given to quality and scale aspects. New technologies should therefore particularly focus on consistency and shelf life of the fish. In addition, institutional innovations focussing on improved supply chain integration between farmers, processors and supermarkets become a major challenge.

At institutional level, the organisation of land and capital markets deserves major attention. Current land right systems need to be improved to guarantee long-term ownership in order to enable farmers to make the necessary investments for aquaculture intensification. Many farmers in South East Asia still rely on sharecropping systems for access to land, usually reducing the incentives for intensification (see Quibria and Rashid, 1986 for an overview). Unless cost-sharing aspects are included and long-term contracts are agreed upon, operators involved sharecropping will be fairly reluctant to invest in improved fishponds.

In a similar vein, large imperfections in rural financial markets may limit the prospects for integrated farming. Most rural credit is still used for consumption smoothing and to mitigate shocks, while only a limited proportion is available for in-depth investment (Kochar, 1999). Fishponds can be, however, considered as suitable collateral for borrowing, and may thus contribute to improved access to credit (to be used for other purposes, like payment of school fees). While many poor rural household have a strong preference for near-liquid savings (in cattle or jewellery), fishponds can also be turned into an appropriate savings and insurance device. If this is indeed the case, then farmers will be more interested in fishponds that provide stable returns and will opt less for profit-optimising aquaculture technologies.

Strategic options

Based on the earlier assessment of appropriate options for integrating fishponds into farming systems - given the specific market and institutional condition - we can finally discuss what the expected contributions of such systems for poverty reduction of rural farm households are. Therefore, it is useful to understand the attractiveness of integrated fish-farm production to different categories of farm-households, and to make a clear difference between various and diverging motives for activity diversification.

Drawing upon a distinction made by Dorward *et al.* (in press), we can identify basically three reasons for activity diversification:

1. Hanging-in: a defensive strategy of passive integration, mainly for defensive purposes and focussing on improved risk management. These farmers will use fishponds mainly as a mere survival strategy, looking particularly at options for more backward integration (e.g. reduced input purchase and improved on-farm food availability) to reduce their vulnerability and dependency.
2. Stepping-up: a strategy of active integration of fishponds into arable cropping and livestock systems, looking for modernisation of agricultural activities through intensive engagement in market exchange (forward integration). These farmers might opt for improved species and external feed purchase, and could finally become more specialised fish producers.
3. Stepping-out: a strategy to invest in different non-farm activities that require start-up capital which may be generated through fish farming. Given the current tight credit constraints, internal funding should be raised. In addition, farmers may invest in children's education to enable better engagement in labour markets and migration activities. Consequently, high-yielding but low labour-intensive fisheries could be selected as a temporary livelihood activity.

Although no formal stratification has been made regarding the proportion of farmers belonging to each of these categories, it seems that the first and the last groups represent the overwhelming majority. This may imply that fishponds serve, either as a defensive device for agricultural diversification, or as a temporary strategy towards activity diversification out of agriculture. The real demand for more advanced fishpond management options might - in perspective - thus be limited to a rather small segment of commercial specialised aquaculture farmers.

Notwithstanding the increasing availability of knowledge regarding appropriate technological regimes for integrated agriculture-aquaculture management, many local farmers appear to be rather reluctant towards further resource use intensification. This can be largely explained due to the simultaneous adjustments in external market conditions and institutional regimes that make fishponds more suitable as a defensive risk management strategy but far less attractive as a pathway towards rural modernisation and poverty alleviation.

References

Achmed, M. and M.H. Lorica, 2002. Improving developing country food security through aquaculture development: Lessons from Asia. Food Policy 27, 125-141.

Adger, W.N., P. Kelly, A. Winkels, Luong Quang Huy and C. Locke, 2002. Migration, remittances, livelihood trajectories, and social resilience. Ambio 31, 358-366.

Allison, E.H. and F. Ellis, 2001. The livelihoods approach and management of small-scale fisheries. Marine Policy 25, 377-388.

Berg, H., 2002. Rice monoculture and integrated rice-fish farming in the Mekong Delta, Vietnam – economic and ecological considerations. Ecological Economics 41, 95-107.

Bosma, R.H., Le Thanh Phong, Uzay Kaymak, J. van den Berg, H.M.J. Udo, M.E.F. van Mensvoort and Le Quang Tri, 2006. Assessing and modelling farmers' decision making on integrating aquaculture into agriculture in the Mekong Delta. NJAS- Wageningen Journal of Life Sciences 53: 281-300.

Bosma, R.H., Cao Quoc Nam, H.H.J. Udo and J.A.J. Verreth, 2007. Assessing farmers' motives for the diversification of livelihoods in the Mekong Delta: household life cycle, virtual farm size, and index of integration. In: A.J. van der Zijpp, J.A.J. Verreth, Le Quang Tri, M.E.F. van Mensvoort, R.H. Bosma and M.C.M. Beveridge (eds.), Fishponds in farming systems. Wageningen Academic Publishers, Wageningen, The Netherlands, pp. 261-270.

Dercon, S., 2005. Vulnerability: a Micro Perspective. Paper presented at Annual Bank Conference on Development Economics, Amsterdam, May 23-24 2005.

Dorward, A., S. Anderson, Y. Nava, J. Pattison, R. Paz, J. Rushton and E. Sanchez Vera (in press). Hanging In, Stepping up and Stepping Out: Livelihood Aspirations and Strategies of the Poor. World Development, in press.

Edwards, P., 2000. Aquaculture, poverty impacts and livelihoods. London: ODI Natural Resource Perspectives No. 56.

Ellis, F., 1998. Peasant economics: farm households and agrarian development. Cambirdge, Cambridge University Press.

Jayaraman, R. and P. Lanjouw, 1999. The Evolution of Poverty and Inequality in Indian Villages, World Bank Research Observer 14, 1-30.

Kochar, A., 1999. Smoothing Consumption by Smoothing Income: Hours-of-Work Responses to Idiosyncratic Agricultural Shocks in Rural India. The Review of Economics and Statistics 81, 50-61.

Kurosaki, T., 2003. Specialization and Diversification in Agricultural Transformation: The Case of West Punjab, 1903–92 American Journal of Agricultural Economics 85, 372-386.

Kuyvenhoven, A. and R. Ruben, 2002. Economic Conditions for Sustainable Intensification. In: M. Uphoff (ed.), Agroecological Innovations: Increasing Food Production with Participatory Development. London, Earthscan, pp. 58-70.

Langeberg, H., M. Rufino, H. Hengsdijk, R. Ruben, J. Dixon, J. Verhagen and K. Giller, 2007. Evaluation of economic and ecological performance of farm household diversification and integration strategies. Rome, FAO (in press).

Lee, D.R., 2005. Agricultural Sustainability and Technology Adoption: Issues and Policies for Developing Countries. American Journal of Agricultural Economics 87, 1325-1334.

Lewis, D., 1997. Rethinking aquaculture for resource-poor farmers: perspectives from Bangladesh. Food Policy 22, 533-546.

Lightfoot, C., M. Prein and J.K. Ofori, 1996. The potential impact of integrated agriculture-aquaculture systems on sustainable farming. Manila: ICLARM Contribution No. 962.

Minot, N. and F. Holetti, 2002. Rice market liberalisation and poverty in Vietnam. Washington D.C., INFPRI Research Report No. 114.

Molini, V., 2002. Food security in Vietnam during the 90s. Mimeo, University of Eastern Piemonte.

Netting, R., 1993. Smallholders, householders: farm families and the ecology of intensive, sustainable agriculture. Stanford: Stanford University Press.

Quibria, M.G. and S. Rashid, 1986. Sharecropping in Dual Agrarian Economies: A Synthesis, Oxford Economic Papers 38, 94-111.

Risseeuw, C., 1988. The fish don't talk about the water: Gender transformation, power and resistance among women in Sri Lanka. Leiden, Brill

Rothuis, A., 1998. Rice-fish culture in the Mekong Delta, Vietnam: Constraint analysis and adaptive research. PhD Thesis Leuven: Katholieke Universiteit Leuven.

Ruben, R. and J. Pender, 2004. Rural Diversity and Heterogeneity in Less-favoured Areas: the Quest for Policy Targeting. Food Policy (29): 303-320.

Ruben, R and L. Clercx, 2003. Rural Finance, Poverty Alleviation and Sustainable Land Use: The Role of Credit for the Adoption of Agroforestry Systems in Occidental Honduras. Journal of Microfinance 5, 77-90.

Sadoulet, E. and A. de Janvry, 1995. Quantitative development policy analysis. Baltimore, The Johns Hopkins University Press.

Sikor, T., 2001. Agrarian Differentiation in Post–Socialist Societies: Evidence from Three Upland Villages in North–Western Vietnam. Development and Change 32, 923-949.

Shoichi I., E.W.F. Peterson and W.R. Grant, 1989. Rice in Asia: Is It Becoming an Inferior Good? American Journal of Agricultural Economics 71, 32-42.

Stokke, K, S. Yapa and H.D. Dias, 1991. Growth Linkages, the Nonfarm Sector, and Rural Inequality: A Study in Southern Sri Lanka. Economic Geography 67, 1-32.

Townsend, R.M., 1995. Consumption insurance: an evaluation of risk-bearing systems in low-income economies. Journal of Economic Perspectives 9, 83-102.

Verdegem, M. (ed.), 2005. Improved resource use efficiency in Asian integrated pond-dike systems. Final Report INCO Project Pond-Live. Wageningen: Wageningen University. Fish Culture & Fisheries Group.

Chapter 7
Strategies to enhance the role of fishponds in farming systems

Strategies to enhance the role of fishponds in farming systems

Johan A.J. Verreth[1], Roel H. Bosma[1], Malcolm Beveridge[5], Le Quang Tri[2], Martinus E.F. van Mensvoort[4] and Akke J. van de Zijpp[3]
[1]Fish culture and Fisheries Group, Wageningen Institute of Animal Sciences, Wageningen University, The Netherlands
[2]Soil Sciences & Land Management Dept., College of Agriculture & Applied Biology, CanTho University, Can Tho City, Vietnam
[3]Animal Production Systems Group, Wageningen University, The Netherlands
[4]Department of Environmental Sciences, Wageningen UR, The Netherlands

Fishponds have a future! At least this seems to be the general conclusion from this book. The research presented in this volume clusters around two objectives: (a) how can constraints be diminished and fish production be improved and (b) how can the outcome of the research contribute to policy development? In relation to this, at the POND symposium the participants reflected on two questions overarching the fish, the pond and the farm level: (1) Are the findings far enough advanced in terms of sustainability, food safety and food security for transfer to extension services? (2) Are the findings far enough advanced to recommend development policies and future research agendas needed to advance the present status of knowledge? In this chapter we will discuss these questions in the context of the conditions and development perspectives for Vietnam, Southeast Asia and Africa.

At the start of the POND programme farmers were consulted to confirm the identified research issues. During these "*State of the System*" workshops the following research priorities were listed by the farmers: the quality of seed and brood stock, the choice of the fish species, the utility of the pond mud, and the improvement of resource use in pond-dike systems integrated with mainly mono-gastric livestock. Some of these questions were subsequently addressed by a combination of on-station and on-farm research. The results were presented in the preceding chapters. The necessity to develop technologies using less pesticides and herbicides was often stressed, as well as a need for training of farmers and extension workers. Main constraints to further develop integrated agriculture aquaculture (IAA) systems were the fluctuating market prices and the lack of capital for investment. This applies as well to Vietnam, Bangladesh as to NE Thailand, but not in a similar way for all sites and social groups (Little *et al.* this volume, p.177-202).

In the following pages we'll discuss the identified problems at the level of the fish, the pond, and the farm. We will further elaborate on food safety, diversification, development of IAA systems, and on interdisciplinary research as a tool in this development. We will conclude by inferring on the development of fishponds in farming systems.

Fish genetics

Selection trials with tilapia are mainly done in high input environments. For aquaculture in developing countries, this has the risk that a sort of Trojan horse is introduced (Hans Komen[11], personal communication). Tilapia are herbivorous fish that do not need feed with protein/oils originating from fish. However their selection is mostly done in pellet-based systems whereby the tilapia are fed with high protein diets. This procedure could lead to an increased dependency on high inputs (and indirectly on Western technologies) to maintain productivity. The selection for growth in high input environments has increased feed intake, but improvements in feed conversion ratio are not evident. The benefit by the realised genetic gain in growth might then be compromised by the increased dependency on feed inputs. In tilapia grow-out enterprises 70% of costs are associated with feed. The dependency on high input technologies is not restricted to feed inputs only: for such high growth varieties, probably mono-sex culture is required to enable the genetic gain to be expressed in increased production.

The results of the current projects showed that there is an alternative. When tilapia was selected in ponds fertilized with manure and without pellets or other inputs, protein retention improved from 1.7% to over 2.5% and weight at harvest increased by more than 50% (Charo, 2006). This is in contrast to the protein retention of 0.8-1.7% in the high input systems. The increased efficiency might be related to the increased gut length found in the selected strains from low-input environments (Beveridge and Baird, 2000). This seems to show us a way out for the resource poor farm households. Further, results from the on-farm trials in Thailand and Bangladesh show that growing fish in low input systems is already profitable. Further research is needed to establish if the use of tilapia selected in low-input environments can improve the economic margins as well.

Reproduction of tilapia in ponds leads generally to erosion of the growth potential. The POND research showed that this is due to multiple paternity and the selective breeding strategy of male and females. Although the heavier males sire more frequently, reproductive success was independent from the magnitude of size difference between males and females. Various breeding strategies are proposed for small hatcheries to prevent the loss of genetic diversity and growth potential of the fingerlings (Ponzoni *et al.*, this volume, p59-67; Fessehaye, 2006). A comparative field study should elucidate the feasibility of these strategies.

Pond nutrient dynamics: feed the pond to grow fish

A pond is essentially a 'hole in the ground with water'. In most cases, the profit margins from pond aquaculture are so small that farmers do not like to invest money in system improvement. Pond production can be enhanced by fertilisation and feeding, improvement of nutrient cycling, mitigation of wastes and sludge discharge. Today, simple and reliable technologies to treat pond effluents are available. Pond effluents can also be used for irrigation, or in case of nutrient-rich bottom sludge, for fertilization of crops (this volume: Yakupitiyage *et al.* p89-96; Shamsul *et al.* p97-104, Muendo *et al.* p 125-133).

[11] Chairgroup Animal Breeding and Genetics, Wageningen University, The Netherlands.

During the last decades little progress has been made in our understanding of the functioning of ponds. Nutrient and energy studies always confirm our present state of knowledge concerning the basic processes occurring in ponds. Sometimes, new insights in the ecology of aquatic systems can be used to fine-tune flows and budgets, but these do not shed light on new pathways to enhance production. A major advantage of pond aquaculture is that it is a low cost technology. In various Asian countries, pond culture significantly contributed to poverty alleviation and income generation (Little *et al.* this volume p177-202). Improvements in extension methodologies and in farming proficiency largely contributed to these successes.

Nevertheless, major challenges are looming. Due to increasing demand for aquaculture products and the limited amounts of land and water available to expand pond acreage, pond production needs to increase 2-3 fold within the next 20-25 years (Verdegem *et al.*, 2006). This needs to be done in a more nutrient-efficient way while polluting less than now. In addition, society demands aquaculture to become more bio-secure, to guarantee animal welfare and to produce healthy food. Doubling to tripling pond productivity, while reducing nutrient losses, will require a shift from our present focus on feeding fishes or shrimps, to fertilising the pond ecosystem in which the fish grow. The use of polyculture ponds and the choice of species for such systems remain a challenge: the positive results obtained in the current projects from mixing carp and rohu (Rahman *et al.*, this volume p 79-88) might be different in polycultures based on tilapia, a species with very plastic feeding habits. Innovative concepts are needed that focus on increasing the natural food availability for farmed animals, producing quality wastes, optimising synergies between species raised in polyculture and reducing water use. Such an approach is needed to guarantee the long-term sustainability of pond aquaculture.

Farm nutrient dynamics

The pond mud may be an important source of fertiliser for poor farmers, especially as flood-protection dikes decrease alluvial deposition. The studies in Bangladesh and Vietnam confirmed that nutrient inputs from pond mud can contribute to maintaining soil fertility and improve farmers' income (this volume: Shamsul *et al.*, p97-104; Nhan *et al.*, p147-157). The pond mud is rich in P and supplementary application of N and K fertiliser is required. Thus in IAAS the ponds can improve on-farm nutrient cycling and the efficiency of nutrient use. However, the nutrient accumulation in the sediments levels off after 6 months, probably due to leaching and precipitation (Yakupitiyage *et al.*, this volume p89-96). Although such a frequent extraction of sediments is not practised, it converges with the short cycle of most culture practices. Efficient tools for the extraction of the mud, used in Vietnam and China, should still be tested in countries that have recently introduced pond-dike farming systems.

The farming systems in the Mekong Delta often have positive nutrient balances (Phong *et al.*, this volume p.135-143; Nhan *et al.*, 2006). The water management of the fishponds in the Delta often leads to excessive discharge of nutrients to the surface water, which may enhance eutrophication of canals and rivers. Nhan *et al.* (this volume p.147-157) demonstrated that changes in management practices can reduce nutrient discharges and improve the efficiency of nutrient use. This is in contrast with large areas of sub-Saharan Africa, for example, where farming systems are nutrient deficient (Stoorvogel, this volume p.107-123). In general the adoption of fishponds in integrated farms by African farmers has been low (Brummett and Haight, 1996), but due to increased market demand and low-input technologies a change is observed in several African countries (Brummett *et al.*, 2004). Examples of aquaculture progress from Burkina Faso (André Kabré[12]), Kenya (Robinson[13]), and Uganda (Rutaisire[14]) were presented at the symposium.

Muendo *et al.* (this volume p.123-133) show that fishponds can improve the N balance of farming systems in Africa. These ideas are based on simulation models of nutrient flows and balances in Kenyan farming systems. In fishponds, N fixation occurs and this reduces the total farm nutrient losses (Muendo, 2006). By further recycling the N contained in the pond sediments and drainage water by using it for crop irrigation, the N balance became less negative in all considered farming systems. N balances become even less negative if the fishponds were constructed on the lower slopes of agricultural fields, thereby trapping Nitrogen from runoff waters. In short, for Kenyan farming systems, where seasonal nutrient deficits occur, fishpond sediments can act as a temporary storage of nutrients that can be used during the nutrient poor seasons.

The question arises here what the value is from the most common research tool in studying the dynamics of IAA systems: calculating nutrient balances. A quick search of the last five years identified 225 published scientific papers on nutrient balances and budgets in farming systems (Van Keulen[15], personal communication). Balances and budgets are quantitative descriptions of the flows of nutrient elements over the boundaries of farms and farming (sub-) systems. They are established using various information sources: farmers, experts, observations, statistics, transfer functions and models. In spite of various limitations associated with uncertainties in the data and the use of subjective knowledge (Schlecht and Hiernaux, 2004), general agreement exists that nutrient balance assessments and exercises are useful instruments to quantify indicators for the sustainability of agricultural systems (Roy *et al.*, 2003). They provide information on management performance, 'weak' links in nutrient cycles and (partial) balances, all of which contribute to increased understanding of the functioning of the systems and provide a basis for action for researchers and farmers involved in participatory action research (Defoer *et al.*, 1998). Nutrient budget analyses can be used as simple devices to stimulate dialogue in a participatory process of negotiating interventions or policies for tackling issues of agricultural sustainability (Scoones and Toulmin, 1998). However, most nutrient balance studies present static pictures, based on short time exercises, and simplify assumptions related to system dynamics. One of the potential problems of such a snapshot approach is the danger of extrapolation and/or up scaling from limited site-specific data sets to wider scales. Moreover, the methods of nutrient balance analyses are fraught with problems such as: limited data availability at different spatial scales, scale-specific spatial variation in nutrient balance input data; non-linearity in up scaling, and lack of reliable techniques for up scaling. Extrapolating the results of models into the future and ensuring their validity presents

[12] Institut du Developpement Rural, Université Polytechnique de Bobo, Bobo-Dioulasso, Burkina Faso.

[13] Kenya Marine and Fisheries Research Institute, Mombasa, Kenya.

[14] Makere University, faculty of Veterinarian Medicine, DWARM, Kampala, Uganda.

[15] Plant Research International, Wageningen, The Netherlands.

practical problems (Roy *et al.*, 2003). Moreover, our understanding of diversity, complexity and uncertainty within smallholder farming systems is still incomplete, as is our insight into many of the socio-economic and institutional factors that influence decision-making at farm level (Scoones and Toulmin, 1998). Although the nutrient budget and balance approach may provide useful information, their application and interpretation require judicious identification of purpose, method and analysis within each particular context (Van Keulen, personal communication).

Food safety and IAA systems

Although IAA systems may have several ecological advantages by turning wastes into fish and reducing the discharge of nutrients and microorganisms in surface water, they also have a potential health risk (Dalsgaard[16], personal communication). Fish-borne zoönotic parasites can become a threat to food safety. Limited information is available on the risks of passive pathogen transfer, genetic changes of viruses, and transmission of chemical residues and antimicrobial resistance (Ghittino *et al.*, 2003; Petersen and Dalsgaard, 2004). Those risks are higher in systems treating wastewater containing excreta from livestock or humans; both the overhung fishpond latrines and the wetlands wastewater treatment systems are balancing on the edge, between benefits and risks. The revised WHO guidelines on the safe use of excreta and wastewater in aquaculture advise avoiding using untreated wastewater and excreta (WHO, 2006). There is a need for risk assessment and for research on systems reducing risk of infections or separating treatment and aquaculture.

Diversification and intensification at farm level

The preliminary results of the POND projects confirm that farm diversification, as it occurs in IAA systems, contributes to increased sustainability at the ecological (Phong *et al.* this volume p.135-143) and economic (Bosma *et al.*, this volume p.261-270) level. In contrast to this, specialised farms demonstrated to be more vulnerable to shocks: for example, in Vietnam many specialised catfish farms went bankrupt after the U.S. import ban on Vietnamese catfish in 2002, and specialised poultry producers suffered more from the Avian Influenza than mixed farms (Jansen, 2006). This confirms the conclusions of Holling *et al.* (1995) and Luu *et al.* (1999), that mixed farming systems have a higher potential to absorb shocks from the natural and economic environment. Being locally adapted the IAA farming systems need less inputs while generating a better income. They are more flexible, thus reducing risk and contributing to the economic sustainability of the family farm. Integrated systems can intensify some of their components while maintaining a high level of nutrient efficiency, thus contributing to ecological sustainability. The high level of labour input can be an advantage in developing economies where industrial wages are still low and the application of low-input technologies in IAA allow obtaining competitive remuneration of relatively low qualified labour (Ruben, this volume p.281-292). However, if this leads to reduced frequenting of school, especially of girls, it impacts negatively on education level and on future options for development.

[16] Royal Veterinary and Agriculture University, Copenhagen, Denmark.

Research and development of IAAS

While various papers in Chapter 5 focused on the People-Planet-Profit aspects of sustainability of the system, Edwards (this volume p.213-227) discussed the factors "required for a given technology to be sustainable for poor farming households". Among the factors he included production technology, socio-economic and environmental aspects of which "the latter invariably determine sustainability". In fact, sustainability of a technology should be looked at from all the aspects mentioned above (Tran Thanh Bê[17], personal communication). A systems researcher equipped with a holistic approach keeps in mind that the criteria for assessing a given technical solution for sustainable development include biological feasibility, technical compatibility, economic viability (including ability to spread risks), socio-cultural acceptability, and suitability for development policy (IRRI, 1986).

The POND programmes demonstrated that high quality research with strong participation of farmers and extension officers can lead to modification of practices. However, there are risks associated with this approach (Koning[18], personal communication). Poverty and unsustainable production may be alleviated at farm level using an interdisciplinary approach on a small scale, but a successful diffusion might well result in a strong increase in the aggregated supply that gluts the market at some higher level. Consequently farm profits are squeezed and farmers remain resource poor, if they have no alternative diversification options, and if market demand does not increase. Thus an agronomic package may not be sufficient to solve the problem, no matter how participatory or interdisciplinary its development has been.

The need for a holistic systems approach is felt even more strongly when considering the dilemma of rural planning. Although central government may consider biodiversity and environmental aspects in its development policy, local communities will inevitably focus on economic development to improve livelihood options for their citizens. The last may infringe on biodiversity and harm the ecology of a region, but that seems to be the trade-off to pay for poverty reduction (Vo Thung Xuan[19], personal communication). We believe that the POND programme showed pathways for animal breeders, fish nutritionists and pond ecologist to contribute to social and economic sustainable development while reducing harmful trade-offs in terms of ecological resources. However, the real challenge for the future remains to strengthen the link between the various research levels and to embed the research more strongly in its' societal context.

Interdisciplinary scientific research and rural development

The POND programme aimed to conduct research that can make a difference either through designing technologies, developing policies, or strengthening human capacities. To what extent did POND contribute to these goals and how could it have done better? Several technologies relevant for development were designed and some continue to be used by farmers in the

[17] Director of the Mekong Delta Development Research Institute, Can Tho University, Vietnam.

[18] International Research and Education Fund (INREF), Wageningen Univerisity, The Netherlands.

[19] Rector of Long Xuyen University, Vietnam; member of the CGIAR advisory board.

research villages. In all three countries we contributed to local policy on aquaculture. The partners' researchers graduated and published peer-reviewed papers, which contributes to human capacity building of the partner institutes.

The obtained results show that high quality research can be combined with a participatory approach to develop technologies, though not all POND studies generated knowledge that can be applied directly by farmers. Thus the broader uptake is very unpredictable, especially as technical research has very long impact cycles (De la Rive Box[20], personal communication). Moreover, measuring impact assumes a good baseline description and a monitoring system measuring or assessing impact during the process, and even then it remains hard to attribute impact to one specific actor (Gunning, 2006). The possibility that individual PhD research contributes to practical development, and thus to solutions of complex problems, is restricted. PhD students also need to explore the boundaries of scientific theory and to analyse the processes behind solutions. Academics should also be able to apply these new insights in other situations and to reflect on fundamental issues and on their own role in the interaction science-society.

The approach of POND-LIVE in both Bangladesh and Thailand showed that engaging national stakeholders in the project pays off through new funding and adjustment of extension policy (Shamsul *et al.*, this volume p.97-104). The policy-relevant discourses within the INREF-POND programme on development issues were mainly restricted to the partner institutions and have hardly had international impact to date. It cannot be expected from individual PhD-candidates to influence policy discourses. Their priority is to finish their research and write their thesis; moreover it may be difficult for them to discuss politically sensitive issues for reasons of status and their future career. Therefore the partner institutions should have both a strong commitment to the project and an interest in the results of the research. Commitment can be achieved through local partners providing staff and financial participation in the project, as that requires the involvement of higher bureaucratic and sometimes even political levels (INREF, 2006). The probability of achieving development impact from a research programme is enhanced through being embedded in networks of partners necessary to scale up and scale out project results. Often this will involve engagement of national policy makers and (inter)national institutes in steering committees. Monitoring and adjustment of programme activities is also essential (Douthwaite *et al.*, 2003).

Conclusions

Some of the POND-LIVE projects were highly successful in establishing linkages between the various levels of production systems, mainly due to a better embedding in the farmers' context and more frequent exchange of experience throughout the project. Although most PhD projects of INREF-POND failed to make these linkages, the symposium facilitated catching up on some of the missed opportunities. We recommend future projects to aim to contribute to the development of farming systems by involving stakeholders in defining the problems and in the identification of solutions. Programme teams should also facilitate a process of group learning throughout the project and allow a continuous adjustment of the programme.

[20] Rector of the Institute of Social Sciences, The Hague, The Netherlands.

To improve the contribution of aquaculture to poverty alleviation, fishes need to be selectively bred in low-input environments and the pond management needs to be based on the paradigm: "feed the pond to grow fish". IAA farming systems deserve the support of research and extension in order to maintain their shock absorbing character, important for both social and economic sustainability. In particular, the medium sized mixed farms showed a high potential to reduce environmental impact and to contribute to ecological sustainability. The management of Integrated Agriculture Aquaculture farming systems requires a high level of farmers' know-how that can be generated through a holistic approach to research and extension.

References

Beveridge, M.C.M. and D.J. Baird, 2000. Diet, feeding and digestive physiology. In: M.C.M. Beveridge and B.J. McAndrew (eds.) Tilapias: Biology and exploitation. Kluwer, Dordrecht, Netherlands, pp. 59-87.

Brummett, R.E. and B.A. Haight, 1996. Research – development linkages. In: FAO Fisheries Report 548 (FIRI/R548). Report of the expert consultation on small-scale rural aquaculture, Appendix 5, pp. 147-169.

Brummett, R.E., D. Jamu, J. Jere and V. Pouomogne, 2004. A farmers' participatory approach to aquaculture technology development & dissemination. Uganda Journal of Agricultural Science 9, 530-536.

Charo-Karisa, H., 2006. Selection for growth of Nile tilapia (*Oreochromis niloticus* L.) in low-input environments. PhD thesis Wageningen University.

Defoer, T., H. de Groote, T. Hilhorst, S. Kanté and A. Budelman,1998. Participatory action research and quantitative analysis for nutrient management in southern Mali: a fruitful marriage? Agriculture, Ecosystems and Environment 71, 215-228.

Douthwaite, B., T. Kuby, E. van der Fliet and S. Shulz, 2003. Impact pathway evaluation: an approach for achieving and attributing Impact in complex systems. Agricultural Systems 78, 243-265

Ghittino, C., M. Latini, F. Agnette, C.F. Panziei, L. Lauro, R. Ciappellloni and G. Petracca, 2003. Emerging pathogens on production and food safety. Veterinary Research Communication 2003 Sept; 27 Suppl 1: 471-479.

Gunning, J.W., 2006. Aid Evaluation: Pursuing development as if evidence matters. Presented at the "Foreign Aid Policy" conference, Economic Council of Sweden, Stockholm, 15 May 2006. Free University Amsterdam.

Fessehaye, Y., 2006. Natural mating in Nile tilapia (*Oreochromis niloticus* L.). Implications for reproductive success, inbreeding and cannibalism. PhD thesis Wageningen University.

Holling, C.S., 1995. Sustainability: The cross-scale dimension. In: Munasinghe, M. and W. Shearer (eds.). Defining and measuring sustainability, the biogeophysical foundations. The International Bank for Reconstruction and Development / World Bank, Washington DC. Ch 4, pp. 65-76.

INREF, 2006. Report of the INREF workshop "Lessons learned from the INREF phase one programmes on interdisciplinary research approaches and organisation". 23 June 2006, Wageningen University.

IRRI, 1986, Farming Systems Socioeconomic Research Training Course – Training Materials.

Jansen, J., 2006. The impact of Avian Influenza on integrated agriculture-aguaculture farming systems in the Mekong Delta, Vietnam. MSc-thesis, Wageningen University, Animal Production Systems.

Luu, L.T., P.V. Trang, N.X. Cuong, H. Demaine, P. Edwards and J. Pant, 1998. Promotion of Small-scale Pond Aquaculture in the Red River Delta Vietnam. In: Edwards, P., Little, D.C. and Demaine, H., Rural Aquaculture, CAB Int.

Muendo, P.N., 2006. The role of fish ponds in the nutrient dynamics of mixed farming systems. PhD thesis Wageningen University.

Nhan, D.K., A. Milstein, M.C.J. Verdegem and J.A.J. Verreth, 2006. Food inputs, water quality and nutrient accumulation in integrated pond systems: a multivariate approach. Aquaculture 261, 160-173.

Petersen, A. and A. Dalsgaard 2003. Antimicrobial resistance of intestinal Aeromonas spp. and Enterococcus spp. in fish cultured in integrated broiler-fish farms in Thailand. Aquaculture 219, 71-82.

Roy, R.N., R.V. Misra, J.P. Lesschen and E.M.A. Smaling, 2003. Assessment of soil nutrient balance. Approaches and methodologies. FAO Fertilizer and Plant Nutrition Bulletin 14, FAO, Rome, Italy. 87 pp.

Schlecht, E. and P. Hiernaux, 2004. Beyond adding up inputs and outputs: process assessment and upscaling in modelling nutrient flows. Nutrient Cycling in Agroecosystems 70, 303-319.

Scoones, I. and C. Toulmin, 1998. Soil nutrient balances: what use for policy? ? Agriculture, Ecosystems and Environment 71, 255-267.

Verdegem, M.C.J., R.H. Bosma and J.A.J. Verreth, 2006. Reducing Water Use for Animal Production through Aquaculture. Water Resources Development 22(1) 101–113.

WHO, 2006. Guidelines for the safe use of wastewater, excreta and greywater. Volume 3: Wastewater and excreta use in aquaculture. World Health Organisation, Geneva. 149 pp.

Keyword index

A

B

C

D

E

F

G

H

I

J

K

L

Q

R

S

T